MOLECULAR
BIOLOGY
INTELLIGENCE
UNIT

THE MOBILE RECEPTOR HYPOTHESIS: THE ROLE OF MEMBRANE RECEPTOR LATERAL MOVEMENT IN SIGNAL TRANSDUCTION

David A. Jans

John Curtin School of Medical Research
Australian National University
Canberra, Australia

CHAPMAN & HALL
ITP An International Thomson Publishing Company

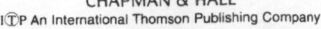

New York • Albany • Bonn • Boston • Cincinnati • Detroit • London • Madrid • Melbourne •
Mexico City • Pacific Grove • Paris • San Francisco • Singapore • Tokyo • Toronto • Washington

LANDES
BIOSCIENCE

AUSTIN, TEXAS
U.S.A.

MOLECULAR BIOLOGY INTELLIGENCE UNIT
THE MOBILE RECEPTOR HYPOTHESIS: THE ROLE OF MEMBRANE
RECEPTOR LATERAL MOVEMENT IN SIGNAL TRANSDUCTION
R.G. LANDES COMPANY
Austin, Texas, U.S.A.

U.S. and Canada Copyright © 1997 R.G. Landes Company and Chapman & Hall
Softcover reprint of the hardcover 1st edition 1997

Please address all inquiries to the Publishers:
R.G. Landes Company, 810 South Church Street, Georgetown, Texas, U.S.A. 78626
Phone: 512/ 863 7762; FAX: 512/ 863 0081

North American distributor:

Chapman & Hall, 115 Fifth Avenue, New York, New York, U.S.A. 10003

CHAPMAN & HALL

U.S. and Canada ISBN 978-1-4757-0682-6   ISBN 978-1-4757-0680-2 (eBook)
DOI 10.1007/978-1-4757-0680-2

While the authors, editors and publisher believe that drug selection and dosage and the specifications and usage of equipment and devices, as set forth in this book, are in accord with current recommendations and practice at the time of publication, they make no warranty, expressed or implied, with respect to material described in this book. In view of the ongoing research, equipment development, changes in governmental regulations and the rapid accumulation of information relating to the biomedical sciences, the reader is urged to carefully review and evaluate the information provided herein.

Library of Congress Cataloging-in-Publication Data
CIP applied for, but not received as of publication date.

PUBLISHER'S NOTE

R.G. Landes Bioscience Publishers produces books in six Intelligence Unit series: *Medical, Molecular Biology, Neuroscience, Tissue Engineering, Biotechnology* and *Environmental.* The authors of our books are acknowledged leaders in their fields. Topics are unique; almost without exception, no similar books exist on these topics.

Our goal is to publish books in important and rapidly changing areas of bioscience for sophisticated researchers and clinicians. To achieve this goal, we have accelerated our publishing program to conform to the fast pace at which information grows in bioscience. Most of our books are published within 90 to 120 days of receipt of the manuscript. We would like to thank our readers for their continuing interest and welcome any comments or suggestions they may have for future books.

Shyamali Ghosh
Publications Director
R.G. Landes Company

CONTENTS

1. **Introduction to the Mobile Receptor Hypothesis** 1
   - A. The Fluid Mosaic Model of Biological Membranes 1
   - B. The Mobile Receptor Hypothesis 3
   - C. Modern Collision Coupling Theory 7
   - D. Summary and Implications 12

2. **Direct Measurement of Lateral Mobility** 17
   - A. Introduction 17
   - B. Fluorescence Microscopy 17
   - C. Confocal Microscopy 18
   - D. Fluorescence Photobleaching Recovery 21
   - E. Lateral Mobility in the Cytoplasm
     and Membranes of Living Cells 26
   - F. Measurements in Artificial and Isolated Membranes 39
   - G. Measurements of Lateral Mobility
     Using Other Methods 40
   - H. Summary 41

3. **Parameters Affecting Plasma Membrane Protein
   Lateral Mobility** 49
   - A. Mechanisms of Protein Immobilization
     in Biological Membranes 49
   - B. Membrane Lipid Mobility 53
   - C. The Cytoskeleton 55
   - D. Anchorage Modulation 56
   - E. Membrane Protein Sequence Motifs 57
   - F. Domain Structure: Regions of Restricted Mobility 65
   - G. Signal Transduction 67
   - H. Summary 72

4. **Lateral Mobility of Polypeptide Hormone Receptors
   and GTP-Binding Proteins** 83
   - A. Introduction 83
   - B. Practical Considerations 83
   - C. Lateral Mobility Measurements of Polypeptide
     Hormone Receptors 88
   - D. Tyrosine Kinase Receptor-Mediated
     Signal Transduction 91
   - E. GTP-Binding Protein Activating Receptor-Mediated
     Signal Transduction 98
   - F. Structural Considerations 101
   - G. Lateral Mobility Measurements
     of GTP-Binding Proteins 107
   - H. Lateral Mobility of Cytokine Receptors 109
   - I. Summary and Implications for Signal Transduction 110

5. **Evidence for the Role of Membrane Receptor Lateral Movement in GTP-Binding Protein-Mediated Signal Transduction** 117
  A. Introduction 117
  B. Kinetic Considerations in GTP-Binding Protein-Mediated Receptor-Effector Systems 118
  C. Indirect Evidence for a Role of Receptor Lateral Movement in GTP-Binding Protein-Mediated Signal Transduction 119
  D. Direct Evidence for a Role of Receptor Lateral Movement in GTP-Binding Protein-Mediated Signal Transduction 122
  E. Gα Signaling in the Cytosolic Phase 126
  F. Stoichiometric Considerations and Trimeric GTP-Binding Protein Immobility 128
  G. Amplification in GTP-Binding Protein-Mediated Receptor-Effector Systems Through Receptor Lateral Movement 129
  H. Summary 130

6. **Evidence for the Role of Receptor Immobilization in Desensitization Subsequent to Hormonal Stimulation** 139
  A. Introduction 139
  B. Receptor Internalization in Desensitization of Response 139
  C. Receptor Immobilization Prior to Internalization 140
  D. Receptor Movement Required for Internalization 145
  E. Kinetic Considerations with Respect to Lateral Mobility Measurements 148
  F. Receptor Phosphorylation 148
  G. Studies with Receptor Antagonists–Receptor Immobilization Is Agonist-Dependent 150
  H. Summary 156

7. **Evidence for the Role of Immobilization of Ligand-Occupied Membrane Receptors in Signal Transduction** 165
  A. Introduction 165
  B. Receptor Immobilization in Tyrosine Kinase Receptor Signaling 165
  C. Receptor Immobilization in Signaling by Fc Receptors ... 167
  D. Receptor Immobilization in Cell-Cell Interaction 173
  E. Receptor Immobilization in Cell-Adhesion to an Extracellular Substratum 180
  F. Summary 182

8. The Mobile Receptor Hypothesis: A Global View 191
   A. Introduction 191
   B. Receptor Lateral Movement in Signal Transduction 191
   C. The Central Role of the Cytoskeleton 193
   D. Signal Transduction: Receptor Lateral Mobility
      Modulation by Heterologous Signaling 199
   E. Potential Pharmacological Applications
      of the Mobile Receptor Hypothesis 205
   F. Concluding Remarks 208

   Index 219

PREFACE

This book seeks to contribute to the understanding of signal transmission at the level of the membrane, which is the primary transducer of extracellular signals represented by hormones, growth factors, etc. The basic tenet of the Mobile Receptor Hypothesis,[1,2] that receptors can diffuse freely within the plane of the membrane and that this is functionally important in signal transduction through bringing about interactions between protein signaling components within the plane of the membrane, has been accepted for over 20 years, despite the fact that there has really been no critical assessment of the evidence for the hypothesis. This book seeks to elaborate the theory behind and experimental evidence for the Mobile Receptor Hypothesis. It concentrates mostly on direct measurements of receptor mobility in a physiological context in living cells and attempts to put the case for the role of membrane receptor lateral movement in signal transduction. Less complete treatments of the theme can be found in review articles (see refs. 3-7).

Although there has been a concerted effort in this book to include the most recent published literature, selectivity has been exercised; e.g., experiments relating to isolated or artificial membranes are only dealt with where deemed appropriate, and apologies are made to colleagues who may feel slighted as a result. Most of the detailed discussion deals with experimental evidence for tyrosine kinase receptors such as those for insulin, and epidermal and nerve growth factors, and GTP-binding protein activating receptors such as the V_1- and V_2-type receptors for vasopressin, although Chapter 7 discusses Fc receptors as well as molecules mediating cell-cell recognition and adhesion interactions such as CD2/LFA-3 and integrins. The technique of fluorescence photobleaching recovery, the method by which most direct measurements of protein lateral mobility have been made, is necessarily treated in some detail and in particular with respect to measurements of the lateral movement of polypeptide hormone receptors. The treatment is not intended to be definitive, and readers are directed to reviews[8-12] which specifically address this topic. It should be stressed that while complex formulations etc. are touched upon, a concerted attempt has been made to make the presentation of the evidence and the arguments for the Mobile Receptor Hypothesis accessible to nonspecialists. It is to be hoped that this has not resulted in oversimplification of what is a complex research field.

If it does nothing else, it is to be hoped that this book may encourage workers in the general field of signal transduction to consider the membrane as a two-dimensional rather than a one-dimensional structure; that is, that the membrane should be seen as a lattice of sites for interaction within its plane, rather than simply as an inert barrier between the extracellular medium (from whence come hormones, growth factors and other ligands) and the cytosol (where "the real signaling" events of phosphorylation, regulation of second messenger concentrations, etc. take place). The importance of this to our understanding of signal transduction at the level of the membrane should not be underestimated.

REFERENCES

1. Cuatrecasas P. Membrane receptors. Annu Rev Biochem 1974; 43:169-214.
2. De Haen C. The non-stoichiometric floating receptor model for hormone sensitive adenylyl cyclase. J Theor Biol 1976; 58:383-400.
3. Helmreich EJM, Elson EL. Protein and lipid mobility. Adv in Cyclic Nucleotide and Prot Phosphor Res 1984; 18:1-62.
4. Chabre M. The G protein connection: is it in the membrane or the cytoplasm. Trends in Biochem Sci 1987; 12:213-215.
5. Peters, R. Lateral mobility of proteins and lipids in the red cell membrane and the activation of adenylate cyclase by β-adrenergic receptors. FEBS Lett 1988; 234:1-7.
6. Jans DA. The mobile receptor hypothesis revisited: a mechanistic role for hormone receptor lateral mobility in signal transduction. Biochim Biophys Acta 1992; 1113:271-276.
7. Jans DA, Pavo I. A mechanistic role for polypeptide hormone receptor lateral mobility in signal transduction. Amino Acids 1995; 9:93-109.
8. Axelrod D, Koppel DE, Schlessinger J et al. Mobility measurement by analysis of fluorescence photobleaching recovery kinetics. Biophys J 1976; 16:1055-1069.
9. Elson EL, Schlessinger J, Koppel DE et al. Measurement of lateral transport on cell surfaces. Prog Clin Biol Res 1976; 9:137-147.
10. Schlessinger J, Elson EL. Fluorescence methods for studying membrane dynamics. In: Ehrenstein G, Lecar H, ed. Methods of Experimental Physics. New York: Academic Press, 1982:197-227.
11. Koppel DE. Fluorescence photobleaching as a probe of translational and rotational motions. In: Shaafi RI, Fernandez SM, ed. Fast Methods in Physical Biochemistry and Cell Biology. Amsterdam: Elsevier-North Holland, 1983:339-367.
12. Peters R. Translational diffusion in the plasma membrane of single cells as studied by fluorescence microphotolysis. Cell Biol Int Rep 1981; 5(8):733-760.

ACKNOWLEDGMENTS

I would like to acknowledge all past and present collaborators for their invaluable contributions over the years; of these, I would particularly like to thank Imre Pavo and Patricia Jans for their intellectual input as well as their friendship. I would also like to thank Jessie Jans for critical reading of parts of the text. The book is dedicated to my understanding family who put up with my being impossible for most of the time of writing, and in particular my daughter Marianna who took it all mostly in her stride and made the moments when I was not writing a sheer pleasure.

CHAPTER 1

INTRODUCTION TO THE MOBILE RECEPTOR HYPOTHESIS

A. THE FLUID MOSAIC MODEL OF BIOLOGICAL MEMBRANES

The idea that proteins float freely within the lipid bilayer of biological membranes gained initial impetus through the formulation of the "fluid mosaic" model in the early 1970s.[1] Based on electron microscopy and freeze-fracture analysis as well as a number of other studies, the structure of biological membranes was purported to be and is still believed to be as shown in Figure 1.1: "dissolved" proteins float within the "sea" of membrane lipid. This includes integral membrane proteins, that is, proteins that completely traverse the lipid bilayer sometimes more than once such as membrane receptors, as well as peripheral membrane proteins such as the guanosine-triphosphate (GTP) binding protein ("G-protein") subunits and cell-cell recognition molecules such as LFA-2 (lymphocyte function-associated antigen-2), which are only associated with the membrane and do not traverse it. They are linked to the lipid bilayer through a variety of covalent modifications, such as glycosyl phosphatidylinositol anchors, myristoyl and palmitoyl fatty acid linkages, etc.

Initial evidence for the fluidity of biological membranes came from "patching" or "capping" experiments (see Fig. 1.2A), where normally diffusely distributed cell surface antigen molecules were induced to aggregate (patch or cap) in one particular region of the membrane of lymphocytes or other cells by divalent antibodies specific to the particular protein.[2] Patching did not occur if the experiments were carried out at 0°C, or if univalent Fab fragments were used instead of bivalent antibodies.[2] Aggregation of surface molecules could clearly occur only if the membrane surface proteins to which the antibodies were bound were able to diffuse within the plane of the membrane.

Further evidence for protein mobility within the plane of the membrane came from experiments such as those performed by Frye and Edidin[3] who derived mouse-human cell heterokaryons by cell fusion and stained for specific antigens using antibodies to assess the distribution of mouse and human proteins on the membrane surface (see Fig. 1.2B). Whereas immediately after cell fusion, mouse and human antigenic components were largely segregated, they were essentially completely intermixed and diffusely distributed after about 30 minutes at 37°C. Such redistribution of membrane proteins clearly implied that the proteins were mobile within the plane of the membrane. The process of intermixing could be inhibited at temperatures below 15°C, but was not affected by inhibitors of ATP or protein

The Mobile Receptor Hypothesis: The Role of Membrane Receptor Lateral Movement in Signal Transduction, by David A. Jans. © 1997 R.G. Landes Company.

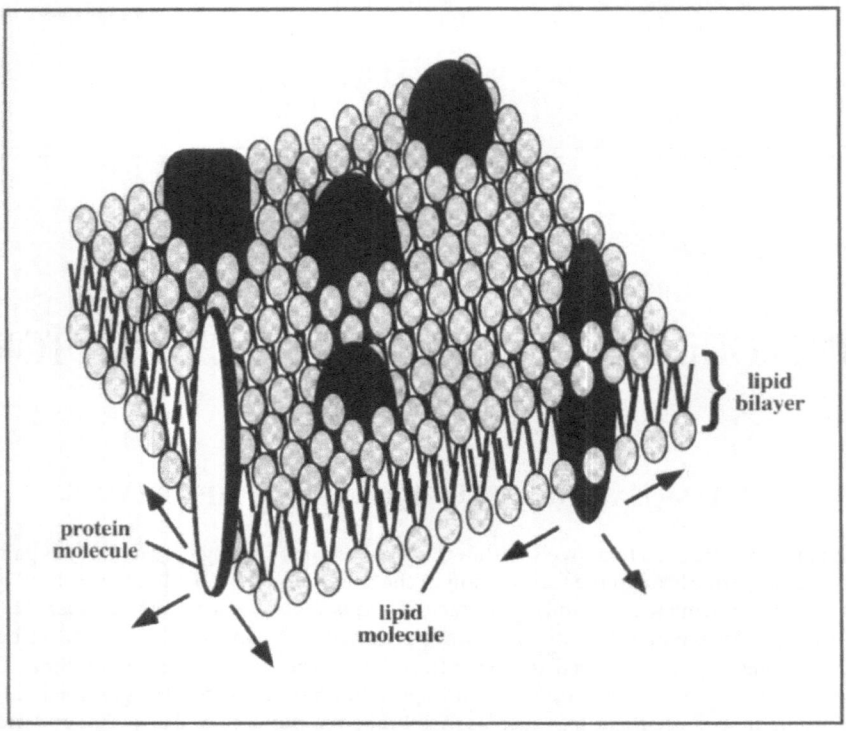

Fig. 1.1. A Schematic Representation of the Fluid Mosaic Model of Biological Membranes. Membrane proteins, either integral (traversing the membrane) or peripheral (associated with the membrane surface), float freely in the two-dimensional "sea" of the lipid bilayer.[1] Proteins can move in any direction within the plane of the membrane.

synthesis. Protein movement in this system was estimated to occur at a rate of about 0.5×10^{-10} cm²/sec,[3] a value not so far removed from results for the estimation of the rate of protein diffusion in the membranes of living cells using direct measurements (see subsequent chapters; e.g., Table 2.3) and more than four orders of magnitude slower than the rate of diffusion of hemoglobin in aqueous solution (7×10^{-7} cm²/sec).[1]

Singer and Nicolson[1] postulated that the physical or chemical perturbation of a membrane might alter or affect a particular membrane protein (e.g., a receptor) or set of proteins; redistribution of membrane proteins could then occur by translational diffusion through the viscous two-dimensional solution, thereby allowing new interactions

among the altered protein components to take place. They suggested that this general mechanism might play an important role in various membrane-mediated cellular phenomena that occur on a time scale of minutes or longer. The basic idea that proteins are intrinsically freely mobile, floating within the lipid layer, led to the formulation of the Collision Coupling or Mobile Receptor Hypothesis,[4,5] where protein lateral movement within the lipid bilayer was proposed to play a critical role in signal transduction at the level of the membrane; i.e., the process by which the extracellular signal represented by a hormone binding to its receptor on the cell surface is converted to an intracellular signal, such as a phosphorylation event or a change in the intracellular

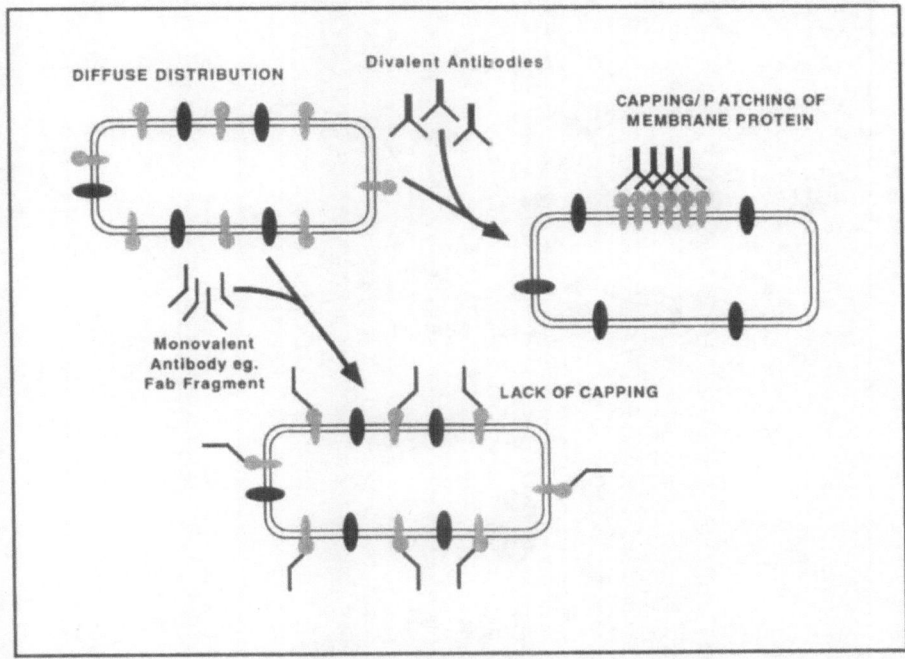

Fig. 1.2A. Lateral Movement of Plasma Membrane Proteins: I. Capping. Normally diffusely distributed cell surface antigen molecules are induced to aggregate (patch or cap) in one particular region of the membrane by divalent (crosslinking) antibodies in contrast to monovalent (non-crosslinking) antibodies specific to the particular surface protein. The proteins aggregate through lateral movement within the plane of the lipid bilayer.[2]

level of a second messenger molecule, such as cAMP (adenosine 3':5'-cyclic monophosphate) or IP_3 (inositol 1,4,5-tris-phosphate).

B. THE MOBILE RECEPTOR HYPOTHESIS

The Mobile Receptor Hypothesis as originally postulated by Pedro Cuatrecasas[4,5] proposed that receptors can diffuse independently in the plane of the membrane and that lateral movement is the mechanism by which, subsequent to hormone binding, membrane receptors are brought into reversible contact with effector molecules, such as adenylate cyclase or phospholipase C, in order to regulate their activity and thereby initiate a biological response. The affinity of the receptor for effector was proposed to be greater when the receptor was hormone-occupied, but the two components

were initially brought into contact through lateral movement. As will be briefly expounded below, the hypothesis and its tenets were based largely on circumstantial evidence and logical deductions founded on the premise "that biological membranes are in a relatively fluid state." This was based on the observations of Singer and Nicolson[1] (above), Gitler,[6] Radda[7] and others, despite the fact that at the time of the hypothesis' conception no direct measurement of the lateral movement of a specific membrane component—either of a lipid or membrane-associated protein—had been performed.

In terms of the Mobile Receptor Hypothesis with particular reference to the adenylate cyclase system, receptor-hormone complexes, once formed, were proposed to diffuse laterally within the plane of the membrane and interact through random

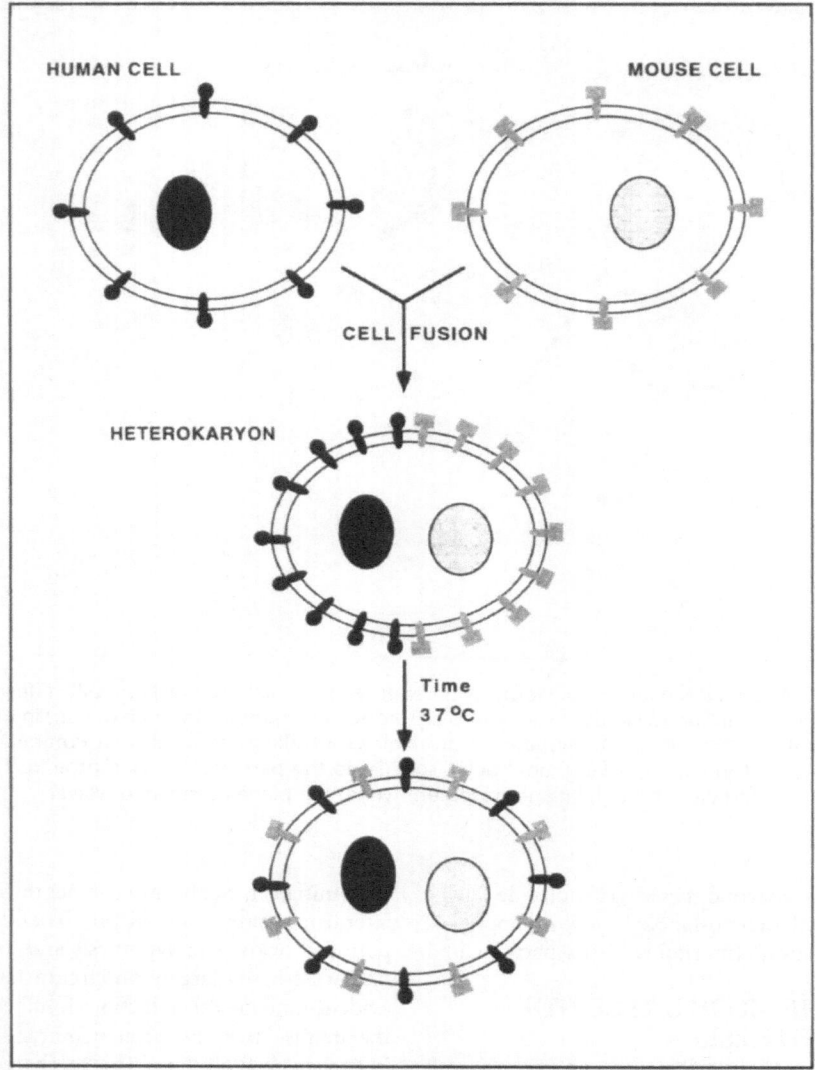

Fig. 1.2B. Lateral Movement of Plasma Membrane Proteins: II. Cell Fusion. Cells of different species containing different surface proteins are fused using polyethylene glycol to create a heterokaryon. Proteins are initially localized in the respective membrane domains of the two cells, but homogeneously distributed throughout the membrane within 30 min at 37°C through protein lateral movement within the plane of the membrane.[3]

encounters with adenylate cyclase, also presumed to be in "a relatively fluid state," "in a way analogous to known interactions in solution."[4] Diffusion-mediated association of the receptor-hormone complex with adenylate cyclase was accordingly proposed to depend on the concentrations of the respective components in the membrane, and their rates of lateral movement, and result from two sequential steps occurring at the level of the membrane:[4,5,8]

$$H + R \Leftrightarrow HR' \qquad [1.1],$$

occurring in the aqueous phase of the outer membrane surface, and

$$HR' + E \Leftrightarrow HR\text{-}E' \qquad [1.2],$$

taking place within the plane of the membrane where H, hormone; R, receptor; E, effector enzyme and indicates "activated state."

It should be remembered that the initial formulation of the Mobile Receptor Hypothesis predated the elucidation of the coupling role of trimeric GTP-binding proteins (namely, the α-, β- and γ-subunits, of which, the α-subunit binds guanine nucleotides and in particular GTP in its activated form) in the case of a number of membrane integral receptors, such as those for the hormones glucagon, vasopressin and isoprotenerol. Cassel and Selinger[9] were among the first to demonstrate that guanine nucleotides were involved in adenylate cyclase activation.

We now know that the receptor-hormone complex in the adenylate cyclase and various other effector systems interacts directly with a trimeric GTP-binding protein complex, with the step of effector activation being mediated by the GTP-binding protein subunits, and the α-subunit in particular in the case of adenylate cyclase. However, as is evident from the equations in the next section, whether a GTP-binding protein complex or effector enzyme such as adenylate cyclase is the signaling component with which the diffusing receptor-hormone complex directly interacts in the initial signaling step within the plane of the membrane does not impinge on the kinetic implications of the hypothesis, movement within the plane

of the membrane is a critical parameter in determining the rate of signal transduction.

Functional evidence for the hypothesis with respect to adenylate cyclase was partly based on experiments with cholera toxin (87 kDa) and in particular the observation that membrane-localized toxin, although not binding directly to adenylate cyclase, was able to activate it.[10-13] Cholera toxin binds with high affinity to the sialo-glycolipid ganglioside GM1, which is normally mobile within the plane of the plasma membrane but can be rendered immobile through treatment with reagents perturbing microtubule or actin microfilament structure.[10] Cholera toxin-mediated activation of adenylate cyclase had been inferred to occur at the level of the membrane based on the observation that 2000 bound toxin molecules per toad erythrocyte were sufficient to effect 50% maximal adenylate cyclase activation, but only 50 of these were found to enter the cell.[11,12] Cholera toxin was not only able to activate adenylate cyclase but also to affect the stimulation of adenylate cyclase by other receptors.[13] It was proposed that membrane lateral diffusion of the cholera toxin-GM1 complex brought it into contact with, and permitted toxin-mediated activation of, adenylate cyclase.[13] The basic features of this scheme were believed to be directly pertinent to the mechanisms by which hormone receptors normally modulate adenylate cyclase activity (e.g., equations [1.1] and [1.2] above).[13] That cholera toxin's action is mediated through ADP ribosylation of the Gs subunit of the trimeric GTP-binding-protein coupling factor complex, and indeed mediated through intracellularly localized toxin (subsequent to GM1-mediated internalization) rather than at the level of the membrane, was not known at the time of these studies.

Cuatrecasas' hypothesis[4] was also based on the view that it was exceedingly unlikely in biological membranes that signal transduction involving two discrete membrane components such as hormone receptors and adenylate cyclase could be explained by large pre-existing complexes of interacting or associated membrane proteins in the normal unstimulated state. This was presumed to be especially true for receptors which were in vast excess relative to the number of adenylate cyclase molecules and in cases where a number of different receptors were capable of activating or inhibiting the same membrane integral effector. In addition, certain receptors, such as those for prostaglandins and growth hormone, which appeared to be able to activate more than one membrane-localized effector (e.g., ion channels, adenylate cyclase, Na^+,K^+-ATPase, etc.), were very unlikely to be in a permanent association with all of the effectors simultaneously. Since more than one receptor could interact with a single effector, and some receptors could interact with more than one effector population, it seemed likely that receptors were not precomplexed with effectors in the basal unstimulated state, but rather, were brought into contact with one another in transient activating interactions through lateral movement of the receptor-hormone complex subsequent to hormonal addition. Neer[14] had isolated peaks of adenylate cyclase activity before and after hormonal activation and indeed found that the size of the cyclase complex subsequent to activation was smaller than that under basal conditions. This implied that hormone-mediated activation of adenylate cyclase did not result in stable association of the cyclase catalytic moiety with the hormone-receptor complex and was consistent with the idea of transient collisionary contacts between hormone-receptor and cyclase being sufficient for activation of the latter.

Craig and Cuatrecasas[4,15] noted that subsequent to hormonal addition there was usually a lag time in terms of the activation kinetics in a number of biological systems, which was not consistent with the idea of preformed receptor/effector enzyme complexes. The temporal lag in activation was temperature-dependent, consistent with activation being brought about through transient contacts effected by lateral movement of the hormone-receptor complex within the plane of the membrane. Again,

the fact that we now know that GTP-binding proteins are the initial components activated by the diffusing receptor-hormone complex, rather than the effector itself, actually alters little in terms of the logic behind the Mobile Receptor Hypothesis. Essentially the same evidence can be used to argue against the existence of preformed receptor-GTP-binding protein complexes and, therefore, for the need for diffusion to bring receptors and GTP-binding proteins together.

Further evidence against precoupled receptor-effector complexes came from the fact that receptor occupancy and biological response did not necessarily appear to be coupled in a linear fashion.[4,5] Maximal biological response in a variety of systems is known to be achieved with only a minority of receptors occupied,[5,8] e.g., maximal stimulation of glucose oxidation by isolated adipocytes occurs when only 2-3% of insulin receptors are occupied.[16-19] Similarly, maximal steroidogenesis in Leydig cells occurs when only 1% of luteinizing hormone-human chorionic gonadotropin receptors are occupied both in cultured cells and in the intact animal,[20] although other biological responses in Leydig and other cells appear to require higher levels of receptor occupancy.[5,21-25] The fact that maximal response can be attained when only a small percentage of total receptors are occupied indicates a high degree of amplification of the signal represented by hormone binding to receptor, the basis of this amplification at the level of the membrane being attributed to lateral movement of the receptor-hormone complex enabling it to activate multiple effector molecules in succession. Importantly in this context, the adenylate cyclase system is distinguished by the fact that one receptor-hormone complex can activate multiple effector (adenylate cyclase) molecules,[26,27] thus amplifying the signal represented by hormone binding to receptor at the membrane surface. Signal amplification is of particular significance in the physiological context of low hormone concentrations and generally low receptor occupancies. This property of the adenylate cyclase system was seen as further evidence for receptor-hormone complex-mediated activation of adenylate cyclase being through transient interactions rather than permanent association. It should be remembered that receptors and effectors are not present in stoichiometric amounts,[28,29] which is particularly relevant in the context of the adenylate cyclase system, where a single adenylate cyclase may be stimulated by more than one distinct hormone although each hormone has its own initial interaction with its specific receptor.

Cuatrecasas stressed the difficulties of experimentation using isolated membranes, where the fluidity and diffusion properties were clearly very different from those in intact biological systems.[4] This is as relevant to current experimentation as it was when the Mobile Receptor Hypothesis was initially formulated. We shall return to this point in the following chapters, but briefly, isolated membranes are severely reduced in terms of the levels of cytoskeletal components (which may play a major role in regulating plasma membrane protein mobility–see chapter 3, section C; chapter 5, section F; chapter 8, section C) and associated domain structures (the "membrane skeleton"–see chapters 3 and 8), and lack all of the proteins, such as soluble intracellular protein kinases and phosphatases, which interact with and regulate the function and mobility of plasma membrane components and the cytoskeleton. Artificial membranes (those reconstituted from lipid/phospholipid fractions isolated from various biological sources as well as from synthetic lipids–see chapter 2, section F) completely lack all relevant components. As a result of this, protein movement within the plane of isolated or artificial membranes is generally much faster than that in the plasma membrane of living cells and hence, much less of a critical parameter or limiting factor in activation kinetics. For this reason, the extensive body of evidence relating to the activation of adenylate cyclase and other effector systems based on isolated and artificial membrane systems will only be treated briefly in the

following chapters. As is generally true for all scientific investigation in biological systems, in vivo *veritas*—only experiments performed in vivo or in well-characterized intact cell systems which approximate the former are truly physiologically relevant. Isolated membrane or other in vitro experiments are valuable only where they corroborate or are consistent with in vivo data–if they do not do so, the in vitro rather than in vivo results must be discarded.

C. MODERN COLLISION COUPLING THEORY

The Mobile Receptor Hypothesis has remained the basis for the rationalization of signal transduction at the level of the membrane to this day. Membrane signal transduction is seen as occurring largely through the capability of membrane integral proteins to diffuse within the plane of the lipid bilayer. The term "collision coupling theory" has been used interchangeably with that of the "Mobile Receptor Hypothesis" to emphasize the idea that the activation of membrane integral and membrane-associated components occurs through transient protein-protein contacts, rather than complexation within the plane of the membrane effected by the lateral diffusion of membrane-integral receptors.

In terms of modern collision coupling theory,[30-34] receptor lateral diffusion within the plane of the membrane is necessary to effect:

I. receptor dimerization in the case of tyrosine kinase receptors such as those for epidermal growth factor (EGF) and platelet-derived growth factor (PDGF) which is essential for signal transduction.[35,36] Upon dimerization, tyrosine kinase receptors undergo an intermolecular tyrosine phosphorylation event which activates their intrinsic tyrosine kinase phosphorylation activity. The receptor tyrosine kinase activity effects the phosphorylation and activation of soluble signaling components, including phospholipase Cγ which associate with the membrane receptors. The sequence of event can be described as follows:

$$H + R \Leftrightarrow HR \qquad [1.1]$$

(as above), which occurs in the aqueous phase of the outer membrane surface,

$$HR + HR \Leftrightarrow (HR)_2' \qquad [1.3]$$

(dimerization), occurring within the plane of the membrane), and

$$(HR)_2' + SSFs \Leftrightarrow (HR)_2'SSFs \Leftrightarrow (HR)_2' + SSFs' \qquad [1.4],$$

which occurs in the aqueous (cytosolic) phase of the inner membrane surface. *SSFs*, soluble signaling factors (see text), other symbols as above.

II. the collisions between receptors and GTP-binding proteins which activate the latter, as in the case of the receptors for vasopressin (type 1 and 2 receptors) and the β-adrenergic receptor.[30,31,37-42] The sequence of events is as follows (compare to that in section B above):

$$H + R \Leftrightarrow HR' \qquad [1.1]$$

(as above), which occurs in the aqueous phase of the outer membrane surface,

$$HR' + G \Leftrightarrow HR.G'' \rightarrow HR' + G' \quad [1.5],$$

occurring within the plane of the membrane), and

$$G' + E \Leftrightarrow E.G' \Leftrightarrow G' + E' \qquad [1.6],$$

which occurs in the aqueous (cytosolic) phase of the inner membrane surface. Cycles of [1.5-1.6] amplify the signal at the level of the membrane. *G* indicates GTP-binding protein; "denotes a transient (millisecond) complex and other symbols as above. *G'* is the GTP-binding protein in its activated (GTP-bound) state.

III. subunit association in the case of multiple subunit receptors, such as those for the cytokines interleukin (IL) 2-6 and granulocyte macrophage colony stimulating factor (GM-CSF), which is necessary for signal transduction (see Fig. 1.3).[43] In the case of the IL-3, IL-5 and GM-CSF receptors, a variety of cytosolic factors, including soluble tyrosine kinases, subsequently associate with the β-receptor subunit to trigger cytosolic signaling events.

Several cytokine receptors such as those for IL-2 and IL-4 possess three receptor subunits, of which the β- and γ-receptor subunits have signaling roles. In the case of the IL-3 receptor, for example, α- and β-subunits have been shown to be separated on the membrane surface in the absence of ligand, while ligand addition induces their dimerization.[44] The sequence of signaling events with respect to the latter is as follows:

$$H + R\alpha \Leftrightarrow HR\alpha \qquad [1.7],$$

comparable to [1.1] above, which occurs in the aqueous phase of the outer membrane surface,

$$HR\alpha + R\beta \Leftrightarrow (HR\alpha R\beta)' \qquad [1.8]$$

(dimerization), occurring within the plane of the membrane, and

$$(HR\alpha R\beta)' + SSFs \Leftrightarrow (HR\alpha R\beta)'SSFs \Leftrightarrow (HR\alpha R\beta)' + SSFs' \qquad [1.9],$$

which occurs in the aqueous (cytosolic) phase of the inner membrane surface. *Rα* and *Rβ* are α/β-receptor subunits (where the α-subunit has hormone binding activity) and other symbols as above.

Figure 1.3 is a schematic representation of the role of receptor lateral movement in signal transduction at the level of the membrane for these three classes of receptor. A further class of receptor/cell surface molecules are those in which receptor immobilization is central to signal transduction. These include: A. receptors which require an immobile aggregated signaling complex for signaling to occur, such as the immunoglobulin (Ig) G Fc receptors of macrophages and neutrophils and IgE Fc receptors of mast cells.[45-47] Soluble factors, including tyrosine kinases, associate with the immobile receptor dimer to effect subsequent signaling events.

$$R \rightarrow L_2 \leftarrow R \qquad [1.10],$$

occurring within the plane of the membrane,

$$R + R + L_2 \Leftrightarrow (RL)_2^{*'} \qquad [1.11],$$

occurring in the aqueous phase of the

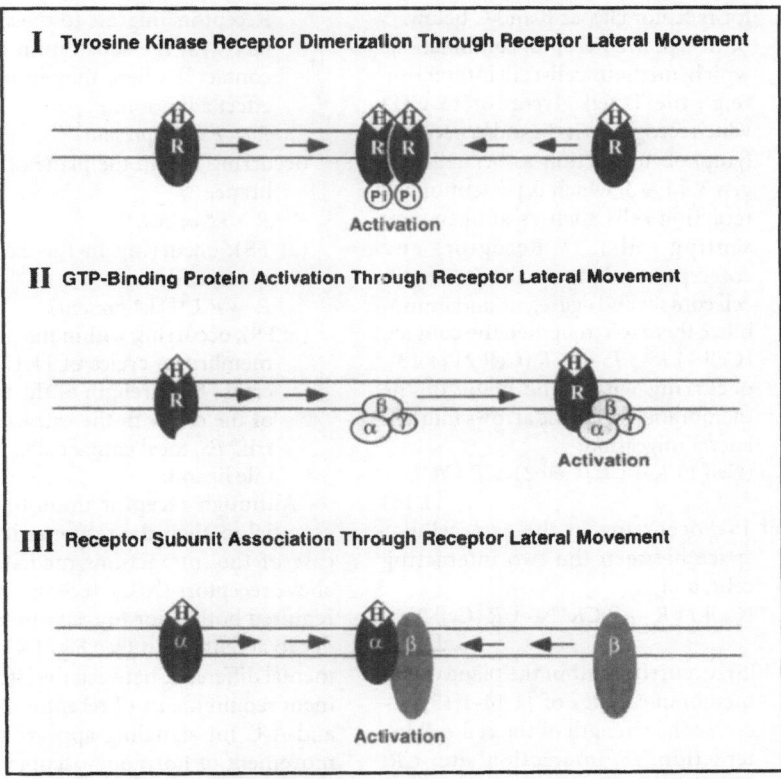

Fig.1.3. A Role for Receptor Lateral Movement in Membrane Signal Transduction. The three basic classes of membrane integral polypeptide hormone or cytokine receptors are shown: I. Tyrosine kinase receptors, such as those for EGF (epidermal growth factor) or PDGF (platelet-derived growth factor)—receptor lateral movement brings about receptor dimerization, thereby permitting the activating intermolecular autophosphorylation event to occur; II. GTP-binding protein coupling receptors, such as those for vasopressin or cholecystokinin—receptor lateral movement brings about transient collisionary contacts with the membrane-associated GTP-binding protein trimeric complex, leading to the latter's activation; III. Multisuburit cytokine receptors such as those for interleukin (IL) 3 or 5 or GM-CSF (granulocyte-macrophage-colony-stimulating-factor)—receptor lateral movement brings the hormone-binding receptor α-subunit into contact with the signal transducing β-subunit to form an active signaling complex. Similarly, lateral movement is required to bring together the three subunits of receptors such as those for IL-2 and -4. H, hormone; R, receptor.

outer membrane surface,

$$(RL)_2*'+ \quad SSFs \quad \Leftrightarrow \quad (RL)_2*'SSFs \Leftrightarrow$$
$$(RL)_2*'+ SSFs' \qquad [1.12],$$

which occurs in the aqueous (cytosolic) phase of the inner membrane surface. L_2, dimeric ligand; * denotes immobilized complex. With respect to [1.11], it is unclear whether the exact series of events is

$$R + R + L_2 \Leftrightarrow RL_2 + R \Leftrightarrow (RL)_2*'$$
or $R + R + L_2 \Leftrightarrow R.R + L_2 \Leftrightarrow (RL)_2*'.$

Higher order aggregation of receptors also occurs in the case of Fc receptors in similar fashion to the cycles of movement to the site of aggregation and immobilization at the site as shown for receptor classes B and C below.

B. receptor/coreceptor combinations which mediate cell-cell interaction (e.g., the T-cell glycoprotein CD2 which recognizes its ligand/coreceptor lymphocyte function-associated antigen 3, LFA-3, which is present on interacting cells such as antigen-presenting cells).[48,49] Receptors and coreceptors migrate to the site of cell-cell contact and aggregate and immobilize there to strengthen the contact.

(Cell 1) $R \rightarrow IS \leftarrow CR$ (Cell 2) [1.13], occurring within the plane of the membrane, where the arrows indicate lateral migration,

(Cell 1) $R + CR$ (Cell 2)$\Leftrightarrow R.CR^{*\prime}$
[1.14]

(at IS), occurring in the extracellular space between the two interacting cells, and

(Cell 1) $R \rightarrow R.CR^{*\prime} \leftarrow CR$ (Cell 2)
[1.15]

(at IS), occurring within the plane of the membrane; cycles of [1.14-1.15] increase the strength of the cell-cell interaction. IS, interaction site; CR, coreceptor.

C. receptors mediating cell adhesion to immobile substrata (e.g., the fibronectin receptor which binds to its extracellular matrix ligand, fibronectin, or Fc receptors of macrophages or neutrophils which mediate phagocytosis of antibody coated particles).[50]

Receptors migrate to the site of contact with the substratum (the "focal contact") where they immobilize to effect adhesion.

$R \rightarrow FS$ $(L^*$ present$)$ [1.16],

occurring within the plane of the membrane,

$R + L^* \Leftrightarrow R.L^{*\prime}$ [1.17]

(at FS), occurring in the extracellular space, and

$R \rightarrow R.L^{*\prime}$ $(L^*$ present$)$ [1.18]

(at FS), occurring within the plane of the membrane; cycles of [1.17-1.18] increase the strength of the interaction of the cell with the extracellular matrix. FS, focal contact site; L^*, immobile ligand.

Although receptor immobilization is essential to elicit the response itself in the case of the interactions mediated by the above receptors (A-C), receptor mobility is required both prior to response and in order to accentuate it (see Fig. 1.4). A fundamental difference between the lateral movement requirements of receptor classes I-III and A-C for signaling appears to be that movement of hormone-occupied receptor is critical to signal transduction in the former case, while receptor movement *prior* to ligand binding is required in the latter case together with subsequent receptor *immobilization* upon ligand binding. Cycles of both movement to and immobilization at a particular site enable the adhesion/recognition interaction to be strengthened and/or amplified.

Fig. 1.4. A Role for Receptor Immobilization in Membrane Signal Transduction. Three basic categories of receptor-mediated interaction are shown: A. Fc receptor activation through lateral movement of receptors to the site of contact with crosslinking dimeric ligand (I), resulting in receptor dimerization, aggregation and immobilization (II), which is a prerequisite for association of cytosolic signaling molecules including soluble tyrosine kinases (K). An example is the high-affinity receptor for IgE (Fcε RI) of mast cells; note the basic similarities to signaling by tyrosine kinase receptors (see Fig. 1.3, receptor class I). B. Cell-cell interaction through receptor lateral movement to the interaction site (I) and subsequent ligation, aggregation and immobilization (II). An example is the CD2 and LFA-3 receptor-ligand/coreceptor combination. C. Cell adhesion through the lateral movement of receptors to immobile ligand at the adhesion site (I) and subsequent ligation, aggregation and immobilization; (II). An example is the fibronectin receptor and its extracellular matrix ligand fibronectin. L, ligand; R, receptor; CR, coreceptor; ECM, extracellular matrix; K, soluble protein kinase.

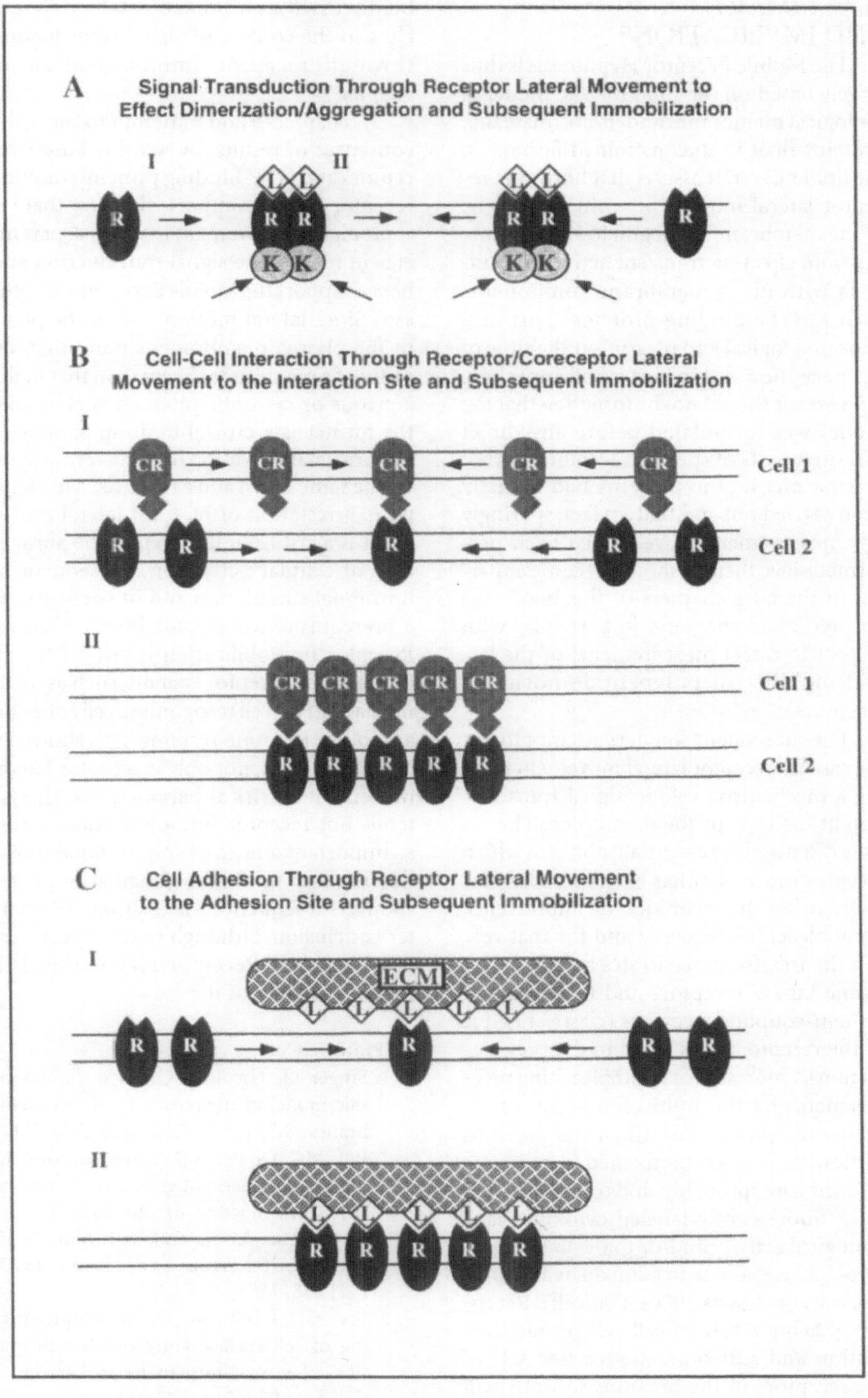

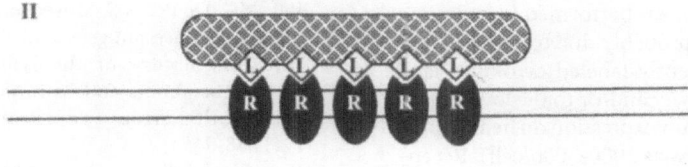

A
Signal Transduction Through Receptor Lateral Movement to Effect Dimerization/Aggregation and Subsequent Immobilization
I
II
L L
R R
K K
R
B
Cell-Cell Interaction Through Receptor/Coreceptor Lateral Movement to the Interaction Site and Subsequent Immobilization
I
CR
R
Cell 1
Cell 2
II
Cell 1
Cell 2
C
Cell Adhesion Through Receptor Lateral Movement to the Adhesion Site and Subsequent Immobilization
I
ECM
L
R
II

Fig. 1.4.

D. SUMMARY
AND IMPLICATIONS

The Mobile Receptor Hypothesis is thus largely based on the fluid mosaic model of biological membranes and its postulate that proteins float in unconstrained fashion in the lipid bilayer. It asserts that hormone receptor lateral movement within the plane of the membrane is integral to signal transduction, effecting transient activating contacts with other membrane components such as GTP-binding proteins. This idea appeared logical and pleasing at the time of its conception and indeed is still appealing. However, it should not be forgotten that the model was formulated before any direct measurements of specific receptor or other plasma membrane proteins had actually been carried out and that, in fact, sparingly few measurements have indeed been performed since then, as should become apparent in the later chapters of this book (see chapter 2 and chapter 4 in particular with respect to direct measurements of the lateral mobility of polypeptide hormone receptors).

The subsequent chapters attempt to put the case for receptor lateral movement playing a mechanistic role in signal transduction at the level of the membrane. The results discussed are essentially those in which receptor movement has been measured directly using the technique of fluorescence photobleaching recovery and for that reason the treatise concentrates largely on tyrosine kinase receptors and GTP-binding protein-coupling receptors (classes I and II of the receptor types listed in the previous section). Fluorescence photobleaching measurements for the multisubunit class of cytokine receptors (class III in the previous section) have been performed in only one isolated case, probably due to the fact that most fluorescently-labeled cytokines lack biological activity and/or to the low level of cytokine receptor expression on hematopoietic cells (as few as 1000-3000/cell). Receptors playing a role in cell-cell contact formation and adhesion, etc. (classes A-C of the receptors in the previous section) will

be discussed as a fundamentally different class in the context of signal transduction through receptor immobilization in chapter 7.

As chapters 5 and 8 attempt to show, the consensus of results for tyrosine kinase receptors and GTP-binding protein-coupling receptors clearly supports the view that receptor lateral movement is a critical parameter in membrane signal transduction and hence support the Mobile Receptor Hypothesis. Since lateral motion within the plane of the plasma membrane is more than an order of a magnitude slower than that in the aqueous or cytosolic phase, it is clear that the former is a crucial limiting parameter in hormonal activation of cellular response. By the same token and as discussed in chapter 6, arrestation of receptor lateral movement is a critical initial step in the abrogation of cellular activation, subsequent to hormonal stimulation and in particular as a prerequisite to receptor internalization. Receptor immobilization is critical to cell activation in receptor systems, such as those mediating cell-cell recognition, cell adhesion and other immune responses, as shown in chapter 7. Thus, not only is receptor lateral movement a critical parameter in all systems, but receptor immobilization is just as important a factor, both in signal transduction and in the downregulation of response subsequent to activation. This latter conclusion, although never a postulate of the Mobile Receptor Hypothesis, is a logical corollary of it.[30,31]

REFERENCES

1. Singer SJ, Nicolson GL. The fluid mosaic model of the structure of cell membranes. Science 1972; 175(23):720-731.
2. Raff MC, De Petris S. Movement of lymphocyte surface antigens and receptors: the fluid nature of the lymphocyte plasma membrane and its immunological significance. Fed Proc 1973; 32(1):48-54.
3. Frye CD, Edidin M. The rapid intermixing of cell surface antigens after formation of mouse-human heterokaryons. J Cell Sci 1970; 7(2):313-335.

4. Cuatrecasas P. Membrane receptors. Annu Rev Biochem 1974; 43:169-214.

5. Jacobs S Cuatrecasas P. The mobile receptor hypothesis and "cooperativity" of hormone binding. Application to insulin. Biochim Biophys Acta. 1976; 433(3):482-495.

6. Gitler C. Plasticity of biological membranes. Annu Rev Biophys Bioeng 1972; 1: 51-92.

7. Radda GK. Enzyme and membrane conformation in biochemical control. Biochem J 1971; 122(4):385-396.

8. Kahn CR. Membrane receptors for hormones and neurotransmitters. J Cell Biol 1976; 70:261-286.

9. Cassel D, Selinger Z. Mechanism of adenylate activation through the β-adrenergic receptor: catecholamine-induced displacement of bound GDP by GTP. Proc Natl Acad Sci USA 1978; 75(9):4155-4159.

10. Craig SW, Cuatrecasas P. Mobility of cholera toxin receptors on rat lymphocyte membranes. Proc Natl Acad Sci USA 1975; 72(10):3844-3848.

11. Bennett V, Cuatrecasas P. Mechanism of activation of adenylate cyclase by Vibrio cholerae enterotoxin. J Memb Biol 1975; 22(1-2):29-52.

12. Sahyoun N, Cuatrecasas P. Mechanism of activation of adenylate cyclase by cholera toxin. Proc Natl Acad Sci USA 1975; 72(9):3438-3442.

13. Bennett V, O'Keefe E, Cuatrecasas P. Mechanism of action of cholera toxin and the mobile receptor theory of hormone receptor-adenylate cyclase interactions. Proc Natl Acad Sci USA 1975; 72(1):33-37.

14. Neer EJ. The size of adenylate cyclase. J Biol Chem 1974; 249(20):6527-6531.

15. Craig SW, Cuatrecasas P. Immunological probes into the mechanism of cholera toxin action. Immunol Commun 1976; 5(5):387-400.

16. Freychet P, Laudat, MH; Laudat G et al. Impairment of insulin binding to the fat cell membrane in the obese hypoglycemic mouse. FEBS Lett 1972; 25:339-342.

17. Kono T, Barham FW. Effects of insulin on the levels of adenosine 3':5'-monophosphate and lipolysis in isolated rat epididymal fat cells. J Biol Chem 1973; 248(21):7417-7426.

18. Megyesi K, Kahn CR, Roth J et al. The NSILA-s receptor in liver plasma membranes. Characterization and comparison with the insulin receptor. J Biol Chem 1975; 250(23):8990-8996.

19. Kono T, Barham FW. The relationship between the insulin-binding capacity of fat cells and the cellular response to insulin. Studies with intact and trypsin-treated fat cells. J Biol Chem 1971; 246(20):6210-6216.

20. Huhtaniemi IT, Clayton RN, Catt KJ. Gonadotropin binding and Leydig cell activation in the rat testis in vivo. Endocrinology 1982; 111(3):982-987.

21. Goldfine ID, Gardner JD, Neville DM Jr. Insulin action in isolated rat thymocytes. I. Binding of 125 I-insulin and stimulation of alpha-aminoisobutyric acid transport. J Biol Chem 1972; 247(21):6919-6926.

22. Hollenberg MD, Cuatrecasas P. Insulin and epidermal growth factor. Human fibroblast receptors related to deoxyribonucleic acid synthesis and amino acid uptake. J Biol Chem 1975; 250:3845-3853.

23. Wicks WD. Induction of hepatic enzymes by adenosine 3',5'-monophosphate in organ culture. J Biol Chem 1969; 244:3941-3950.

24. Hsueh AJ, Dufau ML, Catt KJ. Gonadotropin-induced regulation of luteinizing hormone receptors and desensitization of testicular 3':5'-cyclic AMP and testosterone responses. Proc Natl Acad Sci USA 1977; 74(2):592-595.

25. Moyle WR, Lee EY, Bahl OP et al. New method of quantifying ligand binding based on measurement of an induced response. Am J Physiol 1977; 232(3):E274-E285.

26. Orly J, Schramm M. Fatty acids as modulators of membrane functions; catecholamine-activated adenylate cyclase of the turkey erythrocyte. Proc Natl Acad Sci USA 1975; 72:3433-3437.

27. Brandt DR, Ross EM. Catecholamine-stimulated GTPase cycle; multiple sites of regulation by β-adrenergic recceptor and Mg^{2+} studied in reconstituted receptor-G, vesicles. J Biol Chem 1986; 261:1656-1664.

28. De Haen C. The non-stoichiometric floating receptor model for hormone sensitive adenylyl cyclase. J Theor Biol 1976; 58:383-400.

29. Perkins JP. Adenyl cyclase. Adv Cyclic Nucleotide Res 1973; 3:1-64.

30. Jans DA. The mobile receptor hypothesis revisited: a mechanistic role for hormone receptor lateral mobility in signal transduction. Biochim Biophys Acta 1992; 1113:271-276.

31. Jans DA, Pavo I. A mechanistic role for polypeptide hormone receptor lateral mobility in signal transduction. Amino Acids 1995; 9:93-109.

32. Helmreich EJM, Elson EL. Protein and lipid mobility. Adv in Cyclic Nucleotide and Prot Phosphor Res 1984; 18:1-62.

33. Peters R. Translational diffusion in the plasma membrane of single cells as studied by fluorescence microphotolysis. Cell Biol Int Rep 1981; 5(8):733-760.

34. Jans DA. Nuclear signaling pathways for extracellular ligands and their membrane-integral receptors? FASEB J 1994; 8:841-847.

35. Schlessinger J. Signal transduction by allosteric receptor oligomerization. Trends Biochem Sci 1988; 13:443-447.

36. Schlessinger J. The epidermal growth factor receptor as a multifunctional allosteric protein. Biochemistry 1989; 27:3119-3123.

37. Jans DA, Peters R, Zsigo J et al. The adenylate cyclase-coupled vasopressin V_2-receptor is highly laterally mobile in membranes of LLC-PK$_1$ renal epithelial cells at physiological temperature. EMBO J 1989; 8(9):2431-2438.

38. Jans DA, Peters R, Fahrenholz F. Lateral mobility of the phospholipase-C-activating vasopressin V_1-type receptor in A7r5 smooth muscle cells: a comparison with the adenylate cyclase-coupled V_2-receptor. EMBO J 1990; 9(9):2693-2699.

39. Jans DA, Peters R, Jans P et al. Ammonium chloride affects receptor number and lateral mobility of the vasopressin V_2-type receptor in the plasma membrane of LLC-PK$_1$ renal epithelial cells: role of the cytoskeleton. Exper Cell Res 1990; 191:121-128.

40. Jans DA, Peters R, Jans P et al. Vasopressin V_2-receptor mobile fraction and ligand-dependent adenylate cyclase-activity are directly correlated in LLC-PK$_1$ renal epithelial cells. J Cell Biol 1991; 114(1):53-60.

41. Pavo I, Jans DA, Peters R et al. A vasopressin antagonist that binds to the V_2-receptor of LLC-PK$_1$ renal epithelial cells is highly laterally mobile but does not effect ligand-induced receptor immobilization. Biochim Biophys Acta 1994; 1223:240-246.

42. Zakharova OM, Rosenkranz AA, Sobolev AS. Modification of fluid lipid and mobile protein fractions of reticulocyte plasma membranes affects agonist-stimulated adenylate cyclase. Application of the percolation theory. Biochim Biophys Acta 1995; 1236:177-184.

43. Taga T, Kishimoto T. Cytokine receptors and signal transduction. FASEB J 1992; 6:3387-3396.

44. Stomski FC, Sun Q, Bagley CJ et al. Human interleukin-3 (IL-3) induces disulfide-linked IL-3 receptor alpha- and beta-chain heterodimerization, which is required for receptor activation but not high-affinity binding. Mol Cell Biol 1996; 16(6):3035-3046.

45. Posner RG, Subramanian K, Goldstein B et al. Simultaneous cross-linking by two nontriggering bivalent ligands causes synergistic signaling of IgE Fc epsilon RI complexes. J Immunol 1995; 155(7):3601-3609.

46. Menon AK, Holowka D, Webb WW et al. Clustering, mobility, and triggering activity of small oligomers of immunoglobulin E on rat basophilic leukemia cells. J Cell Biol 1986; 102:534-540.

47. Menon AK, Holowka D, Webb WW et al. Cross-linking of receptor-bound IgE to aggregates larger than dimers leads to rapid immobilization. J Cell Biol 1986; 102:541-550.

48. Liu SJ, Hahn WC, Bierer BE et al. Intracellular mediators regulate CD2 lateral diffusion and cytoplasmic Ca^{2+} mobilization upon CD2-mediated T cell activation. Biophys J 1995; 68(2):459-470.

49. Chan PY, Lawrence MB, Dustin ML et al. Influence of receptor lateral mobility on adhesion strengthening between membranes containing LFA-3 and CD2. J Cell Biol 1991; 115(1):245-255.

50. Duband J-L, Nuckolls GH, Ishihara A et al. Fibronectin receptor exhibits high lateral mobility in embryonic locomoting cells but is immobile in focal contacts and fibrillar streaks in stationary cells. J Cell Biol 1988; 107:1385-1396.

CHAPTER 2

DIRECT MEASUREMENT OF LATERAL MOBILITY

A. INTRODUCTION

This chapter will concentrate on fluorescence photobleaching recovery (FBR), also known as fluorescence recovery after photobleaching (FRAP) or fluorescence microphotolysis,[1-6] the microscopic technique by which plasma membrane lateral movement can be determined directly. It will deal initially with the basis of fluorescence and the microscopic techniques that enable fluorescence to be visualized and quantified, and then concentrate on lateral mobility measurements themselves in cytosol, and biological and artificial membranes. It will become clear that the lateral movement of molecules in biological membranes is at least two orders of magnitude slower than that in cytosol. Protein lateral mobility is much slower than that of membrane lipids, implying that membrane proteins are normally limited in their movement. The fact that protein movement is restricted means that the lateral diffusion of proteins within the membrane lipid bilayer is rate limiting in terms of signal transduction at the level of the membrane.[7,8]

B. FLUORESCENCE MICROSCOPY

The fluorescence photobleaching recovery technique uses fluorescently-labeled molecules such as ligands and antibodies as probes for lipids/phospholipids or membrane proteins. Fluorescent molecules have a special property, which is that when light of a specific wavelength is shone onto a fluorescent molecule, an "excited state" results (this process is called "excitation") which subsequently leads to a fall in energy and emission of light at a different wavelength, and hence of a different color, to that exciting the fluorescence. As an example, the excitation and emission wavelengths for the most commonly used fluorescent label fluorescein iso-thiocyanate (FITC) are shown in Figure 2.1; it is optimally excited at about 490 nm and emits light maximally at about 530 nm. Particular microscope filter combinations for light excitation and emission can be chosen to enable this fluorescence to be visualized specifically and efficiently even in the presence of other fluorescent labels possessing distinct excitation and emission wavelengths, and "noise" due to autofluorescence and nonspecific staining. Autofluorescence is an intrinsic property of all cells, stemming from a variety of molecules such as nucleotides, heme structures, vitamin cofactors, etc. The term "signal-to-noise-ratio" refers to the "specific" fluorescent signal above background fluorescence due to autofluorescence, nonspecific staining, etc.

The Mobile Receptor Hypothesis: The Role of Membrane Receptor Lateral Movement in Signal Transduction, by David A. Jans. © 1997 R.G. Landes Company.

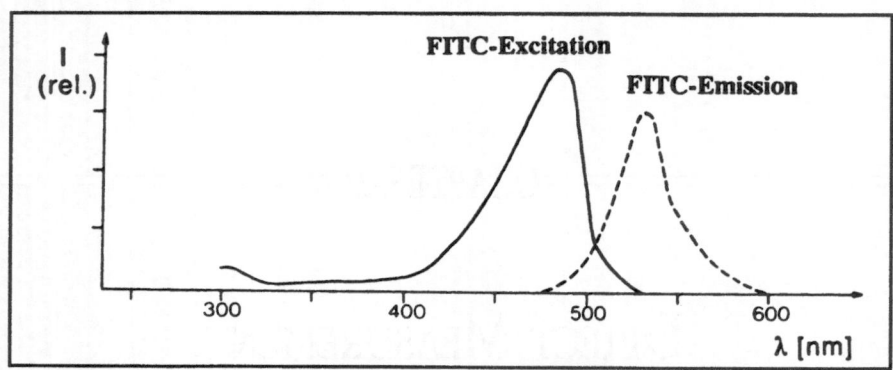

Fig. 2.1. Excitation and emission spectra of the commonly used fluorescent dye fluorescein isothiocyanate (FITC).

C. CONFOCAL MICROSCOPY

The confocal microscope (see Fig. 2.2 and refs. 9-11 for reviews on the topic) has several advantages compared to the conventional fluorescence microscope. A laser is used to excite fluorescence at specific wavelengths, providing high beam powers necessary for certain applications such as fluorescence photobleaching recovery as well as increased sensitivity. The confocal microscope's special optics (see Fig. 2.3) mean that the path of the light exciting fluorescence is exactly the same as that for the fluorescence emitted. This means that only light in the exact plane of focus is collected by the microscope, light from below or above the sample being excluded. These special optics mean that, among other things, resolution is much better than in conventional fluorescence microscopy enabling imaging at the subcellular level with spatial accuracy. This is particularly relevant to applications such as fluorescence photobleaching recovery where spatial selectivity with respect to a specific membrane is required. The confocal optics mean that the in-focus volume is very thin (a "slice" of 0.5-1.5 µm), enabling "optical sectioning" to be performed; i.e., thin slices can be taken through a biological sample in the vertical direction (Z axis) and the slices put together to constitute a three-dimensional picture. Figure 2.4 illustrates how it is possible to exploit the confocal

optics to visualize fluorescence associated with different membranes of a polarized adherent cell labeled with a fluorescently-labeled lipid probe.

Since the in-focus volume is very thin, it is possible to quantitate fluorescence accurately in equivalent volumes in different parts of the same specimen or in different specimens. The fluorescent intensity is proportional to the concentration of fluorescently-labeled molecule. With respect to quantitation of fluorescent intensity, it should be remembered that values are relative, and not at all absolute in the sense of a particular fluorescent signal corresponding directly to a chromophore concentration. It is usually necessary to use appropriate controls such as those of autofluorescence to enable valid quantitation. This is especially true in the case of weak signals where only a few fluorescent molecules are present, which is commonly a problem in studies where plasma membrane receptor lateral mobility is measured (see chapter 4, section B). Other relevant controls include appropriate negative controls (e.g., cell mutants or nonexpressing cell types which lack the molecule of interest such as a surface antigen or receptor) and controls for nonspecific binding in the case of antibody probes or fluorescently-labeled ligands. All negative controls and autofluorescence specimens can and should be processed in the same

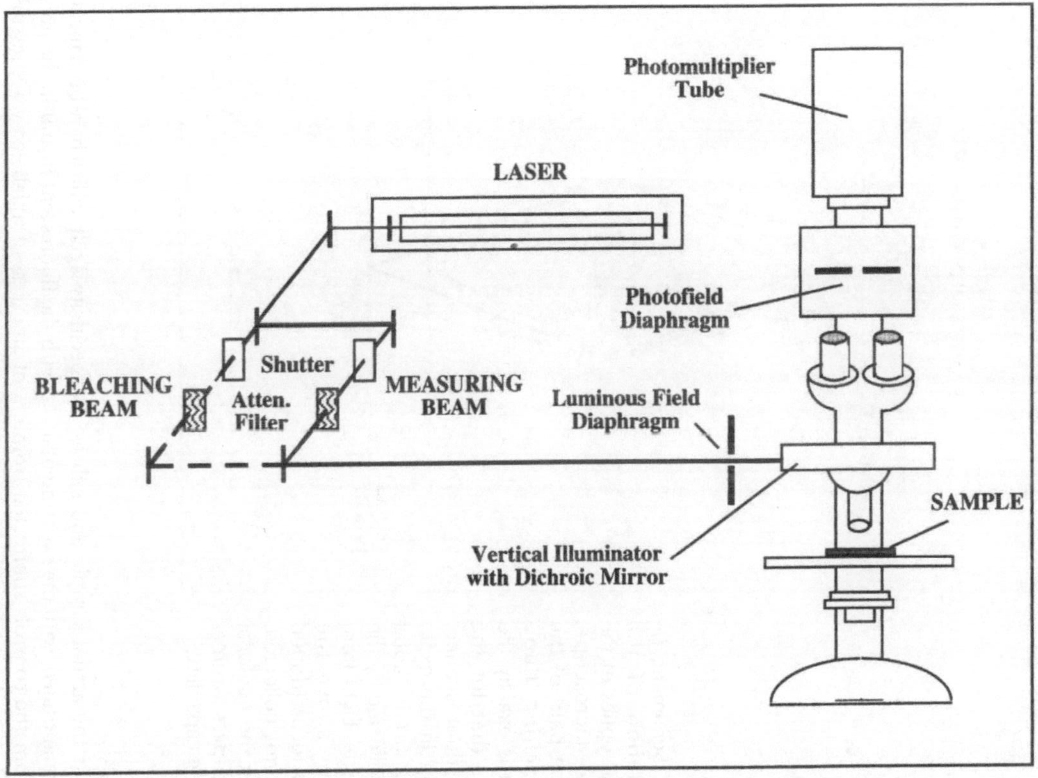

Fig. 2.2. A schematic representation of the confocal microscope as set up to perform fluorescence photobleaching recovery measurements. Lateral diffusion coefficients are measured in a thin layer (about 1 μm in depth) of the biological sample containing a fluorescently-labeled macromolecule which is illuminated by a cylindrical laser beam. The beam is divided into two separate paths, whereby a shutter system enables very rapid transfer from one to the other. The difference between the two is that the bleaching beam is about 10^4 times stronger than the the measuring beam, which is strongly attenuated. Both beam paths are focused onto exactly the same part of the sample through a series of mirrors and filters. The luminous field diaphragm is dimensionally and spatially identical to the photofield diaphragm, providing the confocal microscope's special optics (see Fig. 2.3). Emitted fluorescence is collected and amplified through the photomultiplier tube.

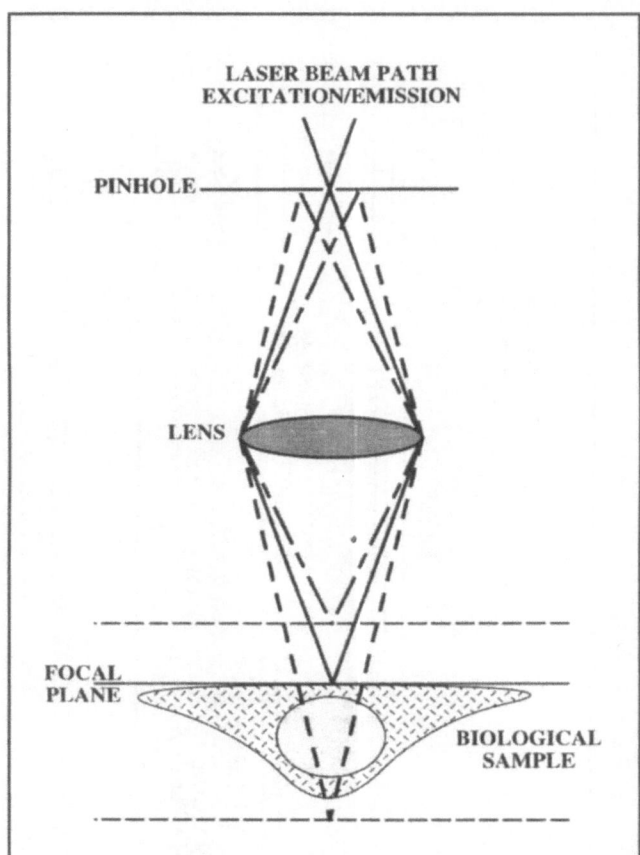

Fig. 2.3. Schematic representation of the particular optics of the confocal microscope. Since the path of the light exciting fluorescence is exactly the same as that for the emitted fluorescence, only light in the exact plane of focus (solid line) is collected by the microscope. Light from below or above the plane of focus (dotted lines) is not collected, providing the confocal microscope's optical sectioning capability.

fashion as the actual sample and subsequently quantitated, with the experimental values from the controls subtracted from those of the sample with specific fluorescent probes to give valid relative estimations of the fluorescence due to specific protein or lipid probes (e.g., see refs. 12,13). In the case of lateral mobility measurements, a further useful negative control is analysis of fixed cells (e.g., see refs. 12, 14) where cell surface molecules are crosslinked, and hence no longer laterally mobile.[12,14]

By way of illustration of some of these principles, an example is shown in Figure 2.5 of studies using fluorescently-labeled vasopressin as a probe for the vasopressin V_2-receptor of renal epithelial cells. A control for

specificity is cells which have been incubated with the fluorescent ligand in the presence of an excess of unlabeled vasopressin[12] (or cells of a low receptor number expressing mutant cell line incubated with the fluorescent ligand).[12,15,16] It is possible to perform binding studies using confocal measurements of fluorescent signals in identical fashion to binding studies with radioligand, using incubations with increasing concentrations of fluorescently-labeled hormone in the absence (total binding) and presence (nonspecific binding) of an excess of unlabeled hormone. Samples are then imaged and image analysis performed, with subtraction of the fluorescent intensity quantified in the nonspecific incubation from that for

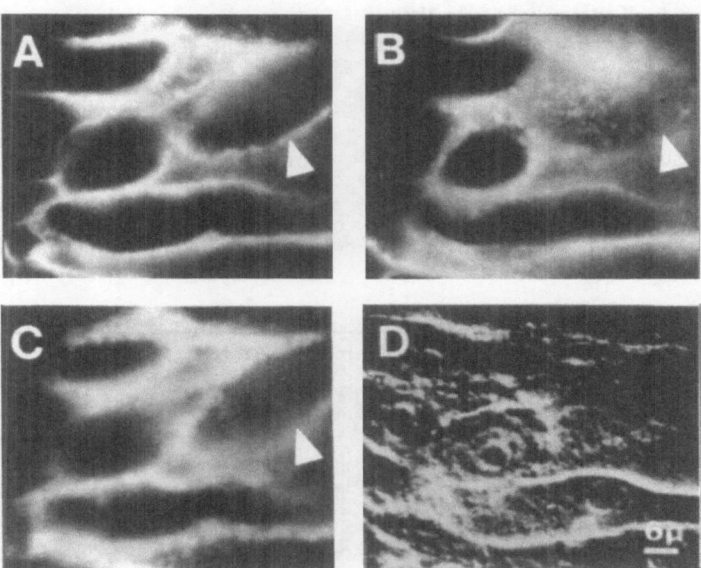

Fig. 2.4. Illustration of use of the confocal microscope's special optics. Renal epithelial cells (the LLC-PK$_1$ porcine line) were labeled with the fluorescent lipid probe DiOC$_{14}$(3) (3,3'-ditetradecyloxacarbocyanine iodide). The arrow indicates a single cell in the monolayer in three focal planes: panel A shows the plane of focus on the middle of the cell lateral membrane, panel B on the apical membrane, and panel C the basolateral membrane. Panel D shows a phase contrast view.

total fluorescence yielding the specific binding curve. From this can be estimated the dissociation constant (K_D) or receptor binding affinity (Fig. 2.5B). Results of confocal measurements using fluorescent hormones agree quite well with those obtained using radioactively labeled ligands (e.g., see refs. 12, 16).

D. FLUORESCENCE PHOTOBLEACHING RECOVERY

Although nonconfocal optics have been used to perform measurement of membrane protein/lipid lateral mobility, the confocal microscope is particularly useful in fluorescence photobleaching recovery because its special optics provide the spatial precision to be able to measure the lateral movement in specific membranes, that is, measurement within the plane of the membrane. Lateral diffusion measurements are made in a thin layer or volume (about 1 µm in depth) containing a fluorescently-labeled macromolecule using an instrumental setup such as that shown diagrammatically in Figure 2.2. The layer is illuminated by a cylindrical laser beam, which is divided or split into two separate beam paths. A shutter system enables very rapid transfer from one beam path (and strength) to the other. The difference between the two is that one is about 10^4 times stronger than the other, which is strongly attenuated. Both beam paths are focused onto exactly the same part of the sample through a series of mirrors and filters.

The idea of the technique,[1-6] shown schematically in Figure 2.6, is as follows: the laser beam is focused onto a small area of a solution, living cell or membrane. A short burst of very high energy light from the laser is used to render all labeled molecules in the irradiated area of the membrane completely nonfluorescent, after which the laser beam energy is reduced by 10^4 times to a normal measuring level. The return of fluorescence in the irradiated area can only occur through the lateral movement of unbleached fluorescent molecules from the surrounding area of the membrane, and this fluorescence recovery can be monitored over time. From the return of fluorescence, two parameters can be calculated: the rate of diffusion of the fluorescent molecule (that is, how fast it moves–the apparent lateral diffusion coefficient–D) and the fraction of mobile molecules (or mobile fraction–f), whereby the complete return of fluorescence

Fig. 2.5. Illustration of the use of appropriate controls to perform quantitative measurements of fluorescent intensity. A. Visualization of binding of an analog, fluorescently-labeled vasopressin (TR-LVP–deamino[Lys⁸(tetramethylrhodamyl-aminothiocarbonyl)]vasopressin), as a specific probe for the vasopressin V_1-receptor of aortic smooth muscle cells (the A7r5 line–I) and the vasopressin V_2-receptor of renal epithelial cells (the porcine LLC-PK₁ line–II). Phase contrast pictures are presented on the left. In I, Binding to A7r5 cells is shown in the absence (Panel B) and presence (Panel D) of an excess of unlabeled vasopressin.[13] In II, binding for LLC-PK₁ cells, which express about 40,000 receptors/cell (Panel B), is compared to that for cells of an LLC-PK₁ mutant cell line (M18), which expresses about 2,000 receptors/cell (Panel D).[12,15]

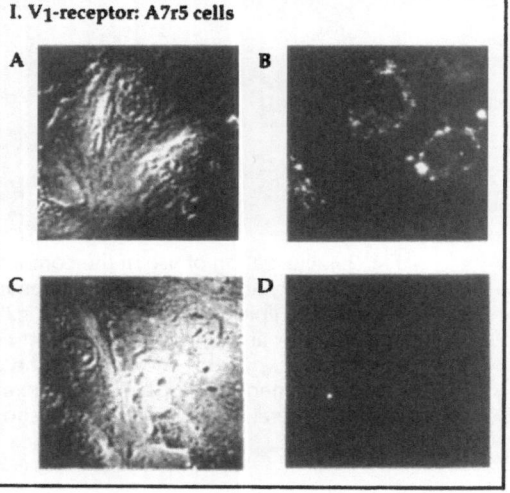

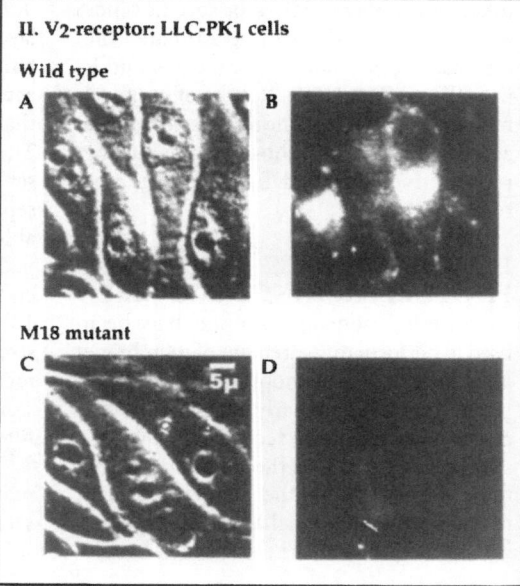

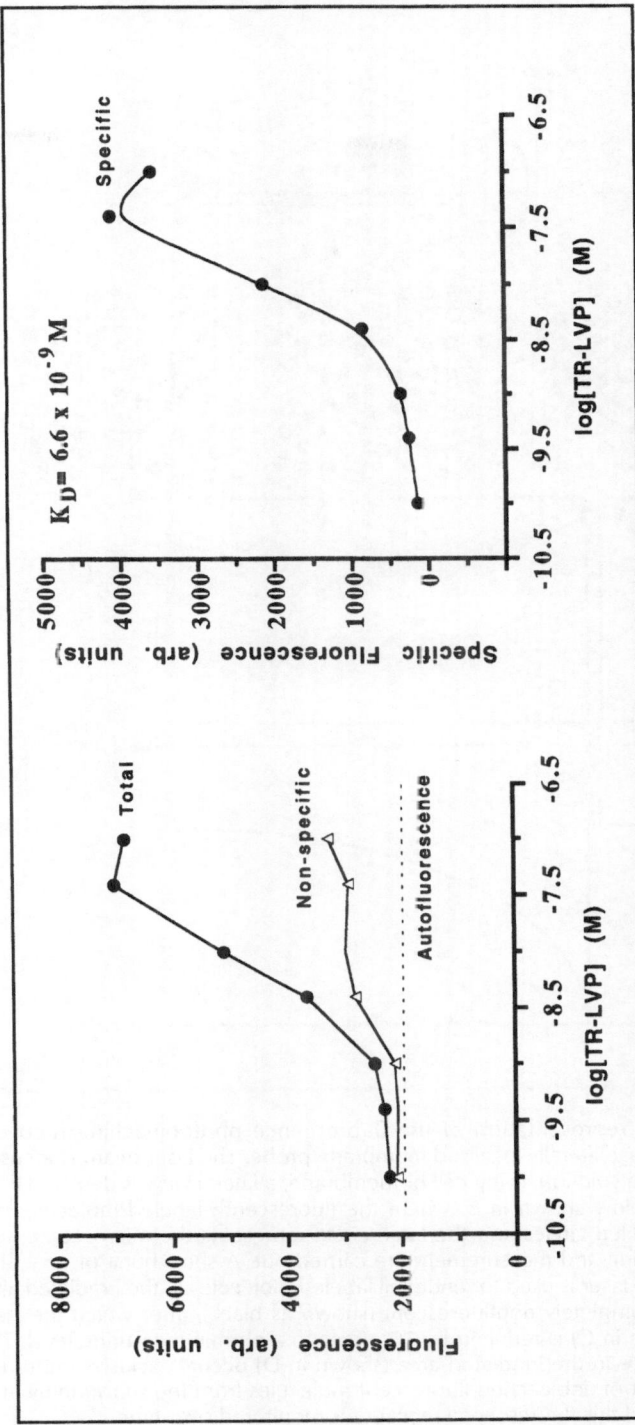

Fig. 2.5B. Quantitation of the receptor binding constant (K_D) using confocal measurements. LLC-PK$_1$ cells were incubated with increasing concentrations of fluorescently-labeled hormone in the absence (total binding) and presence (nonspecific binding) of an excess of unlabeled hormone (left panel). Subtraction of the fluorescent intensity quantified in the nonspecific incubation from that for total fluorescence yields the specific binding curve, from which the dissociation constant can be estimated (right panel). Results of confocal measurement using fluorescent hormones agree well with those obtained using radioactively-labeled ligands.[12,16]

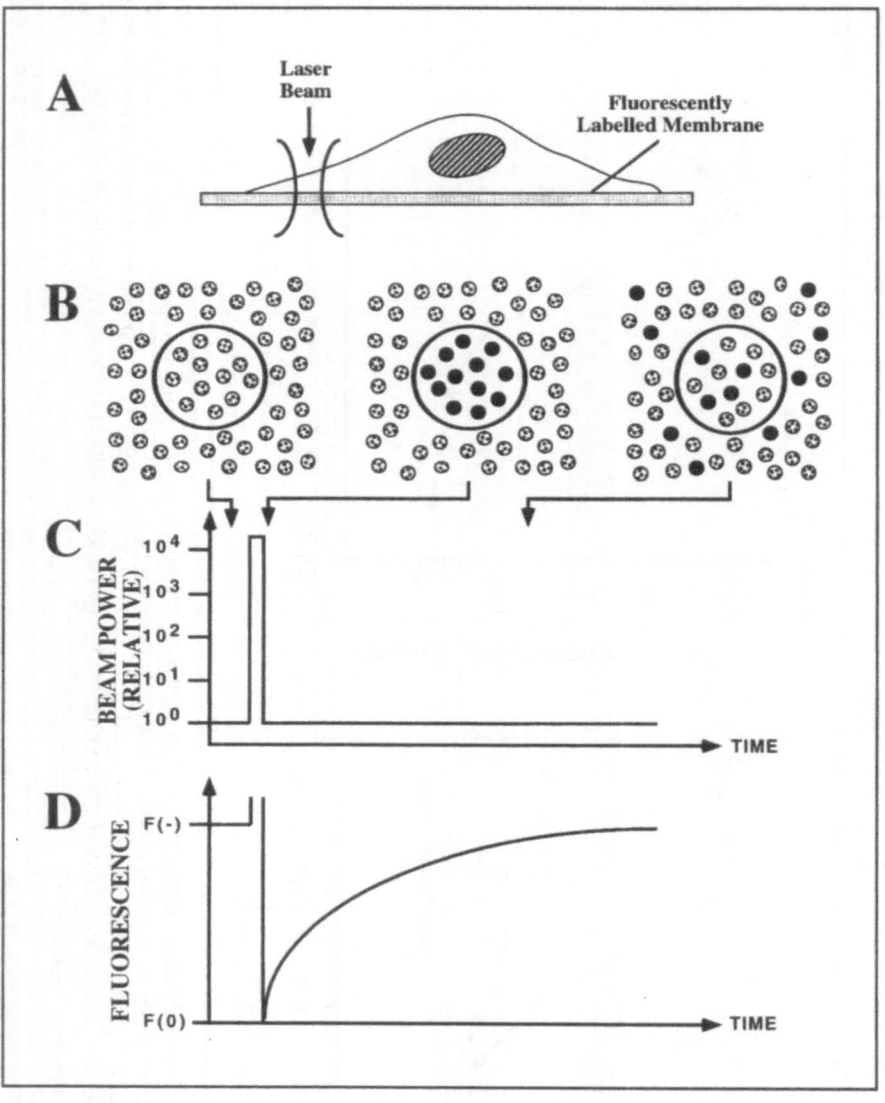

Fig. 2.6. Schematic representation of the fluorescence photobleaching recovery measurement of a basolaterally-localised membrane probe. The laser beam is focused onto a small area of a solution, living cell or membrane; a lateral view is depicted in A, with a view from below shown in B, where the fluorescently-labeled molecules are represented as speckled circles and the large circle represents the microscopic field within which bleaching and measurement are carried out. A short burst of very high energy light from the laser is used to render all labeled molecules in the irradiated area of the membrane completely nonfluorescent (shown as black), after which the laser beam energy (shown in C) is reduced by 10^4 times to a normal measuring level. The return of fluorescence in the irradiated area (shown in D) occurs exclusively through the lateral movement of unbleached fluorescent molecules from the surrounding area of the membrane and this fluorescence recovery is monitored over time.

indicates that 100% of all molecules are mobile (f = 1). Anything less than this indicates that a proportion of the molecules are immobile.

The procedure for fitting experimental data to theoretical curves to calculate the mobile fraction at infinite time has been established for about 20 years[1,17] (see ref. 5). Fluorescence is expressed in absolute terms as F(t) (fluorescence at time t after photolysis) or in fractional form f(t), given by:

$$f(t) = \{F(t)-F(0)\}/\{F(\infty)-F(0)\} \quad [2.1]$$

where F(0) and F(∞) represent fluorescence at t = 0 (immediately after photolysis) and t = ∞, respectively. For a Gaussian laser beam profile, F(t) is given by:

$$F(t) = F(-)vK^{-1}\Gamma(v)P(2K/2v) \quad [2.2]$$

where $v = (1 + 2t/t_D)^{-1}$; $t_D = w^2/4D$, the "characteristic" diffusion time, where D is the apparent lateral diffusion coefficient; G(v) is the gamma function; P(2K/2v) is the c^2-probability distribution (see refs. 1 and 5 and references therein); F(-) is fluorescence before photolysis (i.e., starting fluorescence); w is the exp(-2) beam radius; and K expresses the degree of photolysis and is determined by $F(0)/F(-) = K^{-1}(1-e^{-K})$.

The mobile fraction (f) (extent of fluorescent recovery) is given by:

$$f = \{F(\infty)-F(0)\}/\{F(-)-F(0)\} \quad [2.3]$$

where F(-) is fluorescence before photolysis (i.e., starting fluorescence).

Analysis to estimate the apparent lateral diffusion coefficient (D) can be performed in many cases[1,5] using the simplified equation:

$$D = (w^2/4t_{1/2})\gamma \quad [2.4]$$

where w is the exp(-2) beam radius, $t_{1/2}$ is the half-time of fluorescence recovery and g is a factor depending on the degree of photolysis, which has been tabulated as a function of F(0)/F(-) (see ref. 5). In the case of a uniform circle (disc) profile (the scenario illustrated in Fig. 2.6), $\gamma = 0.88$ independent of F(0)/F(∞).

Among others, Helmreich and Elson[6] have noted the weak dependence of the apparent lateral diffusion coefficient on size, since it varies inversely as the logarithm of the particle radius, as exemplified by the Saffman and Delbruck equation describing the rate of lateral diffusion of proteins in membranes:[18,19]

$$D = \frac{kT}{4\pi\mu h}\left[\log\frac{\mu h}{\mu_w a} - \gamma\right] \quad [2.5]$$

where k is Boltzmann's constant; T, absolute temperature; μ, the viscosity of the lipid bilayer; μ_w, viscosity of the surrounding aqueous solution; h, thickness of the bilayer and length of the particle (molecule whose lateral mobility is being measured); a, radius of the particle; and γ, 0.5772 (Euler's constant).

This expression also indicates that the lateral diffusion coefficients of proteins should not be much less than those of lipids (about 10^{-9} cm^2/sec),[6,20] if the lateral motions of both kinds of molecules are exclusively limited by the viscosity of the membrane bilayer.

That the high laser energies used in the technique of fluorescence photobleaching recovery do not result in photo-induced crosslinking has been shown in various studies.[21-24] Fluorescence photobleaching recovery experiments have been performed, for example, on areas of membrane that have already been photobleached, with results indicating similar mobile fractions and apparent lateral diffusion coefficients before and after a round of bleaching.[22,23] Double label experiments with antibodies enabling the mobility of a second protein in the same membrane to be subsequently measured[24] also indicate that generalized crosslinking does not occur. Since a less than 1°C increase in temperature is measurable subsequent to photobleaching,[25] thermal effects through local heating also do not appear to be a factor.[25,26] Cell permeability (and viability) is also not affected by photobleaching, as shown by trypan blue exclusion experiments.[23] At high laser energies (above 10 W/cm^2), glutathione (2-20 mM) or cystamine (50 mM) have been used to protect against oxidation resulting from the generation of singlet oxygen.[27]

E. LATERAL MOBILITY IN THE CYTOPLASM AND MEMBRANES OF LIVING CELLS

As shown in Table 2.1, the technique of fluorescence photobleaching recovery has been used to measure the lateral mobility of proteins in the cytosol of cells (see ref. 5). The important thing to note is that the rate of diffusion–the lateral diffusion coefficient– is much slower (up to 50 times) in cytosol than in dilute aqueous buffer (results shown in parentheses in Table 2.1). However, as we are about to see, these diffusion rates are about 100 times faster than those of proteins within the plane of the plasma membrane.[6,35,36] Bell,[35,36] for example, observed that while antibody diffuses at a rate of about 5×10^{-8} cm²/sec in solution, membrane-associated antibody diffused at a rate of about 10^{-10} cm²/sec on lymphocytes.

Examples of measurements of the lateral mobility of lipid probes in biological membranes are shown in Table 2.2. Lateral mobility coefficients are generally $10\text{-}100 \times 10^{-10}$ cm²/sec with a mobile fraction of about 1.0, lipid mobility thus being essentially a function of the viscosity of the lipid bilayer, consistent with the Saffman and Delbruck equation ([2.5] above).[6,20,72] Apparent lateral diffusion coefficients increase with increasing temperature, which is also true for membrane proteins (see Table 2.3 below) and molecules in artificial membranes (see Table 2.4 below). Generally speaking, lipid mobilities are constant over different specialized regions of the cell, as well as over various cell types and culture conditions, and in contrast to most membrane proteins, are not influenced by links to the cytoskeleton; lipid mobility, for example, is not affected by agents perturbing the synthesis/assembly of cytoskeletal elements such as actin filaments or microtubules.[14,41] Similarly, measurements in spherocytes, which are spectrin deficient, are essentially the same as those in erythrocytes (Table 2.2). Varying the exact nature of the probe also does not greatly affect the lateral mobility parameters (compare the results for the different lipid probes used in Table 2.2; see also Table 2.4). This is true for photobleaching experiments performed using labeled lipid or phospholipid derivatives spontaneously introduced into membranes, as well as those using fluorescently-labeled compounds such as lipopolysaccharide endotoxin[71] or cholera toxin,[70] which bind to intrinsic membrane lipid components, e.g., the latter binds specifically to the ganglioside GM1[67,70] (see Table 2.2). There are rare exceptions to the general high mobility of lipids in biological membranes; e.g., there are certain specialized membrane regions or domains within which lipids as well as proteins exhibit greatly reduced mobilities–these are discussed in chapter 3 (section F).

The measurements in Table 2.3 imply that protein mobility is not a simple function of the viscosity of the lipid bilayer (see equation [2.5] above),[6,20,72,113] the apparent lateral diffusion coefficient being much less than 10^{-9} cm²/sec. Values are generally of the order of 10^{-10} cm²/sec, about 20-50 times lower than those of membrane lipid probes (compare the values in Tables 2.2 and 2.3). This is clearly instanced by measurements of general protein mobility (see Table 2.3A) in the membranes of living cells labeled:

1. directly with fluorescent dyes such as FITC or TNBS (2,4,6-trinitrobenzene sulfonate) to give general labeling of cell surface molecules,

2. with antibodies prepared to general cell surface proteins, or

3. lectins, which are naturally occurring proteins or glycoproteins usually of plant origin that have high affinity for specific carbohydrate moieties. Since many integral membrane proteins have glycosylated extracellular domains, fluorescently-labeled lectins such as wheat germ agglutinin (WGA) or concanavalin A can be used as general probes for the lateral mobility of N-acetyl-β-D-glucosaminyl- and α-D-mannosyl-/α-D-glucosyl-residue containing membrane proteins respectively.

The mobile fractions of membrane proteins are mostly well below 1.0, again in

Table 2.1. Selected examples of intracellular lateral diffusion coefficients, as determined by the technique of fluorescence recovery after photobleaching

Molecule (Mol. Wt.)	Cell type	Temp.	Apparent Lateral Diffusion Coefficient ($\times 10^{-8}$ cm^2/sec)	Ref.
calmodulin (17 kDa)	mouse 3T3 fibroblasts	22°C	4.0 (110)*	28
lactalbumin (17 kDa)		22°C	6.9 (102)*	
dextran (17.5 kDa)	HTC rat hepatoma	20°C	6.7 (65)*	29
dextran (20 kDa)	mouse 3T3 fibroblasts	22°C	18 (101)*	28
dextran (39 kDa)		22°C	13 (70)*	
actin (43 kDa)	chicken gizzard fibroblasts	22°C	3.4	
G-actin (globular)	amoeba	25°C	92	30
F-actin (filamentous)		25°C	18+	
ovalbumin (43 kDa)	mouse macrophages	22°C	5.9	28
dextran (65.5 kDa)	mouse 3T3 fibroblasts	22°C	10 (52)*	
bovine serum albumin	human fibroblasts	22°C	1.0 (69)*	31
(67 kDa)	mouse 3T3 fibroblasts	22°C	6.8	28
	sea urchin egg	25°C	8.6	32
IgG Fab$_2$ fragment (100 kDa)	human fibroblasts	25°C	1.6 (60)*	33
tubulin (110 kDa)	sea urchin egg	25°C	7.0 (56)*	32
dextran (150 kDa)	HTC rat hepatoma	37°C	2.0^	34
dextran (157 kDa)	HTC rat hepatoma	20°C	1.6° (24)*	29
IgG (160 kDa)	mouse 3T3 fibroblasts	22°C	6.7 (46)*	28
apoferritin (467 kDa)	human fibroblasts	22°C	1.6 (36)*	33
β-galactosidase (476 kDa)	HTC rat hepatoma	37°C	0.9#	34
β-galactosidase-fusion protein (480 kDa)	HTC rat hepatoma	37°C	1.1x	34

Abbreviations: Ig, immunoglobulin; HTC, hepatoma tissue culture
*Numbers in parenthesis represent values for the apparent lateral diffusion coefficient measured in dilute buffer.[5]
+Mobile fraction of about 0.1. The cytotoxic F-actin-binding phalloidin decreases the apparent lateral mobility coefficient to 2×10^{-8} cm^2/sec, but increases the mobile fraction to 0.5.
^Mobile fraction of 0.91.[34]
°Lateral diffusion coefficient in the nucleus of 3.8×10^{-8} cm^2/sec.[29]
#Mobile fraction of 0.8.
xMobile fraction of 0.67. The apparent lateral diffusion coefficient in the nucleus was found to be 0.7×10^{-8} cm^2/sec, with a mobile fraction of 0.68.[34]

Table 2.2. Selected examples of lateral mobility of plasma membrane lipid probes, as measured by the technique of fluorescence recovery after photobleaching

Probe	Cell type	Temp.	Parameter of Lateral Mobility*		Ref.
			D $(10^{-10}\ cm^2/sec)$	f	
FI-PE	human endothelial cell (HEC)	37°C	82	> 0.80	37
	human red blood cell		20-40	0.85-0.95	38
	Epstein barr virus transformed JY B-cells	23°C	49	0.93	39
TMR-PE	K562 human erythroleukemia	21°C	43,44	0.84	40,41
	embryonic chick myotube	22°C	81	0.93	42
	rat spinal cord		72	1.07	43
	CA1 hippocampal neurons	22°C	83	0.89	44
	rat spinal cord neurons: cell body	37°C	73	0.95	45
	: processes		56	0.89	
N-rhodamyl-PE	Trypanosoma brucei	37°C	22	0.92	46
NBD-PE	Chang liver cells	23°C	57	0.87	47
	rat myocyte	22°C	49	0.6	48
	A6 toad kidney		42	0.93	49
	mouse erythrocytes		140		50
	mouse spherocytes#		150		
	mouse splenic lymphocytes		120	0.89	
NBD-PC	L6 rat embryo myoblasts	22°C	44	1.0	51
	transformed 3T3 mouse fibroblasts		35-60	0.6-0.7	52
	GM-2408A human fibroblasts		15	0.90	53
DiOC$_{14}$(3) (3,3'-ditetra-decyloxacarbocyanine iodide)	LLC-PK$_1$ pig kidney	37°C	71	0.97	54
		10°C	18	0.92	12

Table continues on next page.

Table 2.2. (continued)

Probe	Cell / tissue	Temp			Ref
dilC$_{12}$ (1,1'-didodecyl-3,3,3',3'-tetramethylindocarbocyanine iodide)	renal proximal tubule (basolateral membrane)	25°C	51	0.55	55
dilC$_{14}$ (1,1'-ditetradecyl-3,3,3',3'-tetramethylindocarbocyanine iodide)	A6 toad kidney	22°C	119	1.0	49
	acrosome-intact guinea pig sperm		89	0.84	56
	renal proximal tubule (basolateral membrane)	37°C	45	0.60	55
dilC$_{16}$ (1,1'-dihexadecyl-3,3,3',3'-tetramethylindocarbocyanine iodide)	renal proximal tubule (basolateral membrane)	25°C	47	0.60	49
		22°C	44	0.60	
dilC$_{18}$ or diI (1,1'-dioctadecyl-3,3,3',3'-tetramethylindo-carbocyanine iodide)	A6 toad kidney	37°C	102	1.0	49
	3T3 mouse fibroblasts	31°C	100	1.0	57
	chick embryo myotubes	22°C	65	0.93	58
		12°C	37	1.0	
			26	0.85	
	A431 epidermoid carcinoma cells	30°C	64	0.73	59
		25°C	55	0.73	
		15°C	35	0.73	
		5°C	13	0.73	
	rat luteal cells	25°C	104	0.63	60
		22°C	66-78	0.76-0.87	61
		15°C	54	0.66	60
		4°C	49	0.7	
	human red blood cells	25°C	82		62
	mouse red blood cells		130		
	frog red blood cells		53		
	transfected CHO cells	24°C	27	8	63
		10°C			
	rat peritoneal mast cells	23°C	80	0.73	64
	L6 rat embryo myoblasts	22°C##	90	1.0	14
	mouse spleen lymphocytes		95-110	0.91-0.95	65

Table continues on next page.

Table 2.2. (continued)

Probe	Cell type	Temp.	Parameter of Lateral Mobility*		Ref.
			D (10^{-10} cm²/sec)	f	
	NG-108-15 neuroblastoma x glioma cells		40	1.0	66
	C1300 mouse neuroblastoma	20-24°C	61@	1.0	67
5-(N)hexadecanoylamino-fluorescein (HEDAF)	Osteoblasts (F3) (fetal ratcalvaria)	34°C	194	0.90	68
	ROS rat osteosarcoma 17/2.8		164	0.79	
	Osteoblasts (F3) (fetal ratcalvaria)	23°C	50	0.84	
	ROS rat osteosarcoma 17/2.8		43	0.75	
	Madin-Darby canine kidney cells	22°C	20	0.85	69
ganglioside GM1	C1300 mouse neuroblastoma cells	20-24°C	31-62=		67
	HT29 human colonic carcinoma+	23°C	84	0.58	70
lipopolysaccharide	neuroblastoma	25°C	100	0.85-0.90	71
endotoxin	hepatocytes		40	0.60-0.75	
		10°C	18		

Abbreviations: Fl, fluorescein; PE, phosphatidyl-ethanolamine; TMR, tetramethylrhodamine; NBD, N-4-nitrobenzo-2-diazole; PC, phosphatidylcholine; CHO, Chinese hamster ovary.
*D, apparent lateral diffusion coefficient; f, mobile fraction
#Spectrin deficient.
##Pretreatment with colchicine or metabolic inhibitors had no effect on D or f.[14] Fixed cells exhibit six times lower lateral mobility.[14]
@D values varied with the stage of the cell cycle. Results are for cells in mid-S phase of the cell cycle; D was lowest (18 x 10^{-10} cm²/sec) in M phase.[67]
=D values varied with the stage of the cell cycle. Results are for cells in mid-S phase of the cell cycle.[67]
+Measurements were made with fluorescently-labeled cholera toxin.[70]

Table 2.3. Selected examples of lateral mobility of membrane proteins, as measured by the technique of fluorescence recovery after photobleaching%

Protein/probe	Cell type	Temperature	Parameter of Lateral Mobility*		Ref.
			D (10^{-10} cm^2/sec)	f	
A. General					
FITC labeling#	L6 rat embryo myoblasts	22°C	2.2	0.30-0.50	14
FITC labeling	transformed C1 mouse fibroblast		2.6	0.60	73
TNBS labeling^+	L6 rat embryo myoblasts		1.9	0.30-0.50	14
DTAF labeling	mouse erythrocytes		0.45		50
	mouse spherocytes##		25.0		
surface antigens^x=	C1300 mouse neuroblastoma cells	20-24°C	1.8		67
	mouse splenic lymphocytes	23°C	5.3	0.65	51
Lectin binding proteins concanavalin A	rat spinal cord neurons: cell body	37°C	3.5	0.22	45
	: processes		3.1	0.28	
	mouse skeletal muscle fibers		5.3-5.8	0.42-0.51	74
	L185 mouse myoblasts		7.7	0.55-0.60	
	C2 mouse myoblasts		7.6-7.7	0.60-0.61	
	rat luteal cells	27°C	1.3	0.34	49
	rat embryonic cortical neurons: cell body	22°C	4.1	0.39	61
	: axon hillock		0.4	0.13	
	: neuritic terminal		9.2	0.76	
	rat myocyte		4.5	0.26-0.38	48
wheat germ agglutinin	CA1 hippocampal neurons		3.0	0.65	44
	mouse 3T3 fibroblasts	22°C	3.9	0.45	53
	spinal cord neurons		4.5	0.45	43
	rat liver nuclear envelope		3.9	0.40-0.60	75

Table continues on next page.

Table 2.3. (continued)

B. Specific proteins

vesicular stomatitis viral glycoprotein[x]	BHK baby hamster kidney	39°C	6.3	0.73	76
	CEF chicken embryo fibroblasts	37°C	7.2	0.65	
	mouse 3T3 fibroblasts	37°C	5.4	0.68	77
	transfected cos-1 monkey kidney	24°C	0.85	0.75	78
		22°C	3.8	0.50	
Sendai virus envelope proteins					
F (fusion)[x]	red blood cells	22°C	3.1	0.62	79
HN (hemagglutinin-neuraminidase)[x]			3.3	0.60	
influenza virus hemagglutinin[x]	CV-1 monkey kidney	37°C	11.5	0.86	80
		22°C	3.0	0.73	
		10°C	1.3	0.70	
variant surface glycoprotein	Trypanosoma brucei	37°C	1.04	0.82	46
of T. brucei (mVSG)[x]		27°C	0.74	0.79	
		4°C	0.42	0.81	
alkaline phosphatase (AP)	BHK baby hamster kidney		0.66	0.56	
rat AP[x]	Osteoblasts (F3) (fetal ratcalvaria)	23°C	12.8	0.73	68
	UMR106 rat osteosarcoma		17.7	0.83	
	ROS rat osteosarcoma 17/2.8		5.8	0.73	
PLAP (human placental AP)[x]	transfected COS-1 cells	22°C	24.0	0.70	78
cytochrome c	giant mouse mitochondria	22°C	1.6	0.90	81
cytochrome b5	red blood cells	22°C	23	0.52	82
	mouse 3T3 fibroblasts		82	0.56	
rhodopsin	toad rod cell	22°C	35		83-85
	mudpuppy rod cell	22°C	37	c. 0.40-0.60	83
β-adrenergic receptor /β-antagonist	Chang liver cells[&]	23°C	14	c. 0.20	47

Table continues on next page

Table 2.3. (continues)

Receptor / ligand system	Cell type	Temp			Ref.
nicotinic acetylcholine receptor/α-bungarotoxin	chick embryo myotubes	31°C	1.0	0.50	58
α2-macroglobulin receptor@	mouse 3T3 fibroblasts	12°C	0.75	0.18	86
thyroid hormone receptor/5-triiodo-L-thyroxine**	mouse 3T3 fibroblasts	23°C	5.5	0.37	87
	mouse 3T3 fibroblasts	23°C	2.8	0.69	
red blood cell band 3 (anion transporter)	red blood cells	37°C	0.38, 0.1-0.2	0.88, 0.5-0.7	21,38
	sickle cells		0.1-0.2	0.05-0.38	38
glycine receptor/strychnine antagonist	rat spinal cord: cell body	22°C	11.5	0.48	43
	: neuronal processes		5.5	0.30	
Na^+,K^+-ATPase^x	renal proximal tubule epithelial^{xx}	25°C	3.3	0.47	55
	Madin-Darby canine kidney cells	22°C	5.0	0.50	69
Amiloride-sensitive Na^+ channel^x	A6 toad kidney	25°C	0.43, 0.99	0.12, 0.14	88,89
Voltage-dependent Na^+ channel/α-scorpion toxins	rat spinal cord neurons: cell body	37°C	11.2	0.92	45
	: processes		0.09	0.12	
	rat embryonic cortical neurons: cell body	22°C	13.5	0.77	61
	: axon hillock		1.9	0.42	
	: neuritic terminal		2.1	0.20	
	dorsal root ganglion neurons	22°C	0.50-1.3	0.20	90
Voltage-dependent Ca^+ channel/ω-conotoxin	CA1 hippocampal neurons: cell body	22°C	1.1	0.30	44
	: dendrites		8.6	< 0.15	
GABA/benzodiazepine receptor	rat spinal cord neurons: cell body	37°C	2.1, 1.9	0.3, 0.27	45
	: processes		0.45, 0.30	0.19, 0.15	
	chicken cortical neurons: cell body		3.3	0.68	
	: processes		0.90	0.23	
asialoglycoprotein receptor H1 polypeptide^x	transfected NIH 3T3 mouse fibroblasts	12°C	1.0	0.66	91
	Hep G2 human hepatoma		1.0	0.48	

Table continues on next page.

H2 polypeptide[x]	transfected NIH 3T3 mouse fibroblasts		1.1	0.56	
			1.1	0.47	
LDL-receptor/	Hep G2 human hepatoma				
dil(3)-LDL	GM-2408A human fibroblasts[++]	28°C	0.45	0.60	54
		22°C	0.14	0.75	
		10°C	0.05-0.3	0.20	
Transferrin receptor	normal human fibroblasts		0.05-0.2	0.20	
	K562 human erythroleukemia%%	37°C	10.3	0.58	41
Thy-1	AKR1/G1 T-cell lymphoma cells[x]	21°C	5.5,5.0	0.56,0.44	40,41
	AKR1/G1M1 (Thy-1⁻) cells[x$]	23°C	20	0.45	92
	Chicken embryo fibroblasts[x$]		21	0.46	
	COS-1 cells (transfected)[x]		16	0.49	
	COS-1 cells[x$]		27	0.55	
	AKR1/G1M1 (Thy-1⁻) cells[$]		19	0.45	
			32	0.52	
CD2[x]	transfected CHO cells	22°C	1.8	0.86	93
	Jurkat T-cell leukemia		7.2	0.70	94
CD4[x]	recombinant virus infected Sf9 cells	22°C	4.7	0.80	95
	CEM human T-lymphoblasts[^^^]		5.8	0.61	96
CD8 α chain (Ly-2)[x]	transfected T-cell hybridoma	20°C	0.69	0.83	97
major histocompatibility complex (MHC)/human leukocyte antigen (HLA) antigens					
class I MHC proteins[x]	human endothelial cell (HEC)	37°C	c. 30	> 0.80	37
	neonatal human skin fibroblasts (HSF)°	30°C	8.8	0.29	98
		19°C	7.5	0.29	
	Epstein Barr virus transformed JY B-cells	23°C	5.0	0.64	39
	VA-2 transformed human fibroblasts°	19°C	4.9	0.25	98
	WI-38 human lung fibroblast°		4.7	0.24	
	cl Id transformed mouse fibroblasts°		3.4	0.22	
	transfected HEPA-OVA mouse hepatoma[x]	22°C	13	0.43	99
class II MHC protein[x]	Epstein Barr virus transformed JY B-cells	23°C	1.2	0.58	39
HLA class I[x]	HUT-102-B2 human T lymphocytes	30°C	1.8	0.31	100
β2-microglobulin	MDCK Madine Darby canine kidney	23°C	13	0.72	101

Table continues on next page/

Table 2.3. (continued)

Protein	Cell type	Temp	D	R	Ref.
(class I HLA)	HT29 human colon carcinoma[x]		12	0.46	70
HLA-DR (class 2 HLA)[x]			20	0.79	
gp96[x]	human endothelial cell (HEC)	37°C	c. 14	> 0.80	37
IgG Fc receptors (low affinity)					
FcγRIIA (CD32) transfected CHO cells		22°C	16	0.79	102
	transfected mouse macrophage-like P388D1 cells		14	0.54	103
FcγRIIIB[$$] transfected mouse 3T3 fibroblast			34	0.73	
High-affinity IgE receptor					
Fcε RI (IgE monomer)	rat peritoneal mast cells[oo]	22°C	2.1	0.50–0.80	64
	rat basophilic leukemia (RBL-2H3) cells	25°C	2.0	0.85	104
(IgE dimer)			c. 2.0	0.57	
(IgE monomer)		22°C	4.2	0.72	105,106
		10°C	1.3		105
(IgE dimer)		22°C	3.6	0.42	
(IgE ≥ trimer)			6.0	0.30	
(IgE monomer)	transfected COS-7 cells	19°C	2.7	0.59	107
(IgE oligomer)			4.8	0.33	
(IgE monomer)			1.2	0.64	
(IgE oligomer)			1.6	0.34	
SCR complex (IgA dimer)	HT29 human colonic carcinoma	23°C	7.7	0.68	108
(receptor for secretory component)					
Lys6E[x]	tranfected COS-1 cells	22°C	28	0.67	76
PH-20 protein[x]	guinea pig acrosome-intact sperm	22°C	1.8	0.73	56
GP80[x]	mouse fibroblasts	22°C	1.2	0.70	109
fibronectin receptor[x]	avian embryonic neural crest cells (motile)[§]	37°C	2.8	0.84	110
		25°C	2.0	0.66	
		15°C	2.1	0.54	

Table continues on next page

Table 2.3. (continued)

		D	f	ref
avian embryonic somitic fibroblasts (motile)$$	25°C	3.1	0.66	
avian embryonic heart fibroblasts (motile)@@		2.1	0.73	
avian embryonic melanoblasts (motile)		3.3	0.84	
avian embryonic dermal fibroblasts (stationary)		3.4	0.13	
CR1 (C1 complement receptor) transfected CHO cells[x]	37°C	12	0.75	111
	14°C	14	0.75	
glycophorin A (FTSC label) red blood cells	37°C	0.2-0.5	0.40-0.60	
sickle cells	37°C	0.2-0.5	0.18-0.36	38
K562 human erythroleukemia	21°C	4.5,4.7	0.50,0.61	40,41
glycophorin C[x] red blood cells	24°C	0.16	0.44	112

Abbreviations: FITC, fluorescein isothiocyanate; TNBS, 2,4,6-trinitrobenzene sulfonate; DTAF, dichlorotriazinylaminofluorescein; GABA, γ-amino butyric acid; CHO, Chinese hamster ovary; LDL, low density lipoprotein; IgG, immunoglobulin; FTSC, fluorescein thiosemicarbazide.

%Results specifically for polypeptide hormone receptors and trimeric GTP-binding proteins are presented in Tables 4.1 and 4.2, respectively.

*D, apparent lateral diffusion coefficient; f, mobile fraction; #Fixed cells exhibited zero lateral movement at 22°C;[14,22] ^Cytochalasin B pretreatment reduced mobility at 22°C (D of 0.2-0.3);[14] +Measurements 2h after labelling;[72] ##Spectrin-deficient;[50] xMeasurements performed using labeled antibodies; =Measurements performed using antibodies prepared to general cell surface antigens;[67] &30 second labeling with antagonist prior to measurement. 30 minute pretreatment with the agonist propanolol increased both D (c. 3.5×10^{-9} cm^2/sec) and f (c. 0.75);[47] @Measurements after 30 min,[86] **Measurements in the presence of methylamine to prevent receptor clustering prior to endocytosis;[87] xCytochalasin D pretreatment increased the apparent lateral diffusion coefficient seven-fold, as well as retargeting the protein from the apical to the basolateral membrane;[55] ++Internalization defective mutant; %%The mobile fraction fell from 0.58 to < 0.1 in 10 min at 37°C and 0.56 to 0.21 in 30 min at 21°C due to internalization;[41] $Thy-1 spontaneously incorporated into T-cell membranes; ^^The envelope glycoprotein (gp120) of HIV-1 was used as a probe;[96] °D and f values varied with confluence (increased mobility at sparse cell density). Results are for confluent cells;[98] $$Coexpression of the integrin C3b complement receptor (CR3 or CD11b/CD18 or Mac-1) reduces mobility (D = 25×10^{-10} cm^2/sec and f = 0.48);[103] °°Cytochalasin B (but not colchicine) treatment reduces mobility (D < less than 0.34×10^{-10} cm^2/sec and f = 0.2–0.5);[22] §The mobile fraction for stationary cells is 0.16, 0.18 and 0.25 at 37°C, 25°C, and 15°C, respectively;[110] §§The mobile fraction for stationary cells is 0.18;[110] @@The mobile fraction for stationary cells is 0.14.[110]

Table 2.4. Selected examples of lateral mobility of lipid probes and proteins in reconstituted phospholipid model membranes and isolated membranes, as measured by the technique of fluorescence recovery after photobleaching

| Molecule | Membranes^ | Temperature | Parameter of Lateral Mobility* | | |
			D (10⁻⁸ cm²/sec)	f	Ref
A. Artificial membranes					
Proteins					
gramicidin C	DMPC	40°C	6.0		114
		30°C	4.0	1.0	115
		25°C	3.0		114
	EPC	35°C	5.0		
		5°C	2.0		
bacteriorhodopsin	DMPC	32°C	3.4		116
		28.5°C	2.8		
		24.5°C	1.4		
rhodopsin	DMPC⁺	36°C	3.3		117
		24°C	1.6		
acetylcholine receptor	DMPC	37°C	2.3		118
(monomer)		36°C	2.4		117
		27°C	1.6		
		25°C	1.0		118
	DMPC/CHS	36°C	2.0		
		14°C	0.6		
	SBL	37°C	3.3		117
		15°C	1.5		
acetylcholine receptor	SBL	37°C	3.2		118
(dimer)		15°C	1.6		
ATPase (sarcoplasmic	SRTL#	36°C	1.8		117
reticulum)		13°C	1.0		
glycophorin	DMPC	30°C	4.0	1.0	119
		24°C	2.0		120
fibronectin (CSP)	DMPC	22°C	2.9		121
surface immunoglobulin	PC/cholesterol/	22°C	1.4	1.0	122
(anti-DNP IgA)	cardiolipin (2:2:1)				
LFA-3 (GPI-anchored form)	EPC	22°C	0.23	0.73	123
LFA-3 (transmembrane form)			0	0	
Lipids					
NBD-PE	DMPC	38°C	9.3		124
		24°C	6.3		
	EPC	38°C	10.1		
		14°C	4.4		
	EPC	22°C	1.4	0.94	123
NBD-DMPE	DMPC	36°C	8.8		117
		30°C	7.0		118
		25°C	3.7		117

Table continues on next page.

Table 2.4. (continued)

	SBL	37°C	11.0		117,118
		15°C	3.4		
	SRTL	37°C	14.0		117
		10°C	3.5		
DiO-C$_{18}$(3)	DMPC	38°C	9.2		124
		32°C	6.9		116
		24.5°C	5.1		
		24°C	6.7		124
	EPC	38°C	8.9		
		14°C	5.9		
	brain PS	38°C	7.8		
		16°C	3.7		
DiI-C$_{18}$(3)	DMPC	38°C	11.9		124
		24°C	8.4		
	EPC	38°C	10.9		
		14°C	6.9		
	brain PS	38°C	9.8		
		16°C	3.5		
B. Isolated membranes					
Proteins					
General/DTAF labelling	rat reticulocyte	25°C	0.0074	0.63	125
red blood cell band 3	red cell ghosts	26°C	0.53	0.42	126
(anionic transporter)		24°C	0.005		127
	red cell ghosts (spectrin deficient)		0.45		
	red cell ghosts[x]	37°C	0.65	0.83	128
	red cell ghosts	37°C	0.001-0.01		129
	red cell ghosts[x]	30°C	0.14	0.60	128
		21°C	0.067	0.34	
cytochrome c	red cell ghosts	22°C	0.28	0.63	81
Lipids					
NBD-PE	red cell ghosts	45°C	1.90	0.90	130
		25°C	0.50	0.90	
		24°C	1.40		50
dil-C$_{18}$(3)	red cell ghosts	40°C	0.75		131
	red cell ghosts	7°C	0.09		

*D, apparent lateral diffusion coefficient; f, mobile fraction; ^Abbreviations: DMPC, dimyristoylphosphatidylcholine; EPC, egg phosphatidylcholine; DMPC/CHS, DMPC with 45 % cholesterylhemisuccinate; SBL, soybean lipids; SRTL, sarcoplasmic reticulum total lipids; CSP, cell surface protein; DNP, dinitrophenol; IgA, immunoglobulin A; LFA-3, lymphocyte function-associated antigen-3; GPI, glycosyl phosphoinositol; NBD-PE, nitrobenzo-2-diazole phosphatidylethanolamine; NBD-DMPE, nitrobenzo-2-diazole dimyristoylphosphatidylethanolamine; DiO-C$_{18}$(3), 3,3'-dioctadecyloxacarbocyanine iodide; PS, phosphatidylserine; dil-C$_{18}$(3), 1,1'-dioctadecyl-3,3,3,3'-tetramethylindocarbocyanine iodide; DTAF, dichlorotriazinylaminofluorescein.
+Lipid to protein ratio of 3000/1; #Lipid to protein ratio of 5000/1; xIonic strength of 13.3 mM (sodium phosphate, pH 7.4).[127]

contrast to lipid probes, which generally are completely mobile (mobile fraction of about 1.0). The clear implication from the measurements of general membrane protein mobility presented in Table 2.3A is that protein movement is restricted in biological membranes. There do appear to be exceptions to this, however; the photoreceptor rhodopsin, for example, appears to be highly mobile (apparent lateral diffusion coefficient of about 3.5×10^{-9} cm^2/sec) in the rod disc membrane of several species.[83-85]

In contrast to lipid probes, the lateral mobility of particular membrane proteins, as illustrated by the examples in Table 2.3B, varies markedly in terms of both the apparent lateral mobility coefficient and mobile fraction according to:

i. different specialized regions of the membrane of a single cell. This is exemplified by the results in Table 2.3B for the lateral mobility of proteins such as the voltage-dependent Na$^+$ or Ca^{2+} channels or GABA (γ-amino butyric acid)/benzodiazepine receptor measured in either the cell processes or cell bodies of neuronal cells (compare also to results in Table 2.2).

ii. different growth stages and growth conditions, e.g., motile and stationary cells exhibit different fibronectin receptor mobile fractions (see Table 2.3B and legend, and chapter 3).[110]

iii. different cell types. This is instanced by comparing the lateral mobility properties of proteins in sickle cell red blood cells and normal red blood cells,[38] where the anionic transporter band 3 (Table 2.3), in contrast to lipid probes (Table 2.2), shows significantly different mobilities (see ref. 38). Analogously, class I major histocompatability (MHC) antigens show a range of values for the apparent lateral diffusion coefficient (from 3.4 to 30×10^{-10} cm^2/sec) and mobile fraction (from 0.22 to 0.8) in different cell types. As already mentioned, protein mobility in spherocytes (which have defective cytoskeletons) is more than 50-fold higher than in normal erythrocytes (see Table 2.3), whereas lipid probes are unaffected (see Table 2.2).[50]

A further difference between lipid and protein mobilities is that while different lipid probes have very similar lateral mobility properties, it is clear that those of specific proteins vary greatly; e.g., compare the mobility parameters of cytochrome b5, Fcγ RIIIB and α-2 macroglobulin receptors measured in mouse 3T3 fibroblasts, or Sendai virus F and N proteins, cytochrome b5 and glycophorin C measured in red blood cells (Table 2.3). Lateral mobility properties seem to be distinct to or specific for a particular protein. In addition, multiple factors, including hormonal stimulation, differentiation, aging, stages of the cell cycle, cell motility, etc. all appear to affect the lateral mobility of specific membrane proteins.[113] The various factors influencing protein lateral mobility will be discussed in detail in the next chapter.

F. MEASUREMENTS IN ARTIFICIAL AND ISOLATED MEMBRANES

As can be seen from the examples in Table 2.4A, protein lateral mobility measured in biological membranes is markedly lower than that measured in either isolated, or artificial or "model" reconstituted membranes. Whereas isolated membranes are those prepared from living cells through cell fractionation procedures, and contain both integral and peripheral membrane proteins, artificial membranes are reconstituted from purified lipid/phospholipid fractions from sources such as soy bean and/or synthetic lipids, containing essentially no membrane proteins. This provides the possibility to examine the mobility of lipid or protein molecules reconstituted into the bilayer in complete isolation from other membrane proteins and with controlled lipid to protein ratios. Diffusion rates in artificial or isolated membranes can be greater than 10^{-8} cm^2/sec for a range of proteins, including cytochrome P-450,[132] glycophorin, rhodopsin,[116] the nicotinic acetylcholine receptor,[117,11] bacteriorhodopsin[116] and the M-13 phage coat protein (see refs. 6,133). Those of lipid probes are about 2-3 times higher, consistent with theoretical predictions based on viscosity parameters (see refs. 6, 20, 72). The diffusion rates are lower, however, than those in the aqueous cytosolic phase (see Table 2.1). High protein concentrations have been shown to reduce lipid mobility in artificial phospholipid membranes (see also chapter 3, section A; Table 3.1).[72,113,116,133,134] Generally speaking, measurements of protein or lipid probes do not appear to vary significantly between different reconstituted membranes, e.g., either DMPC (dimyristoylphosphatidylcholine) or EPC (egg phosphatidylcholine) membranes yield essentially the same results for gramicidin C and NBD-PE (nitrobenzo-2-diazole phosphatidylethanolamine) (see Table 2.4A). Through their composition (e.g., cholesterol content), however, differences do exist in the phase transition properties of various artificial membranes (see refs. 114-118).

Measurements in isolated membranes (see Table 2.4B) indicate about 10-fold lower rates of lipid lateral diffusion than those in artificial membranes (compare the results for the NBD-PE–nitrobenzo-2-diazole phosphatidylethanolamine–lipid probe), with protein lateral movement generally even further reduced in isolated membranes relative to that in artificial membranes. One reason for this is the fact that increasing the protein concentration increases the membrane viscosity (see Table 3.1) and accordingly reduces the apparent lateral diffusion coefficient of both proteins and lipids. The second reason relates to the lack of membrane skeleton in artificial membranes (see chapter 3).[6,113,127,128,133] Several studies have compared the mobility of proteins, and in particular red blood cell band 3, on red blood cell ghosts directly to that on red blood cells,[135,136] (compare the results in Tables 2.3 and 2.4B), confirming that mobility in isolated membranes generally exceeds that in intact cells.[137] However, it should be noted that results with isolated membranes appear to be quite variable (see Table 2.4B). In the light of the above, it is

clear that although lateral mobility measurements in artificial membranes are useful in terms of testing parameters affecting particular protein lateral mobilities (e.g., the role of the type of membrane linkage–transmembrane regions or glycosyl phosphoinositol anchors–with respect to the lateral mobility of proteins, such as the lymphocyte-function-associated antigen LFA-3—see also chapter 3, section E), they are clearly questionable with respect to relevance to the in vivo situation. This means that functional studies, e.g., of adenylate cyclase activation, etc., using isolated or reconstituted membranes have limited usefulness for exactly the same reason.

G. MEASUREMENTS OF LATERAL MOBILITY USING OTHER METHODS

It is not within the scope of this book to deal with alternative methods of measuring plasma membrane protein and lipid lateral mobility in any depth, but it should be mentioned that there are several other measurement techniques which have been used and that the results obtained using them are essentially the same as those derived using fluorescence photobleaching recovery. Wey et al,[138] for example, used the technique of adsorption microphotolysis to measure the rate of lateral movement of rhodopsin (3-5 x 10^{-9} cm^2/sec), results concurring with those of Poo and Cone and others using fluorescence photobleaching recovery.[83-85] Schlessinger et al[14] derived essentially the same apparent lateral diffusion coefficient for a lipid probe using fluorescence photobleaching recovery and fluorescence spectroscopy measurement methods. Other techniques used to measure plasma membrane protein lateral mobility include those of in situ electromigration and post-field relaxation,[139-143] macromolecular-tagging/ single particle tracking used in combination with fluorescence nanovid and other microscopic techniques[144-146] as well as other approaches.[147]

All measurements provide estimates of the rate of lateral movement of membrane proteins, such as the nicotinic acetylcholine receptor of *Xenopus* embryonic muscle fibers,[143] the high affinity IgE receptor Fcε RI of rat basophilic leukemia cells,[140] the low density lipoprotein-receptor of human fibroblasts,[141] the epidermal growth factor receptor of human epidermoid carcinoma cells,[142] the transferrin receptor of cultured rat kidney fibroblasts,[145] the MHC molecules H-2D^b and Qa2,[146] and the Thy-1 antigen[144] of between 10^{-10} and 10^{-9} cm^2/sec. These values are completely consistent with equivalent measurements using fluorescence photobleaching recovery and/or typical of restricted protein movement in biological membranes. The clear implication is that the results for measurements of protein lateral mobility using the technique of fluorescence photobleaching recovery can be relied upon as having been validated not only by appropriate controls (such as those of fixed cells, etc.; see above) but by concurring measurements using other methods where possible. Results for lipid probe measurements using other techniques also concur with those obtained using fluorescence photobleaching recovery (e.g. see ref. 144).

H. SUMMARY

We have seen that the lateral movement of molecules in biological membranes is at least two orders of magnitude slower than that in cytosol. Protein lateral mobility is much slower than that of membrane lipids and not simply a function of the viscosity of the lipid bilayer, implying that proteins are normally limited in their movement. Lateral mobility properties vary from protein to protein, implying that the factors responsible for the restriction of protein lateral movement, which will be discussed in detail in the next chapter, are specific to particular proteins. The fact that protein movement is restricted means that the lateral diffusion of proteins within the membrane lipid bilayer is potentially rate-limiting in terms of signal transduction at the level of the membrane. Subsequent chapters will expound the evidence supporting this postulate, which constitutes the foundation of the Mobile Receptor Hypothesis.

REFERENCES

1. Axelrod D, Koppel DE, Schlessinger J et al. Mobility measurement by analysis of fluorescence photobleaching recovery kinetics. Biophys J 1976; 16:1055-1069.
2. Elson EL, Schlessinger J, Koppel DE et al. Measurement of lateral transport on cell surfaces. Prog Clin Biol Res 1976; 9:137-147.
3. Schlessinger J, Elson EL. Fluorescence methods for studying membrane dynamics. In: Ehrenstein G, Lecar H, ed. Methods of Experimental Physics. New York: Academic Press, 1982:197-227.
4. Koppel DE. Fluorescence photobleaching as a probe of translational and rotational motions. In: Shaafi RI, Fernandez SM, ed. Fast Methods in Physical Biochemistry and Cell Biology. Amsterdam: Elsevier-North Holland, 1983:339-367.
5. Peters R. Fluorescence microphotolysis to measure nucleocytoplasmic transport and intracellular mobility. Biochim Biophys Acta 1986; 864:305-359.
6. Helmreich EJM, Elson EL. Protein and lipid mobility. Adv in Cyclic Nucleotide and Prot Phosphor Res 1984; 18:1-62.
7. Jans DA. The mobile receptor hypothesis revisited: a mechanistic role for hormone receptor lateral mobility in signal transduction. Biochim Biophys Acta 1992; 1113:271-276.
8. Jans DA, Pavo I. A mechanistic role for polypeptide hormone receptor lateral mobility in signal transduction. Amino Acids 1995; 9:93-109.
9. Wang Y-L, Taylor DL, eds. Fluorescence Microscopy of Living Cells in Culture Part A. Fluorescent Analogs, Labeling Cells, and Basic Microscopy. Methods in Cell Biology Volume 29. San Diego, New York, Boston, London, Sydney, Tokyo, Toronto. Academic Press, 1989.
10. Wang Y-L, Taylor DL, eds. Fluorescence Microscopy of Living Cells in Culture Part B. Quantitative Fluorescence Microscopy–Imagaing and Spectroscopy. Methods in Cell Biology Volume 30. San Diego, New York, Boston, London, Sydney, Tokyo, Toronto. Academic Press, 1989.
11. Matsumoto B, ed. Cell Biological Applications of Confocal Microscopy. Methods in Cell Biology Volume 38. San Di-

ego, New York, Boston, London, Sydney, Tokyo, Toronto. Academic Press, 1993.

12. Jans DA, Peters R, Zsigo J et al. The adenylate cyclase-coupled vasopressin V_2-receptor is highly laterally mobile in membranes of LLC-PK$_1$ renal epithelial cells at physiological temperature. EMBO J 1989; 8(9):2431-2438.

13. Jans DA, Peters R, Fahrenholz F. Lateral mobility of the phospholipase-C-activating vasopressin V_1-type receptor in A7r5 smooth muscle cells: a comparison with the adenylate cyclase-coupled V_2-receptor. EMBO J 1990; 9(9):2693-2699.

14. Schlessinger J, Axelrod D, Koppel DE et al. Lateral transport of a lipid probe and labeled proteins on a cell membrane. Science 1977; 195:307-309.

15. Jans DA, Resink TJ, Wilson E-L et al. Isolation of a mutant LLC-PK$_1$ cell line defective in hormonal responsiveness: a pleiotropic lesion affecting receptor function. Eur J Biochem 1986; 160:407-412.

16. Luzius H, Jans DA, Jans P et al. Isolation and genetic characterization of a renal epithelial cell mutant defective in vasopressin V_2-type receptor binding and function. Exper Cell Res 1991; 195:478-484.

17. Yguerabide J, Schmidt JA, Yguerabide EE. Lateral mobility in membranes as detected by fluorescence recovery after photobleaching. Biophys J 1982; 40(1):69-75.

18. Saffman PG, Delbrueck M. Brownian motion in biological membranes. Proc Natl Acad Sci USA 1975; 72:3111-3113.

19. Saffman PG. Lateral and rotational movement in membranes. Fluid Mech 1976; 73:593-602.

20. Webb WW, Barak LS, Tank DW et al. Molecular mobility on the cell surface. Biochem Soc Symp 1981(46):191-205.

21. Koppel DE, Sheetz MP. Fluorescence photobleaching does not alter the lateral mobility of erythrocyte membrane glycoproteins. Nature 1981; 293(5828):159-161.

22. Schlessinger J, Koppel DE, Axelrod D et al. Lateral transport on cell membranes: mobility of concanavalin A receptors on myoblasts. Proc Natl Acad Sci USA 1976; 73:2409-2413.

23. Jacobson K, Hou Y, Wojcieszyn J. Evidence for lack of damage during photobleaching measurements of the lateral mobility of cell surface components. Biochim Biophys Acta 1978; 433:215-222.

24. Wolf DE, Edidin M, Dragsten PR. Effect of bleaching light on measurements of lateral diffusion in cell membranes by the fluorescence photobleaching recovery method. Proc Natl Acad Sci USA 1980; 77:2043-2045.

25. Axelrod D. Cell surface heating during fluorescence photobleaching recovery experiments. Biophys J 1977; 18:129-131.

26. Wu E-S, Jacobson K, Szoka F et al. Lateral diffusion of a hydrophobic peptide, N-4-nitrobenz-2-oxa-1,3-diazole gramicidin S, in phospholipid multibilayers. Biochemistry 1977; 17:5543-5550.

27. Sheetz MP, Koppel DE. Membrane damage caused by irradiation of fluorescent concanavalin A. Proc Natl Acad Sci USA 1979; 76:3314-3317.

28. Luby-Phelps K, Lanni F, Taylor DL. Behavior of a fluorescent analogue of calmodulin in living 3T3 cells. J Cell Biol 1985; 101(4):1245-1256.

29. Lang I, Scholz M, Peters R. Molecular mobility and nucleocytoplasmic flux in hepatoma cells. J Cell Biol 1986; 102(4):1183-1190.

30. Wang YL, Lanni F, McNeil PL et al. Mobility of cytoplasmic and membrane-associated actin in living cells. Proc Natl Acad Sci USA 1982; 70:4660-4664.

31. Wojcieszyn JW, Schlegel RA, Wu ES et al. Studies on the mechanism of polyethylene glycol-mediated cell fusion using fluorescent membrane and cytoplasmic probes. Proc Natl Acad Sci USA 1981; 78(7):4407-4410.

32. Salmon ED, Saxton WM, Leslie RJ et al. Diffusion coefficient of fluorescein-labeled tubulin in the cytoplasm of embryonic cells of a sea urchin: video image analysis of fluorescence redistribution after photobleaching. J Cell Biol 1984; 99(6):2157-2164.

33. Jacobson K, Wojcieszyn J. The translational mobility of substances within the cytoplasmic matrix. Proc Natl Acad Sci USA 1984; 81(21):6747-6751.

34. Rihs H-P, Peters R. Nuclear transport kinetics depend on phosphorylation-site-containing sequences flanking the karyophilic signal of the Simian virus 40 T-antigen. EMBO J 1989; 8:1479-1484.

35. Bell GI. Models for the specific adhesion of cells to cells: a theoretical framework for adhesion mediated by reversible bonds between cell surface molecules. Science 1978; 200:618-627.

36. Bell GI. Theoretical models for cells-cell interactions in immune responses. Dev Cell Biol 1979; 4:371-392.

37. Stolpen AH, Golan DE, Pober JS. Tumor necrosis factor and immune interferon act in concert to slow the lateral diffusion of proteins and lipids in human endothelial cell membranes. J Cell Biol 1988; 107:781-789.

38. Corbett JD, Golan DE. Band 3 and glycophorin are progressively aggregated in density-fractionated sickle and normal red blood cells. Evidence from rotational and lateral mobility studies. J Clin Invest 1993; 91(1):208-217.

39. Bierer BE, Golan DE, Brown CS et al. A monoclonal antibody to LFA-3, the CD2 ligand, specifically immobilizes major histocompatibility complex proteins. Eur J Immunol 1989; 19(4):661-665.

40. Thatte HS, Bridges KR, Golan DE. ATP depletion causes translational immobilization of cell surface transferrin receptors in K562 cells. J Cell Physiol 1996; 166:446-452.

41. Thatte HS, Bridges KR, Golan DE. Microtubule inhibitors differentially affect translational movement, cell surface expression, and endocytosis of transferrin receptors in K562 cells. J Cell Physiol 1994; 160:345-357.

42. Angelides KJ. Fluorescently labelled Na+ channels are localized and immobilized to synapses of innervated muscle fibres. Nature 1986; 321(6065):63-66.

43. Srinivasan Y, Guzikowski AP, Haugland RP et al. Distribution and lateral mobility of glycine receptors on cultured spinal cord neurons. J Neurosc 1990; 10:985-995.

44. Jones OT, Kunze DL, Angelides KJ. Localization and mobility of ω-conotoxin-sensitive Ca^{2+} channels in hippocampal CAI neurons. Science 1989; 244:1189-1193.

45. Velazquez JL, Thompson CL, Barnes EM Jr et al. Distribution and lateral mobility of GABA/benzodiazepine receptors on nerve cells. J Neurosci 1989; 9(6):2163-2169.

46. Bulow R, Overath P, Davoust J. Rapid lateral diffusion of the variant surface glycoprotein in the coat of Trypanosoma brucei. Biochemistry 1988; 27(7):2384-2388.

47. Henis YI, Hekman M, Elson EL et al. Lateral diffusion of β-receptors in membranes of cultured liver cells. Proc Natl Acad Sci USA 1982; 79:2907-2911.

48. Yechiel E, Barenholz Y, Henis YI. Lateral mobility and organization of phospholipids and proteins in rat myocyte membranes. Effects of aging and manipulation of lipid composition. J Biol Chem 1985; 260:9132-9136.

49. Dragsten PR, Blumenthal R, Handler JS. Membrane asymmetry in epithelia: is the tight junction a barrier to diffusion in the plasma membrane? Nature 1981; 294:718-722.

50. Koppel DE, Sheetz MP, Schindler M. Matrix control of protein diffusion in biological membranes. Proc Natl Acad Sci USA 1980; 78(6):3576-3580.

51. Henis YI, Elson EL. Inhibition of the mobility of mouse lymphocyte surface immunoglobulins by locally bound concanavalin A. Proc Natl Acad Sci USA 1981; 78(2):1072-1076.

52. Tank DW, Wu ES, Webb WW. Enhanced molecular diffusibility in muscle membrane blebs: release of lateral constraints. J Cell Biol 1982; 92(1):207-212.

53. Swaisgood M, Schindler M. Lateral diffusion of lectin receptors in fibroblast membranes as a function of cell shape. Exp Cell Res 1989; 180(2):515-528.

54. Barak LS, Webb WW. Diffusion of low density lipoprotein-receptor complex on human fibroblasts. J Cell Biol 1982; 95:846-852.

55. Paller MS. Lateral mobility of Na,K-ATPase and membrane lipids in renal cells. Importance of cytoskeletal integrity. J Membr Biol 1994; 142(1):127-135.

56. Cowan AE, Myles DG, Koppel DE. Lateral diffusion of the PH-20 protein on guinea pig sperm: evidence that barriers to diffusion maintain plasma membrane domains in mammalian sperm. J Cell Biol 1987; 104:917-923.

57. Schlessinger J, Schechter Y, Cuatrecasas P et al. Quantitative determination of the lateral diffusion coefficients of the hor-

mone-receptor complexes of insulin and epidermal growth factor on the plasma membrane of cultured fibroblasts. Proc Natl Acad Sci USA 1978; 75:5353-5357.

58. Axelrod D, Wight A, Webb W et al. Influence of membrane lipids on acetylcholine receptor and lipid probe diffusion in cultured myotube membrane. Biochemistry 1978; 17(17):3604-3609.

59. Hillman GM, Schlessinger J. The lateral diffusion of epidermal growth factor complexed to its surface receptors does not account for the termal sensitivity of patch formation and endocytosis. Biochemistry 1982; 21:1667-1672.

60. Roess DA, Rahman NA, Kenny N. Molecular dynamics of luteinizing hormone receptors on rat luteal cells. Biochim Biophys Acta 1992; 1137:309-316.

61. Angelides KJ, Elmer LW, Loftus D et al. Distribution and lateral mobility of voltage-dependent sodium channels in neurons. J Cell Biol 1988; 106(6):1911-1925.

62. Bloom JA, Webb WW. Lipid diffusibility in the intact erythrocyte membrane. Biophys J 1983; 42(3):295-305.

63. Roettger BF, Rentsch RU, Hadac EM et al. Insulation of a G protein-coupled receptor on the plasmalemmal surface of the pancreatic acinar cell. J Cell Biol 1995; 130:579-590.

64. Schlessinger J, Webb WW, Elson EL. Lateral motion and valence of Fc receptors on rat peritoneal mast cells. Nature 1976; 264:550-552.

65. Dragsten PR, Henkart P, Blumenthal R et al. Lateral diffusion of surface immunoglobulin, Thy-1 antigen, and a lipid probe in lymphocyte plasma membranes. Proc Natl Acad Sci USA 1979; 76:5163-5167.

66. Kwon G, Axelrod D, Neubig RR. Lateral mobility of tetramethylrhodamine (TMR) labelled G protein α and $\beta\gamma$ subunits in NG108-15 cells. Cellular Signalling 1994; 6(6):663-679.

67. de Laat SW, Van der Saag PT, Elson EL et al. Lateral diffusion of membrane lipids and proteins during the cell cycle of neuroblastoma cells. Proc Natl Acad Sci USA 1980; 77:1526-1528.

68. Noda M, Yoon K, Rodan GA et al. High lateral mobility of endogenous and transfected alkaline phosphatase: a phos-

phatidylinositol-anchored membrane protein. J Cell Biol 1987; 105(4):1671-1677.

69. Jesaitis AJ, Yguerabide J. The lateral mobility of the (Na^+,K^+)-dependent ATPase in Madin-Darby canine kidney cells. J Cell Biol 1986; 102:1256-1263.

70. Magnusson KE, Gustafsson M, Holmgren K et al. Small intestinal differentiation in human colon carcinoma HT29 cells has distinct effects on the lateral diffusion of lipid (ganglioside GM1) and proteins (HLA class 1, HLA class 2, and neoplastic epithelial antigens) in the apical cell membrane. J Cell Physiol 1990; 143(2):381-390.

71. Kilpatrick-Smith L, Maniara G, Vanderkooi JM et al. Cellular effects of endotoxin in vitro: mobility of endotoxin in the plasma membrane of hepatocytes and neuroblastoma cells. Biochim Biophys Acta 1985; 847(2):177-184.

72. Vaz W, Goodsaid-Zalduondo F, Jacobson K. Lateral diffusion of lipids and proteins in bilayer membranes. FEBS Lett 1984; 174:199-207.

73. Edidin M, Zagyansky Y, Lardner TY. Measurement of membrane protein lateral diffusion in single cells. Science 1976; 191:466-468.

74. Zs Nagy I, Zhang X, Kitani K et al. The influence of dystrophin on lateral diffusion of proteins in sarcolemma of L-185 and C2 myoblasts and mature striated muscle cells of rats and mice, as measured by FRAP technique. Biochem Biophys Res Commun 1995; 215(1):67-74.

75. Schindler M, Holland JF, Hogan M. Lateral diffusion in nuclear membranes. J Cell Biol 1985; 100(5):1408-1414.

76. Reidler JA, Keller PM, Elson EL et al. A fluorescence photobleaching study of vesicular stomatitis virus infected BHK cells. Modulation of G protein mobility by M protein. Biochemistry 1981; 20(5):1345-1349.

77. Scullion BF, Hou Y, Puddington L et al. Effects of mutations in three domains of the vesicular stomatitis viral glycoprotein on its lateral diffusion in the plasma membrane. J Cell Biol 1987; 105(1):69-75.

78. Zhang F, Crise B, Su B et al. Lateral diffusion of membrane-spanning and glycosylphosphatidylinositol-linked proteins:

towards establishing rules governing the lateral mobility of membrane proteins. J Cell Biol 1991; 115:75-84.

79. Henis YI, Herman-Barhom Y, Aroeti B et al. Lateral mobility of both envelope proteins (F and HN) of Sendai virus in the cell membrane is essential for cell-cell fusion. J Biol Chem 1989; 264:17119-17125.

80. Fire E, Zwart DE, Roth MG et al. Evidence from lateral mobility studies for dynamic interactions of a mutant influenza hemagglutinin with coated pits. J Cell Biol 1991; 115:1585-1594.

81. Hochman JH, Schindler M, Lee JG et al. Lateral mobility of cytochrome c on intact mitochondrial membranes as determined by fluorescence redistribution after photobleaching. Proc Natl Acad Sci USA 1982; 79(22):6866-6870.

82. George SK, Xu YH, Benson LA et al. Cytochrome b5 and a recombinant protein containing the cytochrome b5 hydrophobic domain spontaneously associate with the plasma membrane of cells. Biochim Biophys Acta 1991; 1066(2):131-143.

83. Poo M, Cone RA. Lateral diffusion of rhodopsin in the photoreceptor membrane. Nature 1974; 247(441):438-441.

84. Gupta BD, Williams TP. Lateral diffusion of visual pigments in toad (Bufo marinus) rods and in catfish (Ictalurus punctatus) cones. J Physiol Lond 1990; 430:483-496.

85. Liebman PA, Entine G. Lateral diffusion of visual pigment in photoreceptor disk membranes. Science 1974; 185(149):457-459.

86. Maxfield FR, Willingham MC, Haigler HT et al. Binding, surface mobility, internalization, and degradation of rhodamine-labeled α_2-macroglobulin. Biochemistry 1981; 20(18):5353-5358.

87. Maxfield FR, Willingham MC, Pastan I et al. Binding and mobility of the cell surface receptors for 3,3',5-triiodo-L-thyronine. Science 1981; 211:63-65.

88. Smith PR, Saccomani G, Joe E-H et al. Amiloride-sensitive sodium channel is linked to the cytoskeleton in renal epithelial cells. Proc Natl Acad Sci USA 1991; 88:6971-6975.

89. Smith PR, Stoner JC, Viggiano SC et al. Effects of vasopressin and aldosterone on the lateral mobility of eptihelial Na+ channels in A6 epithelial cells. J Memb Biol 1995; 147(2):195-205.

90. Joe EH, Angelides KJ. Clustering and mobility of voltage-dependent sodium channels during myelination. J Neurosci 1993; 13(7):2993-3005.

91. Henis YI, Katzir Z, Shia MA et al. Oligomeric structure of the human asialoglycoprotein receptor: nature and stoichiometry of mutual complexes containing H1 and H2 polypeptides assessed by fluorescence photobleaching recovery. J Cell Biol 1990; 111:1409-1418.

92. Zhang F, Schmidt WG, Hou Y et al. Spontaneous incorporation of the glycosyl-phosphatidylinositol-linked protein Thy-1 into cell membranes. Proc Natl Acad Sci USA 1992; 89:5231-5235.

93. Corcao G, Sutcliffe RG, Kusel JR et al. Lateral diffusion of human CD2 wild type and mutants with large deletions in the transmembrane domain. Biochem Biophys Res Commun 1995; 208:1131-1136.

94. Liu SJ, Hahn WC, Bierer BE et al. Intracellular mediators regulate CD2 lateral diffusion and cytoplasmic Ca^{2+} mobilization upon CD2-mediated T cell activation. Biophys J 1995; 68(2):459-470.

95. Grebenkamper K, Tosi PF, Lazarte JE et al. Modulation of CD4 lateral mobility in intact cells by an intracellularly applied antibody. Biochem J 1995; 312(1):251-259.

96. Pal R, Nair BC, Hoke GM et al. Lateral diffusion of CD4 on the surface of a human neoplastic T-cell line probed with a fluorescent derivative of the envelope glycoprotein (gp120) of human immunodeficiency virus type 1 (HIV-1). J Cell Physiol 1991; 147(2):326-332.

97. Letourneur F, Gabert J, Cosson P et al. A signaling role for the cytoplasmic segment of the CD8 α chain detected under limiting stimulatory conditions. Proc Natl Acad Sci USA 1990; 87:2339-2343.

98. Wier M, Edidin M. Effects of cell density and extracellular matrix on the lateral diffusion of major histocompatibility antigens in cultured fibroblasts. J Cell Biol 1986; 103:215-222.

99. Edidin M, Zuniga MC, Sheetz MP. Truncation mutants define and locate cyto-

plasmic barriers to lateral mobility of membrane glycoproteins. Proc Natl Acad Sci USA 1994; 91:3378-3382.

100. Edidin M, Aszalos A, Damjanovich S et al. Lateral diffusion measurements give evidence for association of the Tac peptide of the IL-2 receptor with the T27 peptide in the plasma membrane of HUT-102-B2 T cells. J Immunol 1988; 141:1206-1210.

101. Hochman JH, Shimizu Y, De Marz R et al. Specific association of fluorescent β2-microglobulin with cell surfaces. The affinity of different H-2 and HLA antigens for β2-microglobulin. J Immunol 1988; 140:2322-2329.

102. Zhang F, Yang B, Odin JA et al. Lateral mobility of Fcγ RIIa is reduced by protein kinase C activation. FEBS Lett 1995; 376(1-2):77-80.

103. Poo H, Krauss JC, Mayo-Bond L et al. Interaction of Fc gamma receptor type IIIB with complement receptor type 3 in fibroblast transfectants: evidence from lateral diffusion and resonance energy transfer studies. J Mol Biol 1995; 247(4):597-603.

104. Posner RG, Subramanian K, Goldstein B et al. Simultaneous crosslinking by two nontriggering bivalent ligands causes synergistic signaling of IgE Fc epsilon RI complexes. J Immunol 1995; 155(7):3601-3609.

105. Menon AK, Holowka D, Webb WW et al. Clustering, mobility, and triggering activity of small oligomers of immunoglobulin E on rat basophilic leukemia cells. J Cell Biol 1986; 102:534-540.

106. Menon AK, Holowka D, Webb WW et al. Cross-linking of receptor-bound IgE to aggregates larger than dimers leads to rapid immobilization. J Cell Biol 1986; 102:541-550.

107. Mao SY, Varin-Blank N, Edidin M et al. Immobilization and internalization of mutated IgE receptors in transfected cells. J Immunol 1991; 146(3):958-966.

108. Gustafsson M, Sundqvist T, Magnusson KE. Lateral diffusion of the secretory component (SC) in the basolateral membrane of the human colonic carcinoma cell line HT29 assessed with fluorescence

recovery after photobleaching. J Cell Physiol 1988; 137:608-611.

109. Jacobson K, O'Dell D, August JT. Lateral diffusion of an 80,000-dalton glycoprotein in the plasma membrane of murine fibroblasts: relationships to cell structure and function. J Cell Biol 1984; 99(5):1624-1633.

110. Duband JL, Nuckolls GH, Ishihara A et al. Fibronectin receptor exhibits high lateral mobility in embryonic locomoting cells but is immobile in focal contacts and fibrillar streaks in stationary cells. J Cell Biol 1988; 107:1385-1396.

111. Paccaud JP, Reith W, Johansson B et al. Role of internalization signals and receptor mobility. J Biol Chem 1993; 268(31):23191-23196.

112. Knowles DW, Chasis JA, Evans EA et al. Cooperative action between band 3 and glycophorin A in human erythrocytes: immobilization of band 3 induced by antibodies to glycophorin A. Biophys J 1994; 66(5):1726-1732.

113. Jacobson K, Ishihara A, Inman R. Lateral diffusion of proteins in membranes. Ann Rev Physiol 1987; 49:163-175.

114. Tank DW, Wu ES, Meers PR et al. Lateral diffusion of gramicidin C in phospholipid multibilayers. Effects of cholesterol and high gramicidin concentration. Biophys J 1982; 40(2):129-135.

115. Tank DW, Wu ES, Meers P et al. Lateral diffusion of gramicidin C in phospholipid multibilayers containing 0-50 mole% cholesterol. Biophys J 1981; 33: 109a.

116. Peters R, Cherry RJ. Lateral and rotational diffusion of bacteriorhodopsin in lipid bilayers: experimental test of the Saffman-Delbrück equations. Proc Natl Acad Sci USA 1982; 79(14):4317-4321.

117. Vaz WL, Criado M, Madeira VM et al. Size dependence of the translational diffusion of large integral membrane proteins in liquid-crystalline phase lipid bilayers. A study using fluorescence recovery after photobleaching. Biochemistry 1982; 21(22):5608-5612.

118. Criado M, Vaz WL, Barrantes FJ et al. Translational diffusion of acetylcholine receptor (monomeric and dimeric forms) of Torpedo marmorata reconstituted into

phospholipid bilayers studied by fluorescence recovery after photobleaching. Biochemistry 1982; 21(23):5750-5755.

119. Kapitza HG, Ruppel DA, Galla HJ et al. Lateral diffusion of lipids and glycophorin in solid phosphatidylcholine bilayers. The role of structural effects. Biophys J 1984; 45(3):577-587.

120. Wu E-S, Low PS, Webb WW. Lateral diffusion of glycophorin reconstituted into phospholipid multibilayers. Biophys J 1981; 33:109a.

121. Schlessinger J, Barak LS, Hammes GG et al. Mobility and distribution of a cell surface glycoprotein and its interaction with other membrane components. Proc Natl Acad Sci USA 1977; 74:2909-2913.

122. Peacock JS, Barisas BG. Photobleaching recovery studies of T-independent antigen mobility on antibody-bearing liposomes. J Immunol 1983; 131(6):2924-2929.

123. Chan PY, Lawrence MB, Dustin ML et al. Influence of receptor lateral mobility on adhesion strengthening between membranes containing LFA-3 and CD2. J Cell Biol 1991; 115(1):245-255.

124. Derzko Z, Jacobson K. Comparative lateral diffusion of fluorescent lipid analogues in phospholipid multibilayers. Biochemistry 1980; 19(26):6050-6057.

125. Zakharova OM, Rosenkranz AA, Sobolev AS. Modification of fluid lipid and mobile protein fractions of reticulocyte plasma membranes affects agonist-stimulated adenylate cyclase. Application of the percolation theory. Biochim Biophys Acta 1995; 1236:177-184.

126. Tsuji A, Ohnishi, S. Restriction of the lateral motion of Band 3 in the erythrocyte membrane by the cytoskeletal network: dependence on spectrin association state. Biochemistry 1986; 25:6133-6139.

127. Sheetz MP, Schindler M, Koppel DE. Lateral mobility of integral membrane proteins is increased in spherocytic erythrocytes. Nature 1980; 285:510-512.

128. Golan DE, Veatch W. Lateral mobility of band 3 in the human erythrocyte membrane studied by fluorescence photobleaching recovery: evidence for control by cytoskeletal interactions. Proc Natl Acad Sci USA 1980; 77:2537-2541.

129. Sheetz MP, Febbroriello P, Koppel DE. Triphosphoinositide increases glycoprotein lateral mobility in erythrocyte membranes. Nature 1982; 297:424-425.

130. Henis YL, Rimon G, Felder S. Lateral mobility of phospholipids in turkey erythrocytes. Implications for adenylate cyclase activation. J Biol Chem 1982; 257:1407-1411.

131. Kapitza HG, Sackmann E. Local measurement of lateral motion in erythrocyte membranes by photobleaching technique. Biochim Biophys Acta 1980; 595:56-64.

132. Richter C, Winterthalter, KH, Cherry RJ. Rotational diffusion of cytochrome P-450 in rat liver microsomes. FEBS Lett 1979; 102:151-154.

133. Peters, R. Translational diffusion in the plasma membrane of single cells as studied by fluorescence microphotolysis. Cell Biol Int Reports 1981; 5:733-760.

134. Eisinger J, Halperin B. Effect of spatial variation in membrane diffusibility and solubility on the lateral transport of membrane components. Biophys J 1986; 50:513-521.

135. Nigg EA, Cherry RJ. Anchorage of a band 3 population at the erythrocyte cytoplasmic membrane surface: protein rotational diffusion measurements. Proc Natl Acad Sci USA 1980; 77:4702-4706.

136. Cherry RJ, Buerkli A, Busslinger M et al. Rotational diffusion of band 3 proteins in the human erythrocyte cytoplasmic membrane. Nature 1976; 263:389-393.

137. Bennett V. The molecular basis for membrane-cytoskeleton association in human erythrocytes. J Cell Biochem 1982; 18:49-65.

138. Wey C L, Cone R, Edidin M. Lateral diffusion of rhodopsin in photoreceptor cells measured by fluorescence photobleaching and recovery. Biophys J 1981; 33:225-232.

139. Poo, MM. Mobility and localization of proteins in excitable membranes. Annu Rev Neurosci 1985; 8:369-406.

140. McCloskey MA, Liu ZY, Poo MM. Lateral electromigration and diffusion of Fc epsilon receptors on rat basophilic leukemia cells: effects of IgE binding. J Cell Biol 1984; 99(3):778-787.

141. Tank DW, Fredericks WJ, Barak LS et al. Electric field-induced redistribution and postfield relaxation of low density lipoprotein receptors on cultured human fibroblasts. J Cell Biol 1985; 101(1):148-157.

142. Giugni TD, Braslau DL, Haigler HT. Electric field-induced redistribution and postfield relaxation of epidermal growth factor receptors on A431 cells. J Cell Biol 1987; 104(5):1291-1297.

143. Poo M. Rapid lateral diffusion of functional A Ch receptors in embryonic muscle cell membrane. Nature 1982; 295(5847):332-334.

144. Hicks BW, Angelides KJ. Tracking movements of lipids and Thy1 molecules in the plasmalemma of living fibroblasts by fluorescence video microscopy with nanometer scale precision. J Membr Biol 1995; 144(3):231-244.

145. Sako Y, Kusumi A. Barriers for lateral diffusion of transferrin receptor in the plasma membrane as characterized by receptor dragging by laser tweezers: fence versus tether. J Cell Biol 1995; 129(6):1559-1574.

146. Edidin M, Kuo SC, Sheetz MP. Lateral movements of membrane glycoproteins restricted by dynamic cytoplasmic barriers. Science 1991; 254:1379-1382.

147. Ziomek CA, Schulman S, Edidin M. Redistribution of membrane proteins in isolated mouse intestinal cells. J Cell Biol 1980; 86:849-857.

PARAMETERS AFFECTING PLASMA MEMBRANE PROTEIN LATERAL MOBILITY

A. MECHANISMS OF PROTEIN IMMOBILIZATION IN BIOLOGICAL MEMBRANES

We saw in the last chapter that proteins in biological membranes diffuse both at much lower rates than they do in isolated or artificial membranes, and than membrane lipids in biological membranes. That this is not an artifact of the fluorescence photobleaching recovery measurement technique has been shown by the fact that measurements using other techniques provide similar data (see section G of the previous chapter). Clearly, plasma membrane proteins, and in particular plasma membrane integral receptors, are largely restricted in terms of their lateral mobility.[1]

The mechanisms by which this may occur have been variously speculated upon (e.g., refs. 2-4), but formally demonstrated to be relevant under physiological conditions in only a few cases. Results from experiments using artificial membranes show that high protein concentrations, for example, have been shown to impede diffusion in artificial membranes (e.g. see refs. 2,5-7). Table 3.1 illustrates this, demonstrating that the apparent lateral diffusion coefficient of the protein bacteriorhodopsin, as well as of lipid probes, decreases with increasing amounts of bacteriorhodopsin.[2,5,6] In similar fashion, fusion and dilution of inner mitochondrial membranes with lipid vesicles, thus decreasing the concentration of membrane protein, increases the apparent lateral diffusion coefficient of both lipids and proteins (the mitochondrial complex III) up to 20-fold.[2] Increasing the multiplicity of infection of vesicular stomatitis virus (VSV) leads to a progressive decrease in both the apparent lateral diffusion coefficient and mobile fraction of the VSV glycoprotein (G) spike protein in vivo,[10] consistent with a general immobilizing effect of increased protein concentration.

The effect of high protein concentrations on mobility has been hypothesized to be through increasing overall membrane viscosity and/or restricting diffusion paths through an excluded volume effect.[2] To what extent, however, experiments such as those illustrated in Table 3.1 are relevant to the physiological situation is debatable. In terms of individual proteins, for example, the toad rod disk membrane has a lipid/protein ratio of about 65 (compare to the values in Table 3.1) with respect to rhodopsin (at a density of about $25,000/\mu m^2$–see ref. 11,12), while GTP-binding proteins and phosphodiesterase holomers are at lower densities (of about 2500 and $93/\mu m^2$, respectively; see refs. 11,13,14). Total protein densities in lymphocyte, fibroblast,

Table 3.1. Effects of aqueous phase viscosity and lipid/protein ratio on lateral mobility in artificial membranes as measured by the technique of fluorescence recovery after photobleaching

Parameter	Viscosity°	Apparent lateral diffusion coefficient$ (10^{-8} cm^2/sec)		Ref.
		bacteriorhodopsin	lipid probe	
A. Effect of aqueous phase viscosity^+				
Sucrose (% w/w)	η_w			
0	0.76	3.4		5
30	2.43	3.0		5
40	4.70	2.2		5
48	9.54	1.7		5
B. Effect of phospholipid/protein (L/P) ratio				
L/P ratio	η			
5000		18.0	18.5@	6
2500		7.5	13.0@	6
300		6.9	8.5@	6
210	1.10	3.4^	6.9^*	5
140	1.80	2.3^	4.3^*	5
90	3.50	1.3^	2.5^*	5
30	43.0	0.15^	0.73^*	5

°η_w, viscosity of aqueous phase, and η, membrane viscosity, determined using the Saffman and Delbruck Equation (see chapter 2; equation [2.5]).[8,9] $Measurements made in DMPC (dimyristoylphosphatidylcholine) membranes. ^Measurements made at 32°C. +L/P ratio of 210. @Lipid probe NBD-PE, nitrobenzo-2-oxa-1,3-diazolyl-phosphatidylethanolamine. *Lipid probe DiO-C18(3), dioctadecyloxatricarbocyanine.

erythrocyte plasma membranes are 350-700, 850 (a lipid to protein ratio of about 1000)[5,6] and 4200/μm^2, respectively.[6] In contrast to these "typical" plasma membrane values, local surface densities can be much higher; e.g., bacteriorhodopsin is present at concentrations in the *Halobacterium holobium* membrane sufficient to enable determination of its three-dimensional structure and orientation in the membrane using low intensity microscopy and low angle electron diffraction analysis to a resolution of 0.7 nm[15,16] and is essentially immobile, while the nicotinic acetylcholine receptor immobilized in plaques has been estimated to be at a density of 10,000/μm^2 (see ref. 17). However, despite the fact that rhodopsin is present at concentrations in the rod disc membrane higher than those of the acetylcholine receptor in plaques (see above), it is highly mobile (apparent lateral diffusion coefficient of 35 x 10^{-10} cm^2/sec with a mobile fraction of about 0.5; see Table 2.3),[18-20] which implies that high protein concentration per se does not necessarily result in protein immobilization. The fact that the rod disc membrane lacks cytoskeletal structures (see ref. 5) implies that the cytoskeleton is a more physiologically relevant factor than protein concentration in determining protein lateral mobility (see section B below).

Other parameters which affect protein lateral mobility in artificial or isolated membranes include the ionic strength of the aqueous phase.[21] Experiments with red blood cell ghosts suggest that ionic strength significantly affects the lateral mobility of erythrocyte band 3 anionic transporter; increasing the ionic strength progressively reduces both the apparent lateral diffusion coefficient and the mobile fraction maximally by greater than 13- and 2-fold, respectively (see Fig. 3.1).[21] Above a threshold value of about 26 mM, band 3 is essentially immobile (Fig. 3.1), whereby the stabilization of interacting proteins such as the membrane skeleton-associated protein spectrin may be involved (see ref. 21 and see section C below). Spectrin dissociates from band 3 at low (nonphysiological) ionic strength in a time-dependent manner, implying that it may be involved in immobilizing band 3 through association under physiological conditions (see section C below).[21]

Physiologically relevant mechanisms by which proteins may be immobilized in membranes include:

i. aggregation in localized areas of the membrane. This is shown by a number of observations such as the fact that incubation with crosslinking, aggregation-inducing antibodies results in significantly reduced lateral mobility, particularly with respect to the mobile fraction of their specifically recognized surface molecules, compared to monovalent antibodies.[4,22-24] A number of receptors, such as the nicotinic acetylcholine,[17] glycine and GABA (γ-amino butyric acid)/benzodiazepine receptors (see below)[25,26] and the insulin-responsive glucose transporter,[27] are known to be aggregated and immobile in particular areas of the membrane. Zakharova et al[28] showed directly that general protein aggregation brings about protein immobility by using poly-L-lysine treatment to aggregate protein and bring about a concomitant reduction of protein lateral movement (both the apparent lateral diffusion coefficient and the mobile fraction) in isolated rabbit reticulocyte membranes.

ii. through binding to an extracellular-matrix. One example is the fibronectin receptor which is immobile in areas of attachment to the substratum through binding to its immobile ligand fibronectin.[29]

iii. through binding to an intracellular matrix such as the membrane skeleton/cytoskeleton. As already alluded

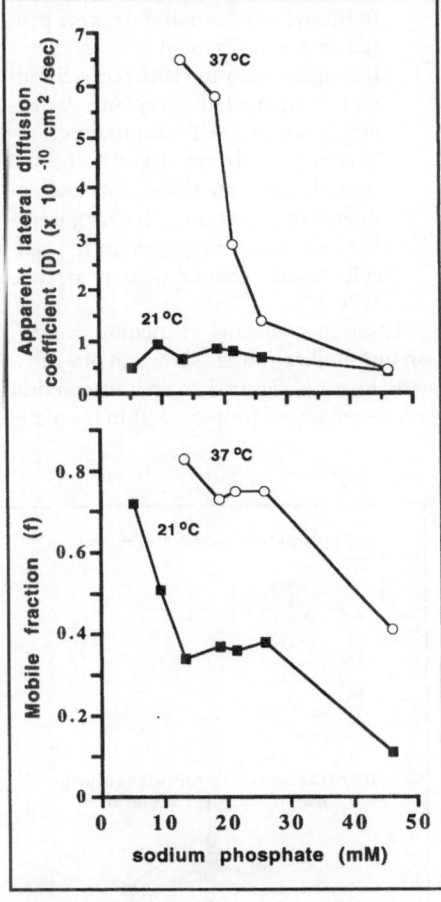

Fig. 3.1. Modulation of lateral mobility of the red blood cell band 3 protein in red blood cell ghosts by ionic strength.[21] Results for the apparent lateral diffusion coefficient (D–top panel) and mobile fraction (f–bottom panel) are shown for measurements at 37° and 21°C.

to above, an example is the erythrocyte band 3 protein, which has specific links with the membrane skeletal components spectrin and ankyrin (see below).[21,30] The lateral mobility of the luteinizing hormone[31] and concanavalin A receptors,[32,33] and membrane proteins in general,[34] can be modulated by treatment with inhibitors of biosynthesis/assembly of cytoskeletal components such as actin filaments and microtubules, supporting the role of the cytoskeleton in restricting protein lateral movement.

iv. through association with cell-cell contact or intercalating regions. An example is the CD2 T cell surface glycoprotein and its ligand LFA-3 (lymphocyte function-associated antigen 3) present on antigen-presenting cells which aggregate in the cell-cell contact area of their respective cells.[35,36]

These mechanisms of membrane protein immobilization are shown in diagrammatic form in Figure 3.2. That interaction with heterologous proteins within the plane of the membrane can play a direct role in mediating the above mechanisms has been shown in a number of studies. The leukocyte integrin adhesion molecule CR3 (C3 complement receptor–CD11b/CD18 or Mac-1), for example, constrains the lateral diffusion of the low-affinity immunoglobulin (Ig) G receptor Fcγ RIIIB, whereby CR3 co-expression reduces both the mobile fraction (from 0.73 in its absence to 0.48 in its presence) and apparent lateral diffusion coefficient (from 3.4 to 2.5 x 10^{-9} cm2/sec).[37] This conclusion is supported by the fact that the heptad sugar N-acetyl-D-glucosamine, known to inhibit Fcγ RIIIB/CR3 cocapping apparently through disruption of lectin-like interactions between the two molecules,[38] reverses the immobilizing effect of CR3 on Fcγ RIIIB (mobile fraction of 0.66).[37] In similar fashion, antibodies to LFA-3, but not to LFA-1 (lymphocyte function-associated antigen-1), effect immobilization of class I major histocompatibility complex (MHC) antigens, but not of CD45 or of a lipid probe, apparently through specific binding interactions of LFA-3 with the former.[39] Similarly, antibodies to glycophorin A effect immobili-

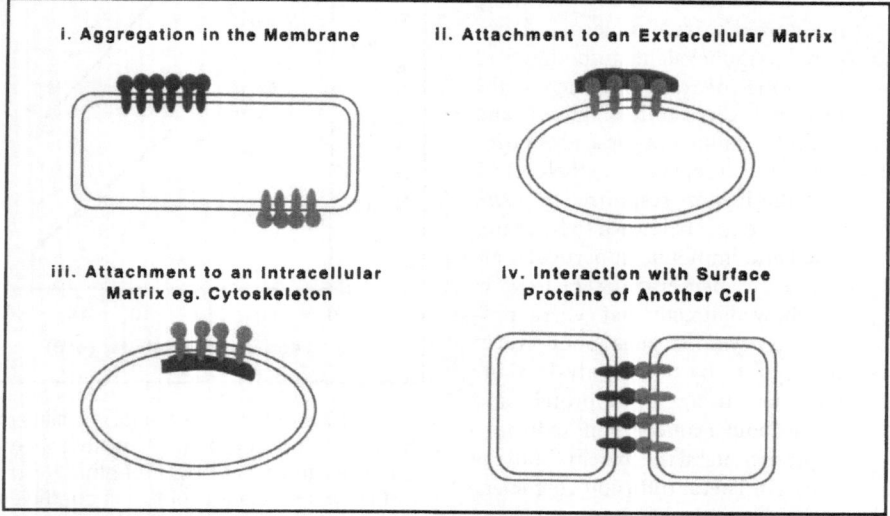

Fig. 3.2. Schematic representation of the various mechanisms by which plasma membrane proteins can be immobilized.

zation,[40] probably through coaggregation,[41] of red blood cell band 3, but not of glycophorin C. A tight association at the level of the membrane between band 3 and red blood cell band 2.1 has also been demonstrated, but this is not the sole mechanism of band 3 immobilization and is estimated only to account for immobilization of 10-15% of band 3 (see ref. 21).[42,43] Experiments with temperature sensitive mutants also imply that the VSV matrix (M) protein influences lateral movement of the VSV G protein in the plasma membrane of VSV infected baby hamster kidney cells.[10]

It should be stressed that the inhibitory effects of protein-protein association on lateral mobility are not likely to result from an effect of increased molecular size. According to the Saffman and Delbruck equation (see Equation [2.5], chapter 2), molecular size only minimally affects the apparent lateral diffusion coefficient and this has been shown in a variety of ways such as using probes for lateral mobility of different sizes e.g., monovalent Fab and (Fab)$_2$ antibody fragments (see ref. 4). Effects are thus probably mediated by vicarious interaction with membrane skeletal and cytoskeletal elements, and/or extracellular matrix components. Henis et al[44] demonstrated that immobilization of one or other of the human asialoglycoprotein receptor subunits using antibody crosslinking led to immobilization of the other, implying that immobile aggregated complexes may include interacting proteins that are not directly crosslinked.

Additional factors that integrally relate to the mechanisms above such as membrane fluidity, primary and secondary sequence motifs of respective membrane proteins,[2,3] and domain structure will be detailed in the following sections. The fact that plasma membrane protein lateral movement can be modulated through various intermolecular mechanisms provides the possibility to regulate mobility. The evidence for a role of signal transduction in doing exactly that is provided in section G, and the importance of this is addressed in detail in the concluding chapter (section D) of this book. Briefly,

since membrane receptor lateral mobility is a crucial limiting factor in hormonal response, signal transduction-mediated regulation of the factors influencing receptor mobility is conceivably central to the coordination of long- and short-term cellular response.

B. MEMBRANE LIPID MOBILITY

Membrane lipid fluidity has been shown to affect membrane protein lateral movement in experiments in which fluidity has been modulated by modifying the fatty acid composition of membranes either in vivo or in vitro through delipidation, alteration of the cellular growth medium or time in culture, etc.[28,45] Reduced membrane fluidity can be effected by a decrease in the percentage of fluid lipids such as *cis*-double bond fatty acids or short chain hydrocarbons, or by an increase in the relative proportion of "non-fluid" lipids, such as cholesterol. Cholesterol molecules orient themselves in the lipid bilayer with their hydroxyl groups close to the polar head groups of the phospholipid molecules, so that their rigid, platelike steroid rings interact with and immobilize the hydrocarbon chain regions that are closest to the polar head groups. While cholesterol makes membranes less fluid, it prevents hydrocarbon chain aggregation and crystallization and thus protects against phase transitions (i.e., conversion to the solid phase). A reduction in membrane fluidity through a decrease in the proportion of cholesterol can directly result in reduced phospholipid as well as protein mobility, as shown directly in artificial lipid bilayers.[46] The lipid acyl chain length can also be critical in determining lipid mobility.[46,47]

Increasing lipid fluidity using some of the techniques mentioned above has been shown to increase the apparent lateral diffusion coefficient of both lipids[48,49] and proteins such as the nicotinic acetylcholine receptor.[48] This is illustrated, by way of example, in Figure 3.3, whereby phospholipase A$_2$ treatment was used on isolated rabbit reticulocyte membranes to demonstrate the direct relationship between membrane fluidity and general protein mobility.[28] Changes in membrane fluidity during the cell cycle have also been shown to affect protein and lipid probe mobility.[50] The real physiological relevance of studies in which membrane fluidity is artificially modulated, however, is questionable; short-term isoprotenerol treatment of cells, for example, appears not to alter lipid mobility although markedly affecting receptor movement (see ref. 51). In contrast, the reduced responsiveness of aging cells to hormonal signals, may relate to the effects of cell aging on membrane fluidity and hence, be physiologically relevant. That aging can result in reduced mobility in terms of the apparent lateral diffusion coefficient has been shown directly for the acetylcholine receptor.[52,53] Both the mobile fraction and the apparent lateral diffusion coefficient of a fluorescent phospholipid probe (NBD-PE–N-4-nitrobenzo-2-diazole phosphatidyl-ethanolamine) and concanavalin A were found to depend on the age of a rat myocyte culture, whereby aged myocytes demonstrated a higher apparent lateral diffusion coefficient and lower mobile fraction compared to young ones.[45] The removal of cholesterol and reintroduction of phosphatidylcholine using liposome

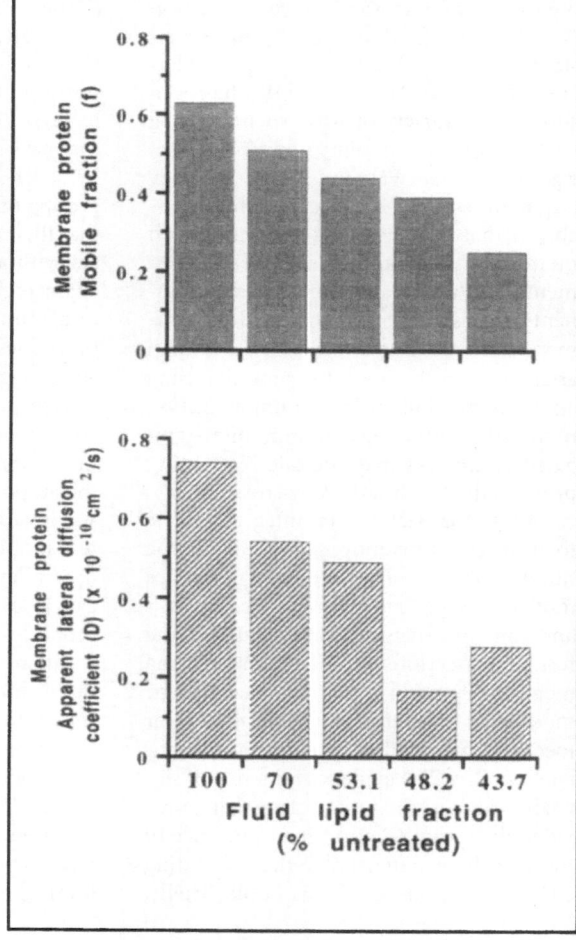

Fig. 3.3. Demonstration of the direct relationship between membrane fluidity and general protein mobility. Phospholipase A_2 treatment was used on isolated rabbit reticulocyte membranes to reduce the membrane fluid lipid fraction to differing extents.[28] General protein lateral mobility (DTAF–dichlorotriazinyl-aminofluorescein labeling) was measured using fluorescence photobleaching recovery. As shown, both the mobile fraction (above) and apparent lateral diffusion coefficient (below) are affected.

treatment reversed the effects of aging on the apparent lateral diffusion coefficient and mobile fraction.[45] Aging has also been shown to result in the reduction of both the apparent lateral diffusion coefficient and mobile fraction of MHC antigens.[54] Certain cell types, such as sickle red blood cells,[40,55] ovalocytic red blood cells[56] and llama elliptocytic red blood cells,[57] appear to possess an intrinsically "rigid" membrane structure and exhibit low plasma membrane mobility in consequence.[41] Specialized membrane regions or domains of reduced lipid and protein mobility will be discussed in section F.

That changes in membrane fluidity, due to aging[58,59] or to delipidation treatments, etc. on isolated membranes,[28] can have significant effects on signal transduction has been shown for several GTP-binding protein coupling hormones and hormone receptors. This will be discussed in detail in chapter 5 (section C).

C. THE CYTOSKELETON

As already alluded to, there is a large body of evidence that the membrane skeleton/cytoskeleton plays a major role in determining membrane protein lateral mobility. It has been hypothesized that steric interactions and/or reversible, low-affinity binding of transmembrane proteins to specific sites on the membrane skeletal/cytoskeletal lattice restrict membrane protein mobility. These interactions may be regulated by phosphorylation or other mechanisms (see sections E and G below and chapter 8, sections C and D).[1,2,4,54] A number of studies[4,32-34,60] have shown that perturbers of cytoskeletal components can influence integral membrane protein lateral mobility. Schlessinger et al[34] for example demonstrated that cytochalasins, fungal metabolites that depolymerize actin filaments and inhibit actin polymerization, could reduce general protein but not lipid lateral movement in the plasma membrane of rat myocytes. In terms of specific proteins, cytochalasin D treatment increases both the apparent lateral diffusion coefficient and

mobile fraction of the Na^+,K^+-ATPase.[61,62] Analogously, the alkaloids colchicine and vinblastin, both of which interact with tubulin and associated proteins—such as Tau and MAP2 (microtubule-associated protein 2), to depolymerize microtubules and inhibit tubulin polymerization, reduce the transferrin receptor mobile fraction.[63] ATP depletion through various treatments also decreases transferrin mobility, presumably through indirect perturbation of microtubule integrity.[64] Treatment with cytochalasin/colchicine also increases mobility of the luteinizing hormone receptor.[31]

Direct links with specific membrane cytoskeletal components have been characterized in some detail in the case of the erythrocyte band 3 protein, which exhibits very low lateral mobility (apparent lateral diffusion coefficient of about $4 \times 10^{-11}/cm^2/sec$)[21,30,65] under normal conditions in red blood cells (see Table 2.3). Its mobility is up to 100 times greater in isolated membranes[66] (where most cytoskeletal-associated and cytoskeletal components are lost) and about 50 times higher in the plasma membrane of spherocytes, which have defective cytoskeletons.[67] Significantly, spherocytes are deficient in the protein spectrin.[67] Studies with red blood cell membranes have demonstrated that increasing the spectrin content of the red blood cell membrane decreases mobility, whereas increasing the content of another cytoskeletally-linked protein ankyrin increases band 3 mobility.[30,65] This indicates that specific protein-mediated association with the membrane skeleton can modulate protein lateral mobility either negatively or positively. The effect of ionic strength on band 3 lateral mobility in red blood cell ghosts alluded to above (see Fig. 3.1 and section A, above) may also be rationalized in this context in terms of association/dissociation of spectrin modulating the band 3 mobile fraction and apparent lateral diffusion coefficient.[21] The lateral movement of concanavalin A receptors in the sarcolemma of myoblasts and mature striated muscle cells similarly appears to be negatively modulated by the cytoskeletally- and extracellular matrix-associated protein dystrophin.[68]

Further evidence for a role of the cytoskeleton in determining plasma membrane protein mobility is provided by studies with specialized cytoskeleton-less areas of the plasma membrane, such as blebs, which are structureless regions which retain integral proteins, but are apparently deficient in membrane skeletal and cytoskeletal components.[69,70] On blebs, nicotinic acetylcholine and concanavalin A receptors as well as other molecules diffuse with a diffusion coefficient of approximately 3×10^{-9} cm^2/sec,[67] which is close to the value for lipid movement in biological membranes (see chapter 2; Table 2.2). This is in stark contrast to the diffusion of these receptors in intact cell membranes, where D is approximately 3×10^{-10} cm^2/sec with a substantial immobile fraction (see Table 2.3).[69] A similar increase in mobility in membrane blebs as opposed to in intact membranes has been observed for the low density lipoprotein (LDL) receptor.[71]

That cell shape also appears to influence protein lateral mobility appears to relate directly to this, through the fact that cell shape is largely a function of the cytoskeletal organization of actin stress fibers and microtubules. Wu et al[72] found that the lateral diffusion coefficient of concanavalin A receptors and receptor complexes on the surface of lymphocytes and lymphoma cells could be enhanced by several orders of magnitude by inducing swelling of the cells to bulbous form using concanavalin A or 7-nitrobenzo-2-oxa-1,3-diazole-phallacidin. The separation of the plasma membrane from most of the F-actin cytoskeleton which results from such treatments (which also induce blebs) releases constraints on lateral diffusion of the cell surface receptors, allowing them to diffuse at rates near to the limit allowed by membrane viscosity.[72] Consistent with this, Swaisgood and Schindler[73] found that attached and flattened cells exhibit significantly reduced mobility of wheat germ agglutinin and concanavalin A receptors compared to spherical shaped cells, whereas phospholipid mobility is unaffected. A mutant cell line largely deficient in its F-actin stress fiber assemblies exhibited increased lectin receptor mobility,[74] implying that the stress fibers, normally increased upon cell adhesion, reduce receptor mobility.

A number of studies[75-78] (see refs. 3 and 4) using inhibitors of cytoskeletal component biosynthesis, such as cytochalasin, have addressed the question of direct participation of the cytoskeleton in signal transduction and in particular that mediated by GTP-binding-protein-coupling receptors. This topic will be discussed in detail in section C of chapter 8 (see also chapter 5, section F), with the conclusion being that since cytoskeletal links may control or modulate membrane protein, and in particular receptor/trimeric GTP-binding protein lateral mobility, the cytoskeleton itself is central to regulation of signal transduction.

D. ANCHORAGE MODULATION

Initial indications that there may a be an underlying membrane structure modulating surface movement through some sort of "anchorage assembly" were provided by the observation that locally-bound concanavalin A can inhibit the formation of patches or capping of crosslinked surface antigen/antibody complexes on B-lymphocytes (e.g., refs. 79-82). This was due to a reduction of the rate of protein, but not lipid, lateral movement.[79-82] Concanavalin A in its dimeric form (succinyl concanavalin A), which has a greatly reduced crosslinking capacity and hence is much less active in agglutination tests, does not show this capability[82] and, in fact, is itself mobile on membrane surfaces (e.g., refs. 32,33,73 and 74; see Table 2.3). Concanavalin A binding does not affect lipid probe mobility.[81] A similar phenomenon may be responsible for the increase of membrane rigidity effected by anti-glycophorin A-antibody-binding to glycophorin A in red blood cells. As alluded to in section A above, the heterologous protein red blood cell band 3 is also immobilized by the anti-glycophorin A-antibody.[40]

Schlessinger et al[80] and Henis et al[81] used localized binding of platelets labeled with concanavalin A to show that the reduction in lateral diffusion effected by concanavalin A did not result from direct crosslinking of surface receptors. Concanavalin A's effect on protein lateral mobility was independent of the vicinity to the actual sites of the bound concanavalin A-platelets, thus demonstrating the propagated and long-range properties of the effect.[80,81] Henis et al[81] also showed that there appeared to be a threshold for the effect of concanavalin A on protein mobility (about 10% occupancy of the cell surface), above which the magnitude of inhibition of receptor mobility was independent of the percentage occupancy.

Significantly, the inhibition of patching/capping by concanavalin A could be reversed by pretreatment with microtubule or actin filament disrupting agents, such as colchicine or cytochalasin B,[81,82] but not by metabolic inhibitors, such as azide or 2-deoxy-D-glucose.[81] The F-actin mutant mentioned in the previous section exhibiting increased membrane protein lateral mobilities is resistant to concanavalin A inhibition of receptor movement.[74] The body of evidence seems to be consistent with the idea that concanavalin A impedes or slows general protein lateral mobility not through immobilizing proteins by directly crosslinking them, but through perturbing the membrane skeleton/cytoskeleton via localized binding of surface components. The implication is that there is a complex network of cytoskeleton and anchored membrane proteins, where not only do events at the level of the cytoskeleton appear to be capable of influencing surface protein movement (as seen in the previous section), but the modifications of surface proteins (e.g., through concanavalin A binding) can influence the cytoskeleton in such a way that other distant membrane surface proteins can be affected. While remaining somewhat dissatisfying in terms of precise molecular mechanisms, the results underline the importance of the cytoskeletal network in modulating membrane protein movement (see also ref. 41).

E. MEMBRANE PROTEIN SEQUENCE MOTIFS

The primary amino acid sequences of membrane proteins contain sequence motifs, some of which appear to be critical in influencing the intrinsic lateral mobility properties of their respective proteins. These include:

i. the nature of the sequence conferring attachment or anchorage to the membrane, including either a transmembrane domain–that is the sequence of amino acids spanning or embedded in the lipid bilayer–or a glycosyl phosphatidylinositol (GPI) anchor, such as those for the lymphocyte surface antigens Thy-1 and LFA-3.

ii. the cytoplasmic domain, i.e., the intracellular domains of the membrane protein that are external to the membrane, and which may contain phosphorylation sites for cytosolic kinases, sequences conferring targeting to coated pits, or motifs conferring interaction with other membrane-associated and cytoplasmic components such as the membrane skeleton/cytoskeleton.[1,2,24]

iii. the extracellular domain or ectodomain, i.e., the domain of the membrane protein external to the cell membrane, which is involved in ligand binding in the case of receptors/coreceptors and may contain glycosylation sequences and/or sequences conferring linkage to the extracellular matrix. The extracellular domain is often the largest domain of a membrane protein and hence potentially, at least likely, to be the most important factor determining lateral mobility properties.

Studies examining the role of these sequence motifs, several of which are highlighted in Table 3.2, have been carried out using modified or wild type derivatives of the cDNAs of membrane proteins expressed in transfected cells, which carries with it certain problems in terms of the physiological relevance of the high levels of expression and

Table 3.2. Lateral mobility of plasma membrane proteins as influenced by particular structural motifs, as measured by the technique of fluorescence recovery after photobleaching.

Protein	Cell type/Membrane	Temperature	Parameter of Lateral Mobility*		Ref.
			D $(10^{-10} cm^2/sec)$	f	
i. Membrane-attachment domain					
CD2§					
wild type (26 a.a. transmembrane domain)	Transfected CHO cells	22°C	1.8	0.87	83
mutant (14 a.a. transmembrane domain)			1.9	0.46	
mutant (12 a.a. transmembrane domain)			3.6	0.86	
Insulin receptor					
wild type	Transfected CHO cells	21°C	6.0	0.58	84
transmembrane domain mutant (Gly^{933}Pro^{934} → AlaAla)			17.1	0.51	
VSV G-protein (vesicular stomatitis virus glycoprotein)**					
wild type (20 a.a. transmembrane domain)	Transfected COS cells	22°C	0.85	0.75	85
wild type (14 a.a. transmembrane domain)			1.9	0.62	
wild type (transmembrane domain)	Transfected COS-1 cells	22°C	3.0	0.42	86,87
chimera with Thy-1 GPI link			5.9	0.62	
LFA-3 (GPI-anchored form)	Artificial EPC^{xx} membranes	22°C	0.23	0.73	35
LFA-3 (transmembrane form)			0	0	
Thy-1**					
wild type (GPI-link)	Transfected COS-1 cells	22°C	27.0	0.67	86,87
chimera with VSV G-protein transmembrane domain^			16.0	0.58	
PLAP (human placental alkaline phosphatase)**					
wild type (GPI-link)	Transfected COS-1 cells	22°C	24.0	0.70	86
chimera with VSV G-protein transmembrane domain			12.0	0.67	

Table continues on next page.

Table 3.2. (continued)

Lys6E**					
wild type (GPI-link)	Transfected COS-1 cells	22°C	28.0	0.67	86
chimera with MHC class I D[b] transmembrane domain			17.0	0.60	
ii. Cytoplasmic domain					
EGF-receptor					
wild type (cytoplasmic domain 542 a.a.)	Transfected COS cells	21°C	1.2	0.61	88
deletion mutant (cytoplasmic domain 479 a.a.)			1.6	0.59	
deletion mutant (cytoplasmic domain 98 a.a.)			1.5	0.70	
deletion mutant (cytoplasmic domain 9 a.a.)			1.2	0.77	
Neu receptor[§]					
wild type (cytoplasmic domain 584 a.a.)	Transfected NIH 3T3 fibroblasts	22°C	2.3	0.57	89
kinase deficient mutant[x] (cytoplasmic domain 584 a.a.)			2.0	0.55	
constitutively active mutant° (cytoplasmic domain 584 a.a.)			2.0	0.28	
constitutively active mutant° (cytoplasmic domain 341 a.a.)			2.6	0.33	
constitutively active mutant° lacking phorphorylation site[#] (cytoplasmic domain 341 a.a.)			2.3	0.68	
H-2L[d] (class I MHC)**					
wild type (cytoplasmic domain 31 a.a.)	HEPA-OVA mouse hepatoma (transfected)	22°C	12.0,13.0	0.31,0.43	90,91
deletion mutant (cytoplasmic domain 7 a.a.)			5.0	0.39	91
deletion mutant (cytoplasmic domain 4 a.a.)	Transfected mouse L-cells		15.0,17.0	0.20,0.40	90,91
deletion mutant (no cytoplasmic domain)			3.0	0.42	91
VSV G-protein**					
wild type (29 a.a. cytoplasmic domain)	Transfected COS cells	22°C	0.85	0.75	85
wild type (1 a.a. cytoplasmic domain)			2.1	0.65	
IgM chimera (3 a.a. cytoplasmic domain)			2.1	0.66	
chimera with E1 spike protein of infectious bronchitis virus (130 a.a. cytoplasmic domain)			1.1	0.64	
chimera with influenza virus haemagglutinin			2.9	0.68	

Table continues on next page.

Table 3.2. (continued)

Protein	Cell type/Membrane	Temperature	Parameter of Lateral Mobility*		Ref.
			D (10⁻¹⁰cm²/sec)	f	
CD4**					
wild type (39 a.a. cytoplasmic domain)	Virus infected Sf9 cells	22°C	4.7	0.80	92
deletion mutant (7 a.a. cytoplasmic domain)			5.7	0.87	
Low-affinity IgG receptor Fcγ RIIa (CD32) (human)					
wild type (79 a.a. cytoplasmic domain)	Transfected mouse macrophage	22°C	14	0.54	93
Tyr → Leu^{252} mutant (79 a.a. cytoplasmic domain) -like P338D1 cells			7.1	0.71	
deletion mutant (59 a.a. cytoplasmic domain)			14	0.51	
deletion mutant (28 a.a. cytoplasmic domain)			9.8	0.66	
deletion mutant (2 a.a. cytoplasmic domain)			8.5	0.62	
High-affinity IgE receptor Fcε RI (human):					
Subunit structure αβγ2; α 20, β 59 and 43, and γ 36 amino acid cytoplasmic domains respectively.					
wild type α20β59;43γ36 monomeric ligand	Transfected COS-7 cells	19°C	1.2	0.64	94
mutant α4β59;43γ36			1.0	0.62	
mutant α20β2;43γ36			1.0	0.70	
mutant α20β59;5γ36			1.3	0.65	
mutant α20β2;5γ36			2.2	0.60	
mutant α20β59;43γ4			1.5	0.56	
mutant α20β59;5γ4			1.3	0.57	
mutant α4β59;43γ4			1.2	0.60	
mutant α4β59;5γ4			1.0	0.64	
mutant α4β2;43γ4			2.7	0.58	
wild type α20β59;43γ36 oligomeric ligand			1.6	0.34	
mutant α4β59;43γ36			2.0	0.33	
mutant α20β2;43γ36			1.8	0.35	

Table continues on next page.

Tablle 3.2. (continued)

mutant α20β59;5γ36			1.7	0.46	
mutant α20β2;5γ36			2.1	0.51	
mutant α20β59;43γ4			2.3	0.34	
mutant α20β59;5γ4			1.5	0.45	0.34
mutant α4β59;43γ4			1.5	0.42	
mutant α4β59;5γ4			1.2	0.36	
mutant α4β2;43γ4			4.2		
complement receptor CR1					
wild type (43 a.a. cytoplasmic domain)	Transfected CHO cells	37°C	12	0.63	95
Glu[2015] → Tyr mutant@ (43 a.a. cytoplasmic domain)			10	0.57	
deletion mutant (1 a.a. cytoplasmic domain)			16	0.77	
chimera with 49 a.a. cytoplasmic domain of LDL receptor$			7.5	0.50	
wild type (43 a.a. cytoplasmic domain)		14°C	14	0.75	
Glu[2015] → Tyr mutant@ (43 a.a. cytoplasmic domain)			13	0.71	
deletion mutant (1 a.a. cytoplasmic domain)			14	0.75	
chimera with 49 a.a. cytoplasmic domain of LDL receptor$			12	0.75	
red blood cell band 3					
intact molecule	red cell ghosts	26°C	0.53	0.37	30,66
trypsin treated			4.0	0.54	
cytochrome b5					
intact molecule	red cell ghosts	22°C	28.0	0.63	24
trypsin treated (intracellular)			25.0	0.45	
iii. Extracellular domain (glycosylation)					
VSV G-protein (462 residue extracellular domain)**					
wild type (2 glycosylation sites)+	Transfected COS-1 cells	22°C	0.85	0.75	85
mutant (Gln → Asn[264]; 3 glycosylation sites)+			1.3	0.55	
mutant (Thr → Ala[338]; 1 glycosylation site)+			1.8	0.65	
cytochrome b5 (4 residue extracellular domain)					

Table continues on next page.

Table 3.2. (continued)

Protein	Cell type/Membrane	Temperature	Parameter of Lateral Mobility* D (10^{-10} cm^2/sec)	f	Ref.
wild type (16 kDa)	red blood cells	22°C	23.0	0.52	24
chimera with β-galactosidase (1023 residue extracellular domain; tetramer - 500 kDa)			62.0	0.41	
wild type (16 kDa)	mouse 3T3 fibroblasts	22°C	11.0	0.46	
chimera with β-galactosidase (1023 residue extracellular domain; tetramer - 500 kDa)			16.0	0.42	

a.a., amino acids; CHO, Chinese hamster ovary; LFA-3, lymphocyte function-associated antigen-3; GPI, glycosyl phosphatidylinositol; MHC, major histocompatability complex; LDL, low density lipoprotein.

*Values are for the apparent lateral diffusion coefficient (D) and mobile fraction (f).

§Measurements using specific monovalent Fab antibodies.

**Measurements using specific divalent Fab antibodies.

xxEPC, egg phosphatidylcholine

^Contains both transmembrane domain and GPI anchor.

xKinase mutant Lys → Ala[758]

°Val → Glu[664]

#Tyrosine phosphorylation site mutant Tyr → Phe[1253]

@Mutation creates a coated pit binding/internalization sequence Asn-Pro-Lys-Tyr[2015].

$The LDL receptor cytoplasmic domain contains the coated pit binding/internalization sequence Asn-Pro-Val-Tyr (see Ref. 95).

+Each glycosylation site is estimated to add 3 kDa to the molecular weight.[85]

cell type chosen for the studies. Such studies have, however, provided some useful information with respect to defining the sorts of sequences which can influence intrinsic lateral mobility properties.[2]

The possibility that transmembrane domains may play a role in determining membrane protein lateral mobility has been addressed in a few mutagenic studies. Corcao et al[83] used the approach of modifying the length of the transmembrane segment of the CD2 lymphocyte antigen (see Table 3.2). Perhaps surprisingly, truncated 12- or 14-amino acid forms of CD2 appear to have lateral mobility properties very similar to those of wild type CD2.[83] Removal of six amino acids from the transmembrane domain of the VSV G protein similarly only marginally increases its lateral mobility.[85] In contrast, mutation of the insulin receptor to effect "structural optimization" of the transmembrane helix increases the apparent lateral mobility coefficient about 3-fold.[84]

An alternative to a transmembrane domain in terms of protein anchoring to or within the membrane is a post-translationally added GPI anchor. The functional significance of this type of protein linkage is not completely understood, although it has been proposed to play a role in protein sorting to the apical surface of the cell, reinsertion into cell membranes and even cell signaling.[96] Several reports suggest that GPI anchors also enhance plasma membrane receptor mobility quite markedly, presumed to be because of the relatively small surface area of interaction between the lipid bilayer and the GPI anchor, compared to that of a transmembrane sequence of about 20 amino acids.[35,96,97] In the case of the adhesion glycoprotein LFA-3, there are two distinct membrane-associated forms of the protein which exhibit radically different lateral diffusion properties in artificial phosphatidyl choline membranes.[35] While the transmembrane domain-containing protein is essentially immobile, the GPI form is highly mobile (see Table 3.2) as well as being functionally more active in promoting

adhesion (see also chapter 7, section D), supporting the idea that GPI anchors may be sequence elements conferring facilitated lateral mobility on the proteins carrying them. Zhang and colleagues[86,87] performed a study of chimeric proteins of normally GPI-linked membrane proteins and the membrane spanning VSV G-protein where transmembrane domains and GPI-links were exchanged to test the effect on lateral mobility (see Table 3.2). The substitution of the GPI anchors of the membrane surface antigen Thy-1 or human placental alkaline phosphatase with the VSV G-protein transmembrane domain decreased the apparent lateral diffusion coefficient 2-fold in both cases,[86,87] while the replacement of the transmembrane domain of VSV G-protein with the GPI-linkage of Thy-1 increased the apparent lateral diffusion coefficient 2-fold and the mobile fraction by 50%. The clear implication is that GPI-anchors confer higher lateral mobility compared to transmembrane domains.

Interestingly, Kooyman et al[96] have shown that GPI-linked proteins can undergo intermembrane transfer in vivo, whereby GPI-linked proteins expressed on the surface of transgenic mouse red blood cells could be transferred to endothelial cells in a functional form. The lateral mobility of the proteins is presumed to be a key factor in this transfer, which may be potentially useful for the delivery of therapeutic proteins to the vascular endothelium of specific target tissues.[96] This is an example not only of the physiological importance of protein lateral movement, but also of the possibility of exploiting protein lateral mobility parameters in clinical and other applications (see also chapter 8, section E).

Evidence that the cytoplasmic domain can determine the lateral mobility properties of membrane proteins has been shown in several studies. As mentioned, among other features within the cytoplasmic domain are phosphorylation sites and other domains conferring the capability to interact with components on the cytoplasmic side of the membrane (see below). One mechanism by which the intracellular domain can affect plasma membrane lateral movement is through conferring or modulating linkage to components of the membrane skeleton/cytoskeleton. Tsuji and Ohnishi,[30,65] for example, showed that proteolysis of erythrocyte band 3 increases its lateral mobility, presumably through impairing interactions of the cytoplasmic domain with spectrin and the cytoskeleton (see section C above).

Molecular approaches have also been used to determine the role of the cytoplasmic domain by varying its primary sequence through point mutation or deletion. Deletions of the 31-amino acid cytoplasmic domain of the mouse class I MHC molecule H-2L^d, for example, have been shown to decrease the apparent lateral diffusion coefficient by a factor of up to 4-fold (see Table 3.2).[91] In similar fashion, successive deletions of the low-affinity IgG receptor Fcγ RIIa (CD32) reduce the apparent lateral diffusion coefficient by up to 40%.[93] Significantly, a point mutation of tyrosine at position 252 within the CD32 cytoplasmic domain increases the mobile fraction by about 30%, as well as reducing the apparent lateral diffusion coefficient by 50% (see Table 3.3 below), implying that tyrosine phosphorylation is important in regulating CD32 lateral mobility, probably through modulating the association of cytoplasmic factors (see also section G below).[93] Multiple modifications of the VSV G-protein 29 amino acid cytoplasmic domain, including almost complete removal of the domain and/or its substitution with the cytoplasmic domains from surface IgM, from the E1 spike protein of the infectious bronchitis virus or from the hemagglutinin protein of the influenza virus, maximally increase the apparent lateral diffusion coefficient about 3-fold.[85]

In contrast, extensive deletions of the 542 amino acid cytoplasmic domain, including the kinase domain, of the epidermal growth factor (EGF) receptor do not significantly affect mobility.[88] Deletion of 32 of the 39 amino acids of the cytoplasmic

domain of the immune receptor CD4 also has a negligible effect on its lateral mobility properties;[92] the cytoplasmic domain, however, appears to be essential for phorbol ester-induced immobilization of the receptor (mobile fraction reduced from 0.8 to 0.5).[92] In contrast, phosphorylation within the cytoplasmic domain of CD32 (above) appears to increase its lateral mobility (see also section G; Table 3.3).[93] Greberkamper et al[92] demonstrated that introduction into the cells of a specific monoclonal antibody directed against the CD4 cytoplasmic domain by microinjection or liposome fusion could effect a reduction of the mobile fraction (to 0.4) of full-length but not truncated CD4. A variety of deletion mutations in each of the total of 5 extensive cytoplasmic domains of the four subunit high-affinity IgE receptor Fcε RI similarly do not markedly affect lateral mobility of the monomeric- or oligomeric-ligand-occupied receptor (see Table 3.2). Mutations truncating the β-subunit amino-terminal cytoplasmic domain, however, in conjunction with truncation mutations in either the α-subunit or β-subunit carboxy-terminal cytoplasmic domain appear to increase the apparent lateral mobility coefficient of monomeric- or oligomeric-ligand-occupied receptor relative to wild type, and to some extent the mobile fraction in the case of oligomer-occupied receptor (see Table 3.2).[94] Other mutations affect the receptor's internalization kinetics (see also chapter 7; Fig. 7.3).[94]

Sequences within the cytoplasmic domain increasing the affinity for structures such as coated pits also play a major role in determining lateral mobility properties. Using the complement receptor CR1 as a model system, for example, Paccaud et al[95] showed that the introduction of sequences conferring binding to coated pits, either through mutation or replacement of the CR1 cytoplasmic domain with that of the LDL receptor, both decreased the lateral mobility (Table 3.2) and concomitantly increased the internalization rate. Deletion of the 43 amino acid cytoplasmic domain increased lateral mobility slightly.[95]

The naturally occurring glycophorin A variant of Miltenberger MiV erythrocytes contains the 6 amino acid cytoplasmic domain of glycophorin B in place of that of glycophorin A (39 amino acids), resulting in a greatly increased mobile fraction (0.69) at 24°C compared to that of normal red blood cells (mobile fraction of less than 0.1).[40] It should, however, be noted that the glycophorin A/B variant also contains the glycophorin B transmembrane domain, so that differences in the mobile fraction cannot formally be attributed solely to the cytoplasmic domain of glycophorin A (see however ref. 40).

The extracellular domain is the largest domain of a number of integral membrane proteins, often containing a number of sites for glycosylation and binding to extracellular matrix components.[85] The extracellular domain of mVSG, the variant surface glycoprotein of *Trypanosoma brucei*, is purported to mediate interaction with the glycocalix extracellular matrix.[98] Consistent with this, mVSG's mobility in mammalian cells, such as baby hamster kidney cells which lack a glycocalix is, lower (mobile fraction of 0.56) than in *T. brucei*.[98] Linkage to the extracellular matrix modulates the mobility of other proteins, such as MHC antigens[54] and receptors such as that for fibronectin.[29] In the case of the VSV G protein, increasing or decreasing the number of glycosylation sites increases the apparent lateral diffusion coefficient maximally up to 2-fold, and decreases the mobile fraction slightly (see Table 3.2).[85]

Thus, while several studies imply that specific sequence elements may be critical in determining protein lateral mobility properties, it is fair to say at this stage that our understanding is insufficient to enable protein lateral mobility to be engineered according to need. Since protein lateral movement appears to be critical in terms of signal transduction and also downregulation of hormonal response (see chapters 5, 6 and 8), it is to be hoped that this situation will be remedied in the near future so that this knowledge can be applied in clinical situations (e.g., see ref. 96).

F. DOMAIN STRUCTURE: REGIONS OF RESTRICTED MOBILITY

The membrane is of course not quite so simply organized as envisaged in the fluid mosaic model,[99] and, in fact, there is clearly a domain structure, probably with many physical restrictions to long range protein lateral movement within the plane of the membrane (e.g., see refs. 21,100). This means that diffusion may only occur within limited areas or domains of the membrane. As seen in the previous chapter and Tables 2.2 and 2.4 in particular, lipids, in contrast to proteins, are highly mobile within the plasma membrane, their movement being essentially a function of the membrane viscosity.[2,4] There are, however, particular membrane domains where lipid mobility is reduced and protein movement similarly impaired. One example appears to be the specialized membrane regions ("macrodomains") of *Xenopus* eggs,[101-103] some of which appear to be inaccessible to lipids or proteins which can freely diffuse in other domains. Even lipid probes within these macrodomains[101] are impaired in lateral mobility; e.g., apparent lateral diffusion coefficients within the animal pole of unfertilized eggs are five times lower than those within the vegetal pole.[102] Subsequent to fertilization, the differences between the membrane domains are further enhanced; lipids at the animal side of the egg are essentially immobilized, whereas the apparent lateral diffusion coefficient at the vegetal side is only slightly reduced (from 7.6 to 4.4 x 10^{-8} cm^2/sec).[102] This apparent immobilization of animal pole plasma membrane lipids may in part constitute the mechanistic basis of polyspermy block, with the sharp transition between the animal and the vegetal domain coinciding with the boundary between the presumptive ecto- and endoderm of the developing embryo.[102] Similar animal-vegetal polarity in the organization of the mollusk egg membrane may play an important role in the process of cell diversification during early development.[103]

Epithelial cell junctions appear to participate in cell polarization through restricting lipid molecule lateral movement, which appears to result in, or be the result of, differences between the apical and basolateral plasma membranes in terms of lipid lateral mobilities.[104] Probes such as 5N-(hexadecanoyl)-aminofluorescein (HEDAF) appear to be largely restricted to the apical membrane in bovine aortic endothelial cells, only localizing to the basal membrane if cell junctions are disrupted.[104] The apparent lateral diffusion coefficient of HEDAF was found to be the same in both apical and basal membranes when junctions were disrupted and regained its initial higher value in the apical poles (compared to the basal membrane) when cell contacts were restored.[104] This strongly suggests that cell junctions can affect overall plasma membrane organization in a reversible manner, defining specific membrane domains with distinctive lipid mobility properties.[101]

A number of proteins, and particularly receptors, display restricted mobility in certain areas or domains of the plasma membrane but not others. Restricted lateral movement may relate to or be the mechanism by which the precise localization of particular receptors in a small area of the membrane is achieved. This is appears to be the case for the PH-20 protein, which is normally restricted to the posterior head of acrosome-intact guinea pig sperm and concomitantly exhibits a significantly lower apparent lateral diffusion coefficient than when it is localized in the inner acrosomal membrane to which it migrates during the exocytotic acrosome reaction.[107] Other examples of proteins exhibiting restricted mobility in particular domains of the plasma membrane include the voltage dependent Na+ channel, which displays restricted mobility on axon hillocks where it appears to be aggregated and immobile but is highly mobile on cell bodies.[106] Similar results have been reported for the nicotinic acetylcholine[26,52] and GABA/benzodiazepine receptors.[26] The fibronectin receptor is immobile in focal contacts and fibrillar streaks but highly mobile in embryonic locomoting cells,[29] which may result from the less rigid

structure of the cytoskeleton in the latter. Similar observations apply to the GP80 integral glycoprotein of murine fibroblasts, whose apparent lateral diffusion coefficient is 3-fold higher near the leading edge of motile cells compared to the trailing region.[107] This difference is presumed to result from weaker coupling of the glycoprotein to the underlying cytoskeleton in the dynamic leading edge region.[107] The latter two examples of increased lateral mobility[29,107] resemble the increased lateral diffusion rates of proteins in damaged cytoskeleton-less membrane regions (blebs) (see section C above).[69-72]

In analogous fashion to the above proteins, glycine receptors are segregated to the neuronal cell body and confined to a domain within which their redistribution is restricted.[25] However, the receptors are highly mobile within this domain, as exemplified by the fact that more than 50% of the receptors have a high apparent lateral diffusion coefficient (1.15×10^{-9} cm^2/sec) characteristic of unrestricted protein lateral diffusion.[25] In contrast, the sparsely distributed glycine receptors on neuronal processes exhibit much lower values for both the mobile fraction and apparent lateral diffusion coefficient compared to those on the neuronal cell body.[25] Accordingly, the mechanism by which glycine receptors are localized to the cell body is not through simple reduction or restriction of receptor lateral mobility. Similarly, certain guinea pig sperm antigens exhibit high lateral diffusion coefficients and mobile fractions in all parts of the membrane, despite being mostly localized to specific domains of the sperm cell surface such as the posterior tail region.[108] Clearly, there are mechanisms other than protein immobilization through which proteins are localized in specific membrane domains.[105,108]

The exact mechanism by which membrane domain structure is achieved is not clear, but may be defined in part through membrane skeletal/cytoskeletal structures, such as the spectrin-ankyrin meshwork of the red blood cell plasma membrane.[21,100]

The cytoskeletal links associated with the tight junctions of cell-cell contact appear to represent a barrier to lateral diffusion, effectively defining the two membrane domains.[61,109] The Na[+],K[+]-ATPase, for example, is normally basolaterally localized in polarized kidney epithelial cells, but the induction of cytoskeletal protein dysfunction enables its localization in the apical membrane, accompanied by an increased rate of lateral diffusion.[61,109] That the cytoskeleton and or other proteinaceous material play a role in defining lipid domain structures is implied by other studies in which trypsin treatment can enhance lipid probe membrane mobility, both in terms of the mobile fraction and apparent lateral diffusion coefficient.[110] The molecular basis of this observation is not as yet understood but may relate to some sort of anchorage modulation model (see section D above)[79-82] where lipids may be linked in some way to the membrane skeleton.

Saxton[111-114] has expanded these and a number of other observations, particularly with respect to the red blood cell, into a "percolation model" (see also refs. 28,115,116), where protein lateral mobility is proposed to be area-limited through various obstacles within the membrane. These include the membrane cytoskeletal network composed of α/β spectrin/ankyrin, etc., large protein aggregates, domains of immobile or "solid" (gel phase) lipids and "boundary lipids," those lipids ordered "stably" around an integral membrane protein in a layer or "lipid annulus," all of which are postulated to constitute a dynamic barrier restricting protein lateral diffusion (see ref. 4). The nature and extent of this barrier can fluctuate according to the stage of cellular development through changes in the membrane skeleton/cytoskeleton; the percolation backbone is the domain of the membrane accessible to protein lateral movement, consisting of lipids in the liquid-crystalline phase ("fluid lipids"). If the fraction of the membrane accessible to movement falls below a critical value–the "percolation threshold"–movement over long distances becomes prohibited.

In this context, although association with the spectrin membrane skeleton appears to immobilize red blood cell band 3 as discussed above,[21,30,61,65-67] it has been estimated that only about 20% of band 3 is actually spectrin-associated at 37°C,[65] perhaps implying that band 3's restricted mobility may also result from the spectrin network's potentially obstructing rather than anchoring properties. The cell-cell tight junctions mentioned above which define and separate apical and basolateral membrane domains, preventing the lateral diffusion of specific proteins between them, can be regarded as representing specific examples where the percolation threshold with respect to long range apical/basolateral exchange has been obtained, e.g., in the case of membrane proteins such as the Na[+],K[+]-ATPase which is basolateral in polarized cells.[61,109] The percolation model has also proved to be relevant with respect to less defined membrane domains, and in rationalizing results for β-adrenergic receptor-mediated activation of adenylate cyclase in isolated membranes[28,115,116] as well as in other systems (see refs. 28, 111-115).

G. SIGNAL TRANSDUCTION

While receptor lateral mobility is important in the signaling context as will be discussed in subsequent chapters at length, hormonal stimulation can also influence the lateral mobility of heterologous membrane proteins, this effect conceivably being a significant but largely ignored mechanistic aspect of signal transduction. Treatment with the cytokines tumour necrosis factor and interferon-γ, for example, slows the lateral diffusion of proteins and lipids in human endothelial cell membranes[117] (see ref. 118 and Table 3.3A). Similarly, the diffusion rates of lipid probes in the apical (as opposed to the basal) cell membrane can be increased about 30% in bovine vascular endothelial cells by the addition of fibroblast growth factor (FGF) or allowing the cells to attach to an extracellular matrix.[122] Clustering through lateral movement of the integrin CR3 (C3 receptor–see section A above) plays

Table 3.3. Selected examples of signal transduction and cell-cycle-mediated regulation of plasma membrane protein lateral mobility.

A. Signal Transduction

Protein	Cell type	Condition /Treatment	Temp.	Parameter of Lateral Mobility*		Ref.
				D (10^{-10} cm^2/sec)	f	
class I MHC protein	human endothelial cell (HEC)[xx]	none	23°C	29-30	> 0.80	117
		TNF		30-35	> 0.80	
		IFNγ		23-27	> 0.80	
		TNF/IFNγ		6.3-8.5	> 0.80	
gp96	HEC[xx]	none	23°C	13-14	> 0.80	117
		TNF		12-18	> 0.80	
		IFNγ		13	> 0.80	
		TNF/IFNγ		5.9-10	> 0.80	
CD2	Jurkat T leukemia cells[xx]	none	22°C	7.2	0.70	36
		EGTA		7.5	0.73	
		EGTA + ionomycin		8.8	0.67	
		DB-cAMP		7.8	0.59	
		activated[§]		ND[1]	< 0.10	
		activated[§] + EGTA		ND[1]	0.16	
		activated[§] + EGTA + ionomycin		7.5	0.35	
		activated[§] + DB-cAMP		7.2	0.29	
CD4	Sf9 insect cells[xx]	none	23°C	4.7	0.80	92
		PMA		3.8	0.53	
		PMA/W7[^]		4.1	0.79	
		PMA/H7[^]		4.0	0.86	

Table continues on next page.

Table 3.3. (continued)

Fcγ RIIa (CD32)[xx] (Low-affinity IgG Fc receptor)	P388D1 cells[@]	none	22°C	14	0.54	93
		PMA		6.1	0.53	
		PMA/calphostin[^]		15	0.71	
acetylcholine receptor[=] (nicotinic)	C2C12 mouse myoblasts agrin[+]	none	22°C	0.07	0.28	119
Amiloride-sensitive Na[+] channel[xx]	A6 toad kidney	pervanadate[+]	25°C	0.99	0.17	62
		none			0.14	
		aldosterone			0.32	
		vasopressin			0.13	

B. Stage of Cell Cycle

Surface antigens[o]	C1300 neuroblastoma cells	M phase	22°C	1.3		50
		early G1 phase		3.1		
		late G1 phase		2.7		
		mid S phase		1.8		
		G2 phase		2.0		
class I MHC antigens	cl ld mouse fibroblasts	Sparse	19°C	7.5	0.47	54
		Confluent		3.4	0.22	
	human skin fibroblasts	Sparse		8.0	0.58	
		Confluent		7.5	0.29	
	VA-2 SV40 transformed human fibroblasts	Sparse		6.5	0.32	
		Confluent		4.9	0.25	
	WI-38 human lung fibroblast	Sparse		9.0	0.36	
		Confluent		4.7	0.24	
β2-microglobulin (class I HLA)	HT29 human colonic carcinoma	Nondifferentiated[$]	22°C	11.5	0.46	120
		Differentiated[$]		15.5	0.84	
PDGF-receptor	human foreskin fibroblasts (AG1523)	None	23°C	3.2	0.60	121

table continues on next page.

Table 3.3. (continued)

Protein	Cell type	Condition /Treatment	Temp.	Parameter of Lateral Mobility*		
				D $(10^{-10}\ cm^2/sec)$	f	Ref.
cl		Starved		5.1	0.73	
		Starved/refed		12.0	0.96	
glycoprotein GP80	C3H/10T1/2 cells mouse fibroblasts	Interphase	24°C	1.4	0.70	107
		M phase		0.86	0.54	
		Go phase		1.4	0.34	

Abbreviations: TNF, tumour necrosis factor; IFN, interferon; MHC, major histocompatibility complex; PMA, phorbol-12-myristate-13-acetate; EGTA, ethylene glycol-bis(β-amino ethyl ether) N,N,N',N'-tetraacetic acid; DB-cAMP, dibutryl cAMP; HLA, human leukocyte antigen; PDGF, platelet-derived growth factor.
*D, apparent lateral diffusion coefficient; f, mobile fraction.
xxMeasurements performed using labeled Fab antibodies.
§Cell activation effected by crosslinking antibody combinations.[36]
¹ND, not able to be determined.
^W7, peptide inhibitor of calmodulin-dependent protein kinase; H7, peptide inhibitor of the Ca^{2+}, phospholipid-dependent protein kinase (PK-C); calphostin, PK-C inhibitor.
@PMA has no effect on CD32 mobility in transfected Chinese hamster ovary cells, implying that the effect is cell-type specific.[93]
=Probe was rhodamine labeled α-bungarotoxin.
+Agrin stimulates tyrosine phosphorylation; pervanadate inhibits tyrosine phosphatases.
°The probe of lateral mobility was rhodamine-labeled Fab antibodies prepared against general lymphoid cell surface antigens.
$Differentiation was induced by replacing glucose in the medium with galactose.[120]

a role in activating macrophages[123] and polymorphonuclear leukocytes[124] for phagocytosis, whereby receptor movement within the plane of the membrane can be facilitated by lymphokines.[123,125,126] Clustering can be effected by phorbol-12-myristate-13-acetate (PMA) treatment,[124,127-129] which activates serine protein kinases such as the Ca^{2+}, phospholipid dependent kinase (PK-C).

PMA treatment also reduces the mobile fraction of the immune receptor CD4[92] as well as the apparent lateral diffusion coefficient of CD32 (see section E above).[93] Phosphorylation by PK-C appears to be directly implicated in the case of CD32, since a PK-C-specific inhibitor (calphostin) reverses the PMA effect,[93] while a peptide inhibitor specific to PK-C but not calmodulin-dependent protein kinase reverses the effect in the case of CD4.[92] Interestingly, PMA has no effect on CD32 mobility in transfected Chinese hamster ovary cells, indicating that the regulatory effect is cell-specific. This is possibly due to the induced association of particular cytoplasmic proteins purported to be important in physiological response as well as endocytosis, which are possibly absent from the transfected cell type or not present in the appropriate form.[93] As mentioned in section E above, mutation of tyrosine at position 252 within the CD32 cytoplasmic domain increases the CD32 mobile fraction by about 30% as well reducing the apparent lateral diffusion coefficient by 50% (see Table 3.2), implying that tyrosine phosphorylation may also be important in regulating CD32 lateral mobility, probably through modulating the association of cytoplasmic factors.[93] Consistent with this, truncated cytoplasmic domain mutants lacking tyrosine at position 252 as well as a second putative tyrosine phosphorylation site are not responsive to PMA in terms of altered lateral mobility properties.[93] Synergistic phosphorylation events mediated by different kinases in response to various stimuli thus probably act to regulate the lateral mobility of CD32. The mobile fraction of Fcγ RIIIB in transfected mouse fibroblasts in the presence of the interacting immobilizing integrin CR3 (see section A above) is increased by the protein kinase inhibitor staurosporine (from 0.48 to 0.59),[37] implying that phosphorylation can modulate membrane protein mobility by regulating protein-protein interactions within the plane of the membrane. The anti-inflammatory agent indomethacin, which affects intracellular cyclo-oxygenase activity, has no significant effect on Fcγ RIIIB lateral mobility in the presence of CR3.[37]

The nicotinic acetylcholine receptor appears to be immobilized in specialized regions on myotubes in analogous fashion to CD32 through tyrosine phosphorylation[119] induced by the extracellular matrix heparin sulfate proteoglycan basal membrane synaptic protein agrin, which effects acetylcholine receptor clustering.[130] Increasing tyrosine phosphorylation through treatment with the phosphatase inhibitor pervanadate also reduces the acetylcholine receptor mobile fraction.[119] In analogous fashion, the mobile fraction of epithelial cell amiloride-sensitive Na^+ channels is increased by aldosterone but not vasopressin treatment.[63] The CD2 mobile fraction seems to be modulated negatively by intracellular calcium (see chapter 8; Fig. 8.4), the mobile fraction being highest at resting state Ca^{2+} concentrations.[36] Activation through crosslinking antibodies in the absence or presence of treatments modulating intracellular Ca^{2+} levels, including the divalent cation chelator EGTA (ethylene glycol-bis[β-amino ethyl ether] N,N,N',N'-tetraacetic acid), the Ca^{2+} ionophore ionomycin or a permeant cAMP analog, affects the CD2 mobile fraction to various degrees due to the differing extent of the Ca^{2+} response (Table 3.3A).[36]

The stage of the cell cycle has also been shown to affect plasma membrane protein lateral mobility (see Table 3.3B). The mobile fraction of the GP80 glycoprotein in mouse fibroblasts, for example, is lower during G_0.[107] Analogously, the mobile fraction of MHC antigens is decreased in confluent as opposed to nonconfluent cells, apparently through association with the extracellular matrix (see Table 3.3B).[54] Differentiation of confluent HT29 colonic carcinoma cells induced by replacing glucose with galactose in the growth medium causes changes in the lateral mobility of the β2-microglobulin of human leukocyte antigen HLA class 1 (Table 3.3B) and neoplastic antigens but not of GM1 or class 2 HLA-DR antigen.[120] A study of lateral mobility of surface proteins and lipid probes demonstrated that the apparent lateral diffusion coefficients of lipid probes are lowest just before and during mitosis and maximal in S-phase, probably as a result of changes in lipid fluidity, whereas protein mobility is highest in G_1.[50] Finally, a further example of modulation of plasma membrane protein lateral mobility properties by the cell or growth stage is that of the platelet-derived growth factor (PDGF) receptor.[121] Starved cells exhibit enhanced lateral mobility with respect to both the rate of lateral diffusion and PDGF-receptor mobile fraction, both of which are further increased when starved cells are refed with serum (see Table 3.3B).[121]

H. SUMMARY

As outlined in this chapter, a number of factors modulate plasma membrane protein lateral mobility. These are summarized in Table 3.4. While our understanding of these processes is incomplete, it is clear from what we do know that protein lateral movement is modulated by a variety of specific mechanisms. Interestingly, the factors modifying membrane protein lateral mobility such as linkages to the cytoskeleton and lipid fluidity appear to be able to be regulated themselves by growth and other signals, as seen in the previous section. This latter observation is most significant, since it means that the lateral mobility of membrane receptors, so integral to membrane signal transduction in response to the ligands they bind, may itself be regulated by hormonal signals. The cytoskeleton would appear to play a central role in mediating these processes, which will be discussed again in the concluding chapter of this book.

Table 3.4. Summary of the factors affecting the lateral mobility of proteins in biological membranes.

Factors Influencing Membrane Protein Lateral Mobility	Examples
Protein concentration	bacteriorhodopsin (high concentration reduces mobility in artificial membranes)[5], VSV G protein (increased multiplicity of infection reduces mobility)[10]
Protein-protein interaction in the plane of the membrane	Fcγ RIIIB* (mobility is reduced by CR3+)[37] red blood cell band 3 anionic transporter (mobility reduced by antibodies to glycophorin A)[40] VSV G protein (mobility is reduced by VSV M protein)[10]

Table continues on next page.

Table 3.4. *(continued)*

Aggregation in localized areas of the membrane (immobilization)	GABA/benzodiazepine,[26] nicotinic acetylcholine receptors,[26,52] the insulin-responsive glucose transporter[27]
Binding to extracellular matrix (immobilization)	fibronectin receptor (binds to extracellular matrix through its ligand fibronectin),[29] GP80 (binds to surface chondroitin glycosaminoglycan and an extracellular proteoglycan),[2] nicotinic acetylcholine receptor (binds to heparan surface proteoglycan - agrin)[131]
Binding to an intracellular matrix such as the membrane skeleton/cytoskeleton	red blood cell anionic transporter band 3 (specific links to membrane skeletal components spectrin and ankyrin),[21,30,65] GP80,[107] LH receptor,[31] transferrin receptor (microtubules),[63] LDL receptor,[71] amiloride-sensitive Na^+ channel (association with the cytoskeleton is hormonally regulated)[62]
Association with cell-cell contact or intercalating regions	CD2 and its ligand LFA-3[35,36]
Membrane Lipid Mobility	nicotinic acetylcholine receptor (increasing membrane lipid fluidity increases mobility)[48,52,53]
Primary Sequence	Membrane anchoring; transmembrane domain (e.g., EGF or insulin receptor) or GPI anchor (e.g., Thy-1 or LFA-3)[35] Cytoplasmic domain; determines interaction with other membrane-associated and cytoplasmic components, including the cytoskeleton (e.g., red blood cell band 3);[2,30] determines interaction with coated pits[95] (e.g., LDL receptor)[71] Cytoplasmic domain; presence of phosphorylation sites (e.g., Neu receptor, CD4)[89,92]
Domain Structure: Regions of Restricted Mobility	glycine receptor (lower mobility on neuronal processes than on the neuronal cell body)[25] PH-20 (lower mobility in the posterior head of acrosome-intact guinea pig sperm than on the inner acrosomal membrane)[105] voltage dependent Na^+ channels,[106] nicotinic acetylcholine[26,52]

Table continues on next page.

Table 3.4. (continued)

	and GABA/benzodiazepine[26] receptors (all show restricted mobility on axon hillocks but high mobility on cell bodies)
Signal Transduction	MHC Class I/glycoprotein gp96 (TNF/IFNγ decrease lateral mobility)[117] CD4 (PMA treatment reduces the mobile fraction; a calmodulin but not a PK-C inhibitor reverses the effect)[92] Fcγ RIIa* (PMA treatment reduces lateral mobility through PK-C phosphorylation)[93] CD2 (Ca^{2+} elevation upon T-cell activation reduces the mobile fraction)[36] CR3+ (receptor mobilization through lymphokine treatment;[125,126] PMA treatment can induce clustering)[124,127-129] nicotinic acetylcholine receptor (agrin-induced tyrosine phosphorylation effects aggregation and reduces mobility)[119] amiloride sensitive Na^+ channel (mobile fraction is increased by aldosterone treatment through modulation of linkage to the cytoskeleton)[62]
Cell cycle	Cell surface proteins (highest mobility during S and G_2)[50] GP80 glycoprotein (lateral mobility reduced during G_0)[107] MHC Class I (mobility is reduced in confluent cells and in cells with increased cell-cell contacts)[54] HLA Class I β2-microglobulin (mobile fraction and apparent lateral diffusion coefficient are increased upon differentiation)[120] PDGF receptor (starved cells exhibit enhanced lateral mobility; mobility is further enhanced by serum)[121] Voltage dependent Na^+ channel (clustering and immobilization during myelination/differentiation)[132]

VSV G protein, vesicular stomatitis virus glycoprotein; GABA, γ-amino butyric acid; GP, glycoprotein; LH, luteinizing hormone receptor; LDL, low density lipoprotein; LFA-3, lymphocyte function-associated antigen-3; EGF, epidermal growth factor; GPI, glycosyl phosphoinositol; MHC, major histocompatibility complex; TNF, tumor necrosis factor; IFNγ, interferon γ; PMA, phorbol-12-myristate-13-acetate; PK-C, Ca^{2+}, phospholipid dependent kinase; PDGF, platelet-derived growth factor; HLA, human leukocyte antigen.
*Low-affinity IgG Fc receptor.
+CR3, C3b complement receptor also known as Mac-1 or CD11b/CD18.

REFERENCES

1. Edidin M, Kuo SC, Sheetz MP. Lateral movements of membrane glycoproteins restricted by dynamic cytoplasmic barriers. Science 1991; 254:1379-1382.
2. Jacobson K, Ishihara A, Inman R. Lateral diffusion of proteins in membranes. Ann Rev Physiol 1987; 49:163-175.
3. Jans DA, Pavo I. A mechanistic role for polypeptide hormone receptor lateral mobility in signal transduction. Amino Acids 1995; 9:93-109.
4. Helmreich EJM, Elson EL. Protein and lipid mobility. Adv in Cyclic Nucleotide and Prot Phosphor Res 1984; 18:1-62.
5. Peters R, Cherry R. Lateral and rotational diffusion of bacteriorhodopsin in lipid bilayers: Experimental test of the Saffman-Delbrueck equations. Proc Natl Acad Sci USA 1982; 79:4317-4321.
6. Vaz W, Goodsaid-Zalduondo F, Jacobson K. Lateral diffusion of lipids and proteins in bilayer membranes. FEBS Lett 1984; 174:199-207.
7. Eisinger J, Halperin B. Effect of spatial variation in membrane diffusibility and solubility on the lateral transport of membrane components. Biophys J 1986; 50:513-521.
8. Saffman PG, Delbrueck M. Brownian motion in biological membranes. Proc Natl Acad Sci USA 1975; 72:3111-3113.
9. Saffman PG. Lateral and rotational movement in membranes. Fluid Mech 1976; 73:593-602.
10. Reidler JA, Keller PM, Elson EL et al. A fluorescence photobleaching study of vesicular stomatitis virus infected BHK cells. Modulation of G protein mobility by M protein. Biochemistry 1981; 20(5):1345-1349.
11. Dratz EA, Miljanich GP, Nemes PP et al. The structure of rhodopsin and its disposition in the rod outer segment disk membrane. Photochem Photobiol 1979; 29(4):661-670.
12. Lamb TD. Stochastic simulation of activation in the G-protein cascade of phototransduction. Biophys J 1994; 67:1439-1454.
13. Hamm HE, Bownds MD. Protein complement of rod outer segments of frog retina. Biochemistry 1986; 25(16):4512-4523.

14. Dumke CL, Arshavsky VY, Calvert PD et al. Rod outer segment structure influences the apparent kinetic parameters of cyclic GMP phosphodiesterase. J Gen Physiol 1994; 103(6):1071-1098.
15. Henderson R, Unwin PN. Three-dimensional model of purple membrane obtained by electron microscopy. Nature 1975; 257(5521):28-32.
16. Henderson R, Baldwin JM, Ceska TA et al. Model for the structure of bacteriorhodopsin based on high-resolution electron cryo-microscopy. J Mol Biol 1990; 213:899-929.
17. Salpeter MM, Loning RH. Nicotinic acetylcholine receptor in vertebrate muscle: properties, distribution and neural control. Prog Neurobiol 1985; 25:297-325.
18. Poo M, Cone RA. Lateral diffusion of rhodopsin in the photoreceptor membrane. Nature 1974; 247(441):438-441.
19. Gupta BD, Williams TP. Lateral diffusion of visual pigments in toad (Bufo marinus) rods and in catfish (Ictalurus punctatus) cones. J Physiol Lond 1990; 430:483-496.
20. Liebman PA, Entine G. Lateral diffusion of visual pigment in photoreceptor disk membranes. Science 1974; 185(149):457-459.
21. Golan DE, Veatch W. Lateral mobility of band 3 in the human erythrocyte membrane studied by fluorescence photobleaching recovery: evidence for control by cytoskeletal interactions. Proc Natl Acad Sci USA 1980; 77:2537-2541.
22. Pal R, Nair BC, Hoke GM et al. Lateral diffusion of CD4 on the surface of a human neoplastic T-cell line probed with a fluorescent derivative of the envelope glycoprotein (gp120) of human immunodeficiency virus type 1 (HIV-1). J Cell Physiol 1991; 147(2):326-332.
23. Posner RG, Subramanian K, Goldstein B et al. Simultaneous cross-linking by two nontriggering bivalent ligands causes synergistic signaling of IgE Fc epsilon RI complexes. J Immunol 1995; 155(7):3601-3609.
24. George SK, Xu YH, Benson LA et al. Cytochrome b5 and a recombinant protein containing the cytochrome b5 hydrophobic domain spontaneously associ-

ate with the plasma membrane of cells. Biochim Biophys Acta 1991; 1066(2):131-143.

25. Srinivasan Y, Guzikowski AP, Haugland RP et al. Distribution and lateral mobility of glycine receptors on cultured spinal cord neurons. J Neurosci 1990; 10(3):985-995.

26. Velazquez JL, Thompson CL, Barnes EM Jr et al. Distribution and lateral mobility of GABA/benzodiazepine receptors on nerve cells. J Neurosci 1989; 9(6): 2163-2169.

27. Lange K, Brandt U. Insulin-responsive glucose transporters are concentrated in a cell surface-derived membrane fraction of 3T3-L1 adipocytes. FEBS Lett 1990; 261(2):459-463.

28. Zakharova OM, Rosenkranz AA, Sobolev AS. Modification of fluid lipid and mobile protein fractions of reticulocyte plasma membranes affects agonist-stimulated adenylate cyclase. Application of the percolation theory. Biochim Biophys Acta 1995; 1236:177-184.

29. Duband JL, Nuckolls GH, Ishihara A et al. Fibronectin receptor exhibits high lateral mobility in embryonic locomoting cells but is immobile in focal contacts and fibrillar streaks in stationary cells. J Cell Biol 1988; 107:1385-1396.

30. Tsuji A, Ohnishi, S. Restriction of the lateral motion of Band 3 in the erythrocyte membrane by the cytoskeletal network: dependence on spectrin association state. Biochemistry 1986; 25:6133-6139.

31. Roess DA, Niswender GD, Barisas BG. Cytocholasins and colchicine increase the lateral mobility of human chorionic gonadotropin-occupied luteinizing hormone receptors on ovine luteal cells. Endocrinol 1988; 122:261-269.

32. Elson EL, Schlessinger J, Koppel DE et al. Measurement of lateral transport on cell surfaces. Prog Clin Biol Res. 1976; 9:137-147.

33. Schlessinger J, Koppel DE, Axelrod D et al. Lateral transport on cell membranes: mobility of concanavalin A receptors on myoblasts. Proc Natl Acad Sci USA 1976; 73(7):2409-2413.

34. Schlessinger J, Axelrod D, Koppel DE et al. Lateral transport of a lipid probe and labeled proteins on a cell membrane. Science 1977; 195(4275):307-309.

35. Chan PY, Lawrence MB, Dustin ML et al. Influence of receptor lateral mobility on adhesion strengthening between membranes containing LFA-3 and CD2. J Cell Biol 1991; 115(1):245-255.

36. Liu SJ, Hahn WC, Bierer BE et al. Intracellular mediators regulate CD2 lateral diffusion and cytoplasmic Ca^{2+} mobilization upon CD2-mediated T cell activation. Biophys J 1995; 68(2):459-470.

37. Poo H, Krauss JC, Mayo-Bond L et al. Interaction of Fc gamma receptor type IIIB with complement receptor type 3 in fibroblast transfectants: evidence from lateral diffusion and resonance energy transfer studies. J Mol Biol 1995; 247(4):597-603.

38. Zhou MJ, Todd RF, van de Winkel JGJ et al. Co-capping of the leukoadhesion molecules complement receptor type 3 and lymphocyte function-associated antigen-1 with Fcγ receptor III on human neutrophils: possible role of lectin-like interactions in their association. J Immunol 1993; 150:3030-3041.

39. Bierer BE, Golan DE, Brown CS et al. A monoclonal antibody to LFA-3, the CD2 ligand, specifically immobilizes major histocompatibility complex proteins. Eur J Immunol 1989; 19(4):661-665.

40. Knowles DW, Chasis JA, Evans EA et al. Cooperative action between band 3 and glycophorin A in human erythrocytes: immobilization of band 3 induced by antibodies to glycophorin A. Biophys J 1994; 66(5):1726-1732.

41. Corbett JD, Golan DE. Band 3 and glycophorin are progressively aggregated in density-fractionated sickle and normal red blood cells. Evidence from rotational and lateral mobility studies. J Clin Invest 1993; 91(1):208-217.

42. Bennett V, Stenbuck PJ. The membrane attachment protein for spectrin is associated with band 3 in human erythrocyte membranes. Nature 1979; 280(5722):468-473.

43. Bennett V, Stenbuck PJ. Identification and partial purification of ankyrin, the high affinity membrane attachment site

for human erythrocyte spectrin. J Biol Chem 1979; 254(7):2533-2541.

44. Henis YI, Katzir Z, Shia MA et al. Oligomeric structure of the human asialoglycoprotein receptor: nature and stiochiometry of mutual complexes containing H1 and H2 polypeptides assessed by fluorescence photobleaching recovery. J Cell Biol 1990; 111:1409-1418.

45. Yechiel E, Barenholz Y, Henis YI. Lateral mobility and organization of phospholipids and proteins in rat myocyte membranes. Effects of aging and manipulation of lipid composition. J Biol Chem 1985; 260:9132-9136.

46. Fahey PF, Koppel DE, Barak LS et al. Lateral diffusion in planar lipid bilayers. Science 1977; 195(4275):305-306.

47. Vaz WL, Clegg RM, Hallmann D. Translational diffusion of lipids in liquid crystalline phase phosphatidylcholine multibilayers. A comparison of experiment with theory. Biochemistry 1985; 24(3):781-786.

48. Axelrod D, Wight A, Webb W et al. Influence of membrane lipids on acetylcholine receptor and lipid probe diffusion in cultured myotube membrane. Biochemistry 1978; 17(17):3604-3609.

49. Schootemeijer A, Gorter G, Tertoolen LG et al. Relation between membrane fluidity and signal transduction in the human megakaryoblastic cell line MEG-01. Biochim Biophys Acta 1995; 1236(1):128-134.

50. de Laat SW, van der Saag PT, Elson EL et al. Lateral diffusion of membrane lipids and proteins during the cell cycle of neuroblastoma cells. Proc Natl Acad Sci USA 1980; 77(3):1526-1528.

51. Henis YI, Hekman M, Elson EL et al. Lateral diffusion of β-receptors in membranes of cultured liver cells. Proc Natl Acad Sci USA 1982; 79:2907-2911.

52. Stya M, Axelrod D. Mobility and detergent extractability of acetylcholine receptors on cultured rat myotubes: a correlation. J Cell Biol 1983; 97(1):48-51.

53. Axelrod D, Ravdin P, Koppel DE et al. Lateral motion of fluorescently labeled acetylcholine receptors in membranes of developing muscle fibers. Proc Natl Acad Sci USA 1976; 73(12):4594-4598.

54. Wier M, Edidin M. Effects of cell density and extracellular matrix on the lateral diffusion of major histocompatibility antigens in cultured fibroblasts. J Cell Biol 1986; 103:215-222.

55. Havell TC, Hillman D, Lessin LS. Deformability characteristics of sickle cells by microelastimetry. Am J Hematol 1978; 4:9-16.

56. Mohandas N, Winardi R, Knowles D et al. Molecular basis for membrane rigidity of hereditary ovalocytosis: a novel mechanism involving the cytoplasmic domain of band 3. J Clin Invest 1992; 89:686-692.

57. Smith JE, Mohandas N, Clark MR et al. Deformability and spectrin properties in three types of elongated red cells. Am J Hematol 1980; 8:1-13.

58. Moscona-Amir E, Henis YI, Yechiel E et al. Role of lipids in age-related changes in the properties of muscarinic receptors in cultured rat heart myocytes. Biochemistry 1986; 25(24):8118-8124.

59. Moscona-Amir E, Henis YI, Sololovsky M. Aging of rat heart myocytes disrupts muscarinic receptor coupling that leads to inhibition of cAMP accumulation and alters the pathway of muscarinic-stimulated phosphoinositide hydrolysis. Biochemistry 1989; 28(17):7130-7137.

60. Edelman GM. Surface modulation in cell recognition and cell growth. Science 1976; 192:218-226.

61. Paller MS. Lateral mobility of Na,K-ATPase and membrane lipids in renal cells. Importance of cytoskeletal integrity. J Membr Biol 1994; 142(1):127-135.

62. Smith PR, Stoner JC, Viggiano SC et al. Effects of vasopressin and aldosterone on the lateral mobility of eptihelial Na⁺ channels in A6 epithelial cells. J Memb Biol 1995; 147(2):195-205.

63. Thatte HS, Bridges KR, Golan DE. ATP depletion causes translational immobilization of cell surface transferrin receptors in K562 cells. J Cell Physiol 1996; 166:446-452.

64. Thatte HS, Bridges KR, Golan DE. Microtubule inhibitors differentially affect translational movement, cell surface expression, and endocytosis of transferrin receptors in K562 cells. J Cell Physiol 1994; 160:345-357.

65. Tsuji A, Kawasaki S, Ohnishi, S et al. Regulation of band 3 mobilities in erythrocyte ghost membrane by protein association and cytoskeletal network. Biochemistry 1988; 27:7447-7452.

66. Koppel DE, Sheetz MP. Fluorescence photobleaching does not alter the lateral mobility of erythrocyte membrane glycoproteins. Nature 1981; 293(5828):159-161.

67. Sheetz MP, Schindler M, Koppel DE. Lateral mobility of integral membrane proteins is increased in spherocytic erythrocytes. Nature 1980; 285(5765):510-511.

68. Zs Nagy I, Zhang X, Kitani K et al. The influence of dystrophin on lateral diffusion of proteins in sarcolemma of L-185 and C2 myoblasts and mature striated muscle cells of rats and mice, as measured by FRAP technique. Biochem Biophys Res Commun 1995; 215(1):67-74.

69. Tank DW, Wu ES, Webb WW. Enhanced molecular diffusibility in muscle membrane blebs: release of lateral constraints. J Cell Biol 1982; 92(1):207-212.

70. Webb WW, Barak LS, Tank DW et al. Molecular mobility on the cell surface. Biochem Soc Symp 1981(46):191-205.

71. Barak LS, Webb WW. Diffusion of low density lipoprotein-receptor complex on human fibroblasts. J Cell Biol 1982; 95:846-852.

72. Wu ES, Tank DW, Webb WW. Unconstrained lateral diffusion of concanavalin A receptors on bulbous lymphocytes. Proc Natl Acad Sci USA 1982; 79(16): 4962-4966.

73. Swaisgood M, Schindler M. Lateral diffusion of lectin receptors in fibroblast membranes as a function of cell shape. Exp Cell Res 1989; 180(2):515-528.

74. Swaisgood M, Schindler M. Clonal selection by fluorescence redistribution after photobleaching (FRAP)–a "fast" lateral mobility fibroblast mutant (E7G1). Exp Cell Res 1989; 180(2):529-536.

75. Rasenick MM, Stein PJ, Bitensky MW. The regulatory subunit of adenylate cyclase interacts with cytoskeletal components. Nature 1981; 294(5841):560-562.

76. Rudolph SA, Greengard P, Malawista SE. Effects of colchicine on cyclic AMP levels in human leukocytes. Proc Natl Acad Sci USA 1977; 74(8):3404-3408.

77. Rudolph SA, Hegstrand LR, Greengard P et al. The interaction of colchicine with hormone-sensitive adenylate cyclase in human leukocytes. Mol Pharmacol. 1979; 16(3):805-812.

78. Simantov R, Sachs L. Cytoskeleton regulates beta-adrenergic hormonal stimulation in normal and leukemic white blood cells. FEBS Lett 1978; 90(1):69-75.

79. Yahara I, Edelman GM. Modulation of lymphocyte receptor mobility by locally bound concanavalin A. Proc Natl Acad Sci USA 1975; 72:1579-1583.

80. Schlessinger J, Elson EL, Webb WW et al. Receptor diffusion on cell surfaces modulated by locally bound concanavalin A. Proc Natl Acad Sci USA 1977; 74(3):1110-1114.

81. Henis YI, Elson EL. Inhibition of the mobility of mouse lymphocyte surface immunoglobulins by locally bound concanavalin A. Proc Natl Acad Sci USA 1981; 78:1072-1076.

82. Edelman GM, Yahara I, Wang JL. Receptor mobility and receptor-cytoplasmic interactions in lymphocytes. Proc Natl Acad Sci USA 1973; 70:1442-1446.

83. Corcao G, Sutcliffe RG, Kusel JR. Lateral diffusion of human CD2 wild type and mutants with large deletions in the transmembrane domain. Biochem Biophys Res Commun 1995; 208(3):1131-1116.

84. Goncalves E, Yamada K, Thatte HS et al. Optimizing transmembrane domain helicity accelerates insulin receptor internalization and lateral mobility. Proc Natl Acad Sci USA 1993; 90:5762-5766.

85. Scullion BF, Hou Y, Puddington L et al. Effects of mutations in three domains of the vesicular stomatitis viral glycoprotein on its lateral diffusion in the plasma membrane. J Cell Biol 1987; 105(1):69-75.

86. Zhang F, Crise B, Su B et al. Lateral diffusion of membrane-spanning and glycosylphosphatidylinositol-linked proteins: towards establishing rules governing the lateral mobility of membrane proteins. J Cell Biol 1991; 115:75-84.

87. Zhang F, Crise B, Su B et al. The lateral mobility of some membrane proteins is determined by their ectodomains. Biophys J 1992; 62:92-94.

88. Livneh E, Benveniste M, Prywes R et al. Large deletions in the cytoplasmic kinase domain of the epidermal growth factor receptor do not affect its lateral mobility. J Cell Biol 1986; 103:327-331.

89. Gilboa L, Ben-Levy R, Yarden Y et al. Roles for a cytoplasmic tyrosine and tyrosine kinase activity in the interactions of Neu receptors with coated pits. J Biol Chem 1995; 270:7061-7067.

90. Edidin M, Zuniga MC. Lateral diffusion of wild-type and mutant L^d antigens in L-cells. J Cell Biol 1984; 99:2333-2335.

91. Edidin M, Zuniga MC, Sheetz MP. Truncation mutants define and locate cytoplasmic barriers to lateral mobility of membrane glycoproteins. Proc Natl Acad Sci USA 1994; 91:3378-3382.

92. Grebenkamper K, Tosi PF, Lazarte JE et al. Modulation of CD4 lateral mobility in intact cells by an intracellularly applied antibody. Biochem J 1995; 312(1):251-259.

93. Zhang F, Yang B, Odin JA et al. Lateral mobility of Fcγ RIIa is reduced by protein kinase C activation. FEBS Lett 1995; 376(1-2):77-80.

94. Mao SY, Varin-Blank N, Edidin M et al. Immobilization and internalization of mutated IgE receptors in transfected cells. J Immunol 1991; 146(3):958-966.

95. Paccaud JP, Reith W, Johansson B et al. Role of internalization signals and receptor mobility. J Biol Chem 1993; 268(31):23191-23196.

96. Kooyman DL, Byrne GW, McClellan S. In vivo transfer of GPI-linked complement restriction factors from erythrocytes to the endothelium. Science 1995; 269(5220):89-92.

97. Woda B, Gilman S. Lateral mobility and capping of rat lymphocyte membrane proteins. Cell Biol Int Rep 1983; 7:203-209.

98. Bulow R, Overath P, Davoust J. Rapid lateral diffusion of the variant surface glycoprotein in the coat of *Trypanosoma brucei*. Biochemistry 1988; 27(7):2384-2388.

99. Singer SJ, Nicolson GL. The fluid mosaic model of the structure of cell membranes. Science 1972; 175(23):720-731.

100. Sako Y, Kusumi A. Compartmentalized structure of the plasma membrane for lateral diffusion of receptors as revealed by nanometer-level motion analysis. J Cell Biol 1994; 125:1251-1264.

101. Tetteroo PA, Bluemink JG, Dictus WJ et al. Lateral mobility of plasma membrane lipids in dividing *Xenopus* eggs. Dev Biol 1984; 104(1):210-218.

102. Dictus WJ, van Zoelen EJ, Tetteroo PA et al. Lateral mobility of plasma membrane lipids in *Xenopus* eggs: regional differences related to animal/vegetal polarity become extreme upon fertilization. Dev Biol 1984; 101(1):201-211.

103. Speksnijder JE, Dohmen MR, Tertoolen LG et al. Regional differences in the lateral mobility of plasma membrane lipids in a molluscan embryo. Dev Biol 1985; 110(1):207-216.

104. Tournier JF; Lopez A; Gas N et al. The lateral motion of lipid molecules in the apical plasma membrane of endothelial cells is reversibly affected by the presence of cell junctions. Exp Cell Res 1989; 181(2): 375-384.

105. Cowan AE, Myles DG, Koppel DE. Lateral diffusion of the PH-20 protein on guinea pig sperm: evidence that barriers to diffusion maintain plasma membrane domains in mammalian sperm. J Cell Biol 1987; 104:917-923.

106. Angelides KJ, Elmer LW, Loftus D et al. Distribution and lateral mobility of voltage-dependent sodium channels in neurons. J Cell Biol 1989; 106:1911-1925.

107. Jacobson K, O'Dell D, August JT. Lateral diffusion of an 80,000-dalton glycoprotein in the plasma membrane of murine fibroblasts: relationships to cell structure and function. J Cell Biol 1984; 99(5):1624-1633.

108. Myles DG, Primakoff P, Koppel DE. A localized surface protein of guinea pig sperm exhibits free diffusion in its domain. J Cell Biol 1984; 98:1905-1909.

109. Jesaitis AJ, Yguerabide J. The lateral mobility of the (Na^+,K^+)-dependent ATPase in Madin-Darby canine kidney cells. J Cell Biol 1986; 102:1256-1263.

110. Ladha S, Mackie AR, Clark DC. Cheek cell membrane fluidity measured by fluorescence recovery after photobleaching and steady-state fluorescence anisotropy. J Membr Biol. 1994; 142(2):223-228.

111. Saxton MJ. The membrane skeleton of erythrocytes. A percolation model. Biophys J 1990; 57:1167-1177.

112. Saxton MJ. The spectrin network as a barrier to lateral diffusion in erythrocytes: a percolation analysis. Biophys J 1989; 55:21-28.

113. Saxton MJ. Lateral diffusion in an archipelago: distance dependence of the diffusion coefficient. Biophys J 1989; 56:615-622.

114. Saxton MJ. Single-particle tracking: effects of corrals. Biophys J 1995; 69(2):389-398.

115. Sobolev AS, Rosenkranz AA, Kazarov AR. Interaction of proteins of the adenylate cyclase complex: area-limited mobility of movement along the whole membrane? Analysis with the application of the percolation theory. Biosc Rep 1984; 4:897-902.

116. Sobolev AS, Kazarov AR, Rosenkranz AA. Application of percolation theory principles to the analysis of interaction of adenylate cyclase complex proteins in cell membranes. Mol Cell Biochem 1988; 81(1):19-28.

117. Stolpen AH, Golan DE, Pober JS. Tumor necrosis factor and immune interferon act in concert to slow the lateral diffusion of proteins and lipids in human endothelial cell membranes. J Cell Biol 1988; 107:781-789.

118. Polgar K, Yacono PW, Golan DE et al. Immune interferon gamma inhibits translational diffusion of a plasma membrane protein in preimplantation stage mouse embryos: a T-helper 1 mechanism for immunologic reproductive failure. Am J Obstet Gynecol 1996; 174(1/1):282-287.

119. Meier T, Perez GM, Wallace BG. Immobilization of nicotinic acetylcholine receptors on mouse C2 myotubes by agrin-induced protein tyrosine phosphorylation. J Cell Biol 1995; 131(2):441-451.

120. Magnusson KE, Gustafsson M, Holmgren K et al. Small intestinal differentiation in human colon carcinoma HT29 cells has distinct effects on the lateral diffusion of lipid (ganglioside GM1) and proteins (HLA class 1, HLA class 2, and neoplastic epithelial antigens) in the apical cell membrane. J Cell Physiol 1990; 143(2):381-390.

121. Ljungquist P, Wasteson A, Magnusson K-E. Lateral diffusion of plasma membrane receptors labelled with either platelet-derived growth factor (PDGF) or wheat germ agglutinin (WGA) in human leukocytes and fibroblasts. Bioscience Reports 1989; 9:63-73.

122. Tournier JF, Lopez A, Tocanne JF. Effect of cell substratum on lateral mobility of lipids in the plasma membrane of vascular endothelial cells. Exp Cell Res 1989; 181(1):105-115.

123. Griffin Jnr FM, Mullinax PJ. Augmentation of macrophage complement receptor function in vitro. III. C3b receptors that promote phagocytosis migrate within the plane of macrophage plasma membrane. J Med Exp 1981; 154:291-305.

124. Detmers PA, Wright SD, Olsen E et al. Aggregation of complement receptors on human neutrophils in the absence of ligand.

125. Griffin FM Jr, Mullinax PJ. Effects of differentiation in vivo and of lymphokine treatment in vitro on the mobility of C3 receptors of human and mouse mononuclear phagocytes. J Immunol 1985; 135(5):3394-3397.

126. Griffin FM Jr, Mullinax PJ. High concentrations of bacterial lipopolysaccharide, but not microbial infection-induced inflammation, activate macrophage C3 receptors for phagocytosis. J Immunol 1990; 145(2):697-701.

127. Pryzwansky KB, Wyatt T, Reed W et al. Phorbol ester induces transient focal concentrations of functional, newly expressed CR3 in neutrophils at sites of specific granule exocytosis. Eur J Cell Biol 1991; 54(1):61-75.

128. Hermanowski-Vosatka A, Detmers PA, Gotze O et al. Clustering of ligand on the surface of a particle enhances adhesion to receptor-bearing cells. J Biol Chem 1988; 263(33):17822-17827.

129. Ross GD, Reed W, Dalzell JG et al. Macrophage cytoskeletal association with CR3 and CR4 regulates receptor mobility and phagocytosis of iC3b-opsonized erythrocytes. J Leukoc Biol 1992; 51(2):109-117.

130. Ferns M, Deiner M, Hall Z. Agrin-induced acetylcholine receptor clustering in

mammalian muscle requires tyrosine phosphorylation. J Cell Biol 1996; 132(5):937-944.

131. Anderson MJ, Fambrough DM. Aggregates of acetyl choline are associated with plaques of a basal lamina, heparan sulfate proteoglycan on the surface of skeletal muscle fibers. J Cell Biol 1983; 97:1396-1411.

132. Joe EH, Angelides KJ. Clustering and mobility of voltage-dependent sodium channels during myelination. J Neurosci 1993; 13(7):2993-3005.

LATERAL MOBILITY OF POLYPEPTIDE HORMONE RECEPTORS AND GTP-BINDING PROTEINS

A. INTRODUCTION

As we saw in the previous chapter, the membrane, and in particular its integral and peripheral membrane protein components, are not organized as simplistically, or at least functionally, as envisaged in the fluid mosaic model. Firstly, there is clearly a domain structure, with many physical restrictions to long range protein lateral movement within the plane of the membrane; that is, diffusion may only occur within limited areas or domains of the membrane. Secondly, many membrane proteins have reduced or essentially no mobility through a variety of specific mechanisms, such as linkage to the cytoskeleton, aggregation, etc. Accordingly, it is crucial to any critical examination of the tenets of the Mobile Receptor Hypothesis to examine the evidence for the actual mobility of plasma membrane integral receptors. This chapter will thus concentrate on lateral mobility measurements for plasma membrane integral receptors for polypeptide hormones (see refs. 1-3) and draw general conclusions with respect to receptor movement in the context of the possible relevance to signal transduction. The succeeding chapter (chapter 5) will then deal specifically with the direct and indirect evidence for a role for polypeptide hormone receptor lateral movement in signal transduction, while chapter 6 will discuss the role of receptor lateral movement in the desensitization of response subsequent to hormonal stimulation. Chapter 7 rounds the picture by concentrating on the specific role of receptor immobilization in signaling events relating particularly to immune responses and cell adhesion.

B. PRACTICAL CONSIDERATIONS

Fluorescence photobleaching recovery measurements of membrane receptors have several additional complications which do not necessarily apply to protein lateral mobility determinations in general (see also chapter 2, section D). The most physiologically relevant measurements are those performed using fluorescently-labeled ligands, where an important prerequisite is that covalent modification with a chromophore affects neither the high affinity of ligand binding nor its biological activity. In contrast, measurements made using

The Mobile Receptor Hypothesis: The Role of Membrane Receptor Lateral Movement in Signal Transduction, by David A. Jans. © 1997 R.G. Landes Company.

fluorescently-labeled monovalent or divalent antibodies to specific receptors or receptor subunits should be treated with caution, since antibodies are generally much larger than most polypeptide hormone ligands and their binding may induce gross conformational changes in receptor structure as a result. This is a particularly important consideration in the case of small polypeptide ligands such as the nonapeptide vasopressin, cholecystokinin (33 amino acids), insulin (4-6 kDa), etc. Measurements of lateral mobility of the platelet-derived growth factor (PDGF), for example, yield quantitatively quite different results if either Fab antibody fragment or ligand are used as the fluorescent receptor probe (see Table 4.1 below and legend).[4,5] Further to this, ligands in a biological context are of course almost exclusively agonists, activating the receptors they bind; since most antibodies do not function agonistically, their use as probes of receptor lateral mobility does not approximate the physiological situation.

It should be stated at this point that measurements made with hormone analogs which are high-affinity antagonists (see also chapter 6, section G) need to be treated with caution, since such molecules are essentially alien to the in vivo situation and conclusions based on measurements using hormone antagonists alone can be questionable for this reason. While agonists activate receptors upon binding, hormone antagonists bind but do not activate receptors and, hence, do not trigger any of the physiological responses, both stimulatory and downregulatory, normally associated with hormone binding.

Typical additional difficulties of receptor lateral mobility measurements include those of low receptor densities and the accompanying problem of low signal-to-noise ratios; i.e., low-specific fluorescent signals above that of fluorescence due to autofluorescence and nonspecific binding. One approach to overcome this problem is the use of highly-fluorescent ligands, e.g., the biologically-active nerve growth factor (NGF) derivative used by Levi et al[6] con-

tained 8-10 molecules of rhodamine. In a comparable approach, Schlessinger et al[7] coupled either epidermal growth factor (EGF) or insulin to α-lactalbumin containing 7-8 molecules of covalently attached rhodamine. Another way to get around the problem of normally low receptor densities is the use of cell lines which either express abnormally high numbers of receptors (such as the A431 human epidermoid carcinoma line which possesses c. 2,000,000 EGF receptors per cell; see also below),[8,9] or those transfected with cDNAs encoding the receptor of interest.[10-12] The physiological relevance of such studies is questionable, however, since receptor expression is nonphysiologically high and, particularly in the latter case, the appropriate cell context is lacking because the specific signal transduction apparatus apposite to the particular receptor may normally either be absent or not in keeping with the high receptor numbers.

A strategy to optimize low fluorescent signals is to perform measurements of fluorescence at certain intervals (e.g., every 20-40 seconds) rather than in the standard continuous monitoring approach. The latter is illustrated in Figure 4.1, where temperature dependence of lateral mobility of a lipid probe is shown.[13] Measurement of fluorescence intermittently rather than continuously also reduces bleaching during performance of the measurement, which can occur even at low beam intensity.[13-18] This is illustrated in Figure 4.2, which also demonstrates several other aspects of the means by which the fluorescence photobleaching recovery measurement at low signal-to-noise ratios can be carried out and specific fluorescent signals quantitated. As already alluded to in chapter 2 (section C), it is of paramount importance at low receptor densities to perform parallel photobleaching measurements for nonspecific binding of the fluorescent ligand (that is, measurement subsequent to incubation of the cells with fluorescent ligand in the presence of a 100-fold excess of unlabeled ligand). The results for the recovery of fluorescence due to nonspecific binding can be subtracted from

Table 4.1. Lateral mobility of polypeptide hormone plasma membrane receptors, as measured by the technique of fluorescence recovery after photobleaching.

Receptor	Cell type	Temp.	Parameter of Lateral Mobility*		Ref.
			D (10^{-10}cm²/sec)	f	
i. Tyrosine-kinase receptors					
Insulin receptor	mouse 3T3 fibroblasts$@	37°C	0.1 - 1.0	< 0.10	7
		23°C	4.8	0.5 - 0.8	
	transfected CHO cells	21°C	6.0	0.58	10
EGF-receptor	mouse 3T3 fibroblasts$@^	37°C	0.1 - 1.0	< 0.10	7
		23°C	3.4	0.5 - 0.9	
	A431 human epidermoid carcinoma^	37°C×	8.5	0.45	8,9
		23°C	6.0	0.45	8
		4°C	2.8		
		30°C	7.0	0.90	9
		25°C	6.0 c.	0.60	
		20°C	4.7		
		15°C	4.2		
		5°C	2.8		
high-affinity receptor+	A431 human epidermoid carcinoma	22°C	2.3	0.49	20
low-affinity receptor		3-37°C	< 0.01	< 0.05	21
			2.6	0.78	
	transfected COS cells	22°C	1.2	0.61	20
	rat luteal cells	27°C	4.2	c. 0.54	22
	transfected CHO cells	21°C	6.0	0.58	10

Table continues on next page.

Table 4.1. (continued)

Receptor	Cell type	Temp			Ref
Neu receptor	transfected NIH 3T3 mouse fibroblasts§	22°C	2.3	0.57	11
Neu receptor (const. act.)°			2.0	0.28	
PDGF-receptor	human foreskin fibroblasts (AG1523)	23°C	3.2	0.60	4
	human polymorphonuclear leukocytes (PMNL)		3.7	0.59	
PDGF-β receptor	human foreskin fibroblasts (AG1523)ˣˣ		11.0	0.91	5
NGF-receptor	PC12 rat pheochromocytoma**	23°C	6.8	0.35	6
	immature chicken embryonal sensory cells⁺⁺	23°C	5.6	0.22	
	differentiated chicken embryonal sensory cells	37°C	0.05	0	
		4°C	4.6	0.40	
NGF-receptor (high affinity gp75)	transfected cos cells (nonresponsive)	23°C	5.0	0.40	23
	A875 human melanoma (nonresponsive)		5.0	0.52	
	human medulloblastoma MED/NGFR		42	0.12	
	PC12 rat pheochromocytoma (responsive)		40	0.26	
	nnr5 (nonresponsive PC12 derivative)		5.1	0.45	
ii. GTP-binding protein activating receptors					
V₁-receptor	A7r5 rat smooth muscle	37°C	5.1	0.36	14
		23°C	3.6	0.50	
		13°C	2.9	0.44	
V₂-receptor (agonist)	LLC-PK₁ pig kidney	37°C#	2.8	0.91	13,15-18
		23°C	1.5	0.65	
		10°C	UD¹	0.10	
Luteinizing hormone receptor	rat luteal cells	37°C	1.9	0.38	22,24,25
		29°C	1.7		26
		27°C	1.7	0.46	22,24,25
		15°C	UD¹	<0.20	22,24,25
	mouse Leydig cells	4°C	UD¹	<0.20	22,24,25
	sheep luteal cells	29°C	5.8	0.35	26
			1.9		

Table continues on next page.

Table 4.1. (continued)

			D	f	
(human chorionic gonadotropin)°°	rat luteal cells	29°C	<0.2	<0.10	22
(human chorionic gonadotropin)		27°C	0.3	0.10	
(human chorionic gonadotropin)	sheep luteal cells	29°C	<0.2	<0.10	
(human chorionic gonadotropin)	mouse Leydig cells		2.9	0.48	
Cholecystokinin receptor	acinar pancreatic cells	37°C^^	UD[I]	0.04	12
(CCK_A)		24°C	2.2	0.65	
		10°C	1.7	0.17	
	transfected CHO cells	24°C	3.1	0.92	
		10°C	1.4	0.88	
N-formyl peptide receptor	human neutrophils	14°C##	5.5	0.40	27
glucagon receptor	rat hepatocyte	22°C	7.0	0.74	28
iii. Cytokine receptors					
interleukin-2	HUT-102-B2 human T-cell	30°C	2.6	0.37	29
p55 α-subunit@@					

Abbreviations: CHO, Chinese hamster ovary; EGF, epidermal growth factor; PDGF, platelet-derived growth factor; NGF, nerve growth factor.
*Values are for the apparent lateral diffusion coefficient (D) and mobile fraction (f). $Measurements were made using ligands linked to α-lactalbumin, to which 7-8 molecules of rhodamine were covalently attached.[7] @Measurements were made over 20-60 min in the presence of the metabolic inhibitor sodium azide.[7] ^Comparative EGF Measurements for the insulin receptor in the absence of azide yielded values of 4.0×10^{-10} cm^2/sec and 0.4 - 0.8 for D and f, respectively.[7] ^^The high-affinity EGF receptor may be preaggregated and immobilized, possibly through weak association with the cytoskeleton, as implicated by measurements of rotational mobility (see refs. 8,21). $Measurements using a specific monovalent Fab antibody. Measurements using divalent crosslinking antibody result in D values of 2.4 and 2.0 and f values of 0.3 and 0.28 for nonactivated and constitutively activated Neu receptor, respectively.[11] °Constitutively-activated receptor mutant.[11] ××Measurements using a specific monovalent Fab antibody; addition of PDGF to Fab antibody-labeled receptor reduces the mobile fraction to 0.74 with $D = 10 \times 10^{-10}$ cm^2/sec.[5] **Results at 23°C in the presence of methylamine, inhibiting receptor aggregation prior to endocytosis, were 2.1×10^{-10} cm^2/sec and 0.67, respectively.[5] ++Results at 23°C in the presence of methylamine were 8×10^{-10} cm^2/sec and 0.58 for D and f, respectively.[6] #Measurements 10-30 min. IUnable to be determined. °°Luteinizing hormone receptor occupied by human chorionic gonadotropin. ^^Measurements after 15 min.[12] @@Measurements made using a labeled monoclonal antibody (anti-"Tac peptide" antibody) to the α-subunit (p55). The α-subunit has a low-affinity IL-2 binding site; the signaling β-subunit (p95) has an even lower-affinity IL-2 binding site, while the signaling γ-subunit (p75) has no binding site. The high-affinity binding site is constituted by the three subunit receptor.[30]

those for the recovery of fluorescence due to total binding (that is, measurement subsequent to incubation of the cells with fluorescent ligand alone) to yield the recovery of fluorescence due to specific binding of the fluorescent ligand by the receptor (Fig. 4.1).[7,13-18] With this approach, one can perform lateral mobility measurements at low receptor densities with assurance that the results truly reflect movement of the specific receptor in question. A useful control mentioned in previous chapters is that of using cells incubated with ligand that have subsequently been fixed (e.g., using p-formaldehyde) prior to lateral mobility measurements; molecules are crosslinked in fixed cells, so that they are no longer laterally mobile (see refs. 13, 19).

C. LATERAL MOBILITY MEASUREMENTS OF POLYPEPTIDE HORMONE RECEPTORS

Direct measurements of polypeptide hormone receptor lateral mobility, illustrated in Table 4.1, have been performed for receptors of both the tyrosine kinase and GTP-binding-protein activating receptor categories. With one exception, measurements for cytokine receptors do not exist largely because the numbers of cytokine receptors possessed by hematopoietic cell lineages is very low (of the order of 500-3000 receptors per cell) compared to the former on their appropriate cells (e.g., renal epithelial cells possess about 40,000 G_s- and adenylate cyclase-activating vasopressin V_2-recep-

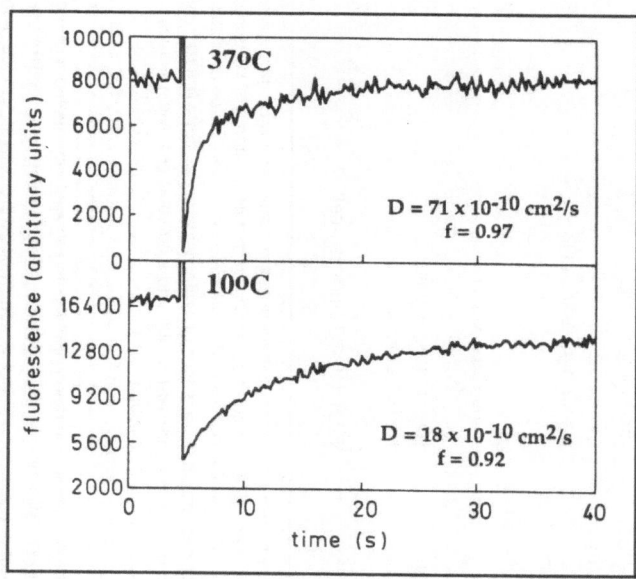

Fig. 4.1. Measurement of lateral mobility of a lipid probe using continuous monitoring. Renal epithelial cells (the LLC-PK$_1$ line) were incubated with the fluorescent lipid probe DiOC$_{14}$(3) (3,3'-ditetra-decyloxacarbocyanine iodide; see also Fig. 2.4) and fluorescence photobleaching recovery measurements carried out at the temperatures indicated using "continuous" measurement. The apparent lateral diffusion coefficients (D) and mobile fractions (f) are indicated. Compare to Fig. 4.2, with respect to the time scale of the measurement and mag-nitude of D and f.

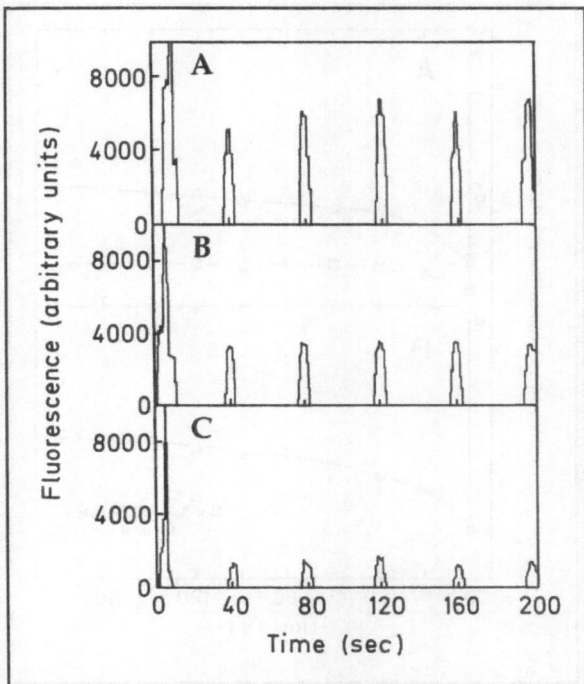

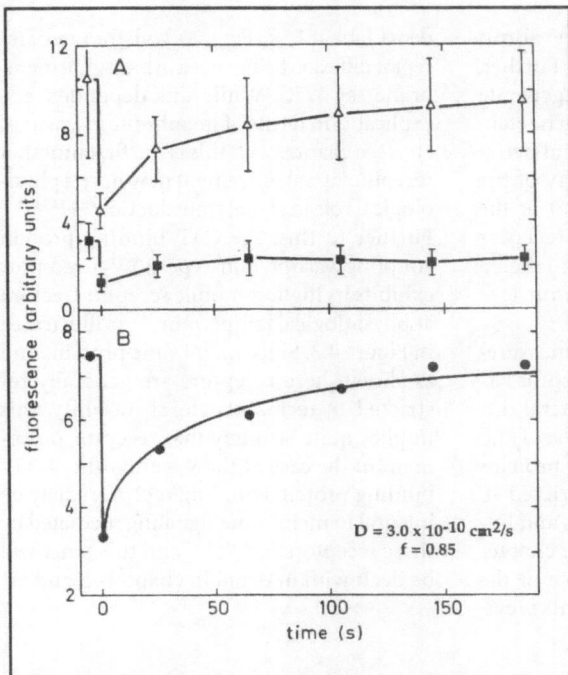

Fig. 4.2. Measurement of lateral mobility of vasopressin receptors at low signal-to-noise ratios. Measurements were performed using the fluorescence photobleaching recovery technique with intermittent monitoring (A) and subtraction of fluorescence due to nonspecific binding (B, bottom and C, next page).

A. (top) Renal epithelial cells (the LLC-PK$_1$ line expressing vasopressin V$_2$-type receptors) were labeled with a fluorescently-labeled vasopressin agonist (deamino-[Lys[8](tetramethylrhodamyl-aminothiocarbonyl)] vasopressin in the absence (panel A-top) or presence (fluorescence due to nonspecific binding - panel B - middle) of a 100-fold excess of unlabeled vasopressin and fluorescence photobleaching recovery measurements performed at 37°C at 40 second intervals. Panel C (bottom) shows identical measurements for autofluorescence.

B. (bottom) Results from measurements such as those illustrated in A for LLC-PK$_1$ cells and the vasopressin V$_2$-type receptor (40 second interval measurements). The results for fluorescence due to nonspecific binding (squares) are subtracted from those for total binding (triangles) (panel A - above) to yield the recovery curve for fluorescence due to specific binding (circles - panel B - below, which can then be fitted to yield estimations of the apparent lateral diffusion

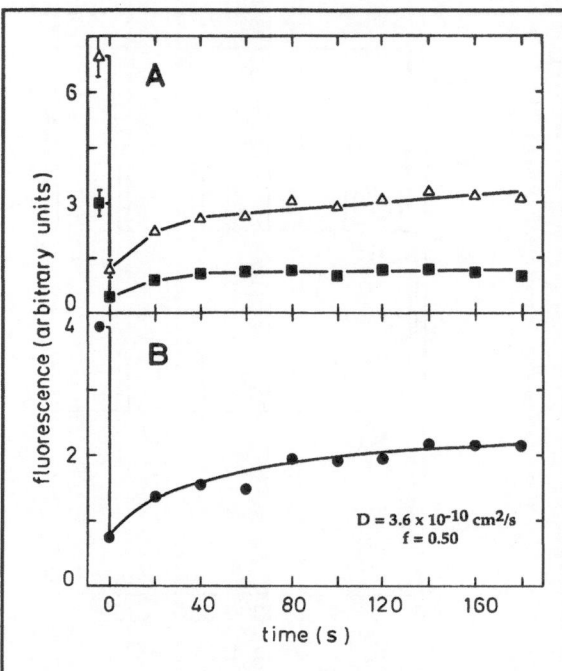

Fig. 4.2C. Results from measurements such as those illustrated in A for smooth muscle aortic cells (the A7r5 line, expressing vasopressin V_1-type receptors), incubated with the vasopressin agonist, with measurements performed at 23°C at 20 second intervals. Results for D and f were obtained as per Fig. 4.2B.

tors/cell, while fibroblasts have about 100,000 EGF receptors/cell).[13,31-33] Further, it appears to be quite difficult to generate fluorescently-labeled cytokines which retain both binding affinity and biological activity. Measurement of lateral mobility of the 55 kDa α-subunit (or Tac-peptide) of the three subunit interleukin (IL) 2 receptor was achieved using fluorescently-labeled monoclonal antibody to the subunit (see also section H below).[29]

At face value, the results for the measurement of polypeptide hormone receptor lateral mobility indicate apparent lateral diffusion coefficients ($1-5 \times 10^{-10}$ cm²/sec) and mobile fractions (0.3-0.9) typical of proteins in physiological membranes restricted in their movement. As for all proteins and lipids in all types of membranes (see chapter 2), a clear temperature dependence of the apparent lateral diffusion coefficients is evi-

dent (Table 4.1; see Fig. 4.3), with the rate of receptor diffusion being fastest in biological membranes at 37°C. While this dependence is explicable in terms of membrane viscosities, etc. (see chapter 2),[34] this is the first hint that receptor lateral movement may have a physiological role in signal transduction.[2,3,13,14,17,18] Further to this, the GTP-binding protein coupling vasopressin type-2 (V_2) receptor exhibits its highest mobile receptor fraction at physiological temperature[13] as illustrated in Figure 4.3. Since membrane proteins and as shown here receptors are generally restricted in terms of lateral mobility, this implies quite strongly that receptor movement in the case of the V_2- and other GTP-binding protein coupling receptors may be integral to membrane signaling mediated by these receptors,[2,3,13,15,17,18] and this tenet will be dealt with in detail in chapters 5 and 6.

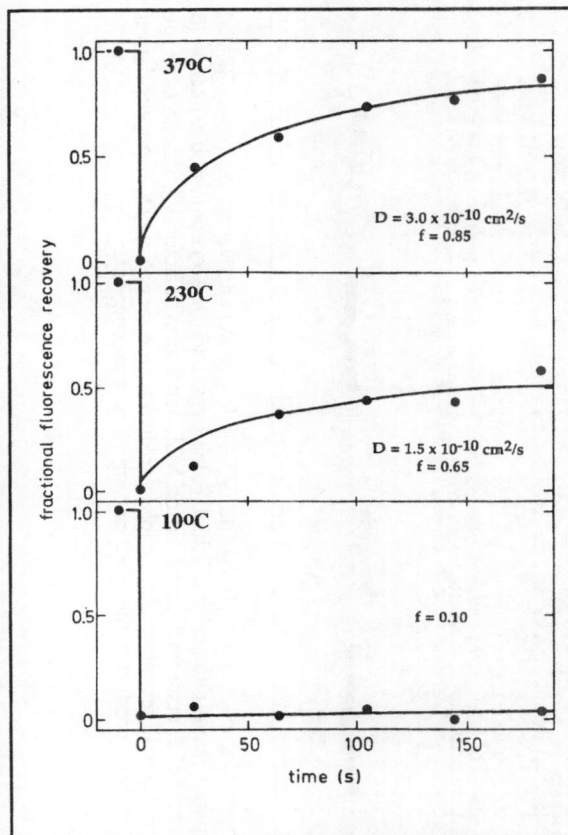

Fig. 4.3 Temperature dependence of lateral mobility of the vasopressin type-2 receptor. Measurements were performed by the fluorescence photobleaching recovery technique in renal epithelial cells (the LLC-PK$_1$ line) essentially as outlined in Figure 4.2. Values for the apparent lateral diffusion coefficient (D) and mobile fraction (f) are indicated.[13]

D. TYROSINE KINASE RECEPTOR-MEDIATED SIGNAL TRANSDUCTION

Growth factors and their tyrosine kinase receptors are ubiquitously involved in proliferative responses by many sorts of fibroblast and other types of cells, although insulin has quite specialized additional roles associated with the regulation of glucose homeostasis. GTP-binding protein activating receptors, on the other hand, appear to mediate more diverse, often short-term systemic and tissue responses, such as control of blood pressure through vasodilation and constriction, osmoregulation including water,

Ca^{2+} and Na$^+$ homeostasis, and stress responses. As evident from Table 4.1, there appear to be differences between tyrosine kinase and GTP-binding-protein coupling receptors with respect to lateral mobility parameters, particularly pertaining to the receptor mobile fraction.

The fundamental differences are highlighted in Figure 4.4, where the apparent lateral diffusion coefficient and receptor mobile fraction are shown for three temperatures. While selected results are shown, the general trend is valid for most of the other receptors in Table 4.1, although there are exceptions as will be discussed below. It

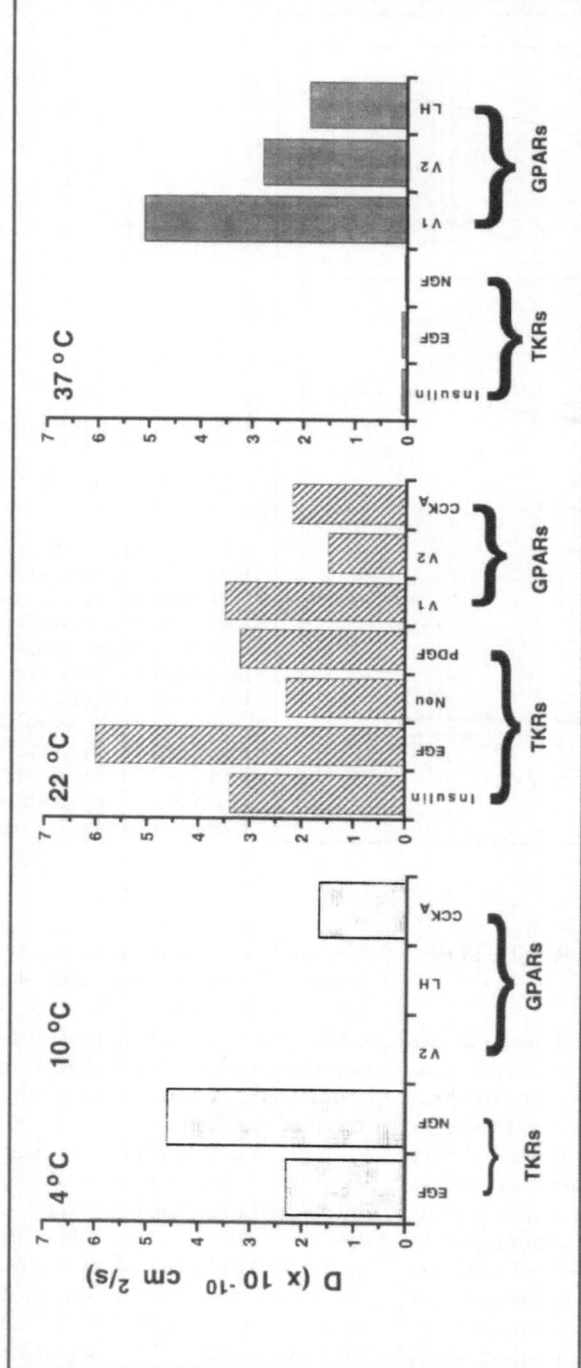

Fig. 4.4. Differences in lateral mobility properties of tyrosine kinase receptors (TKRs) and GTP-binding protein activating receptors (GPARs). Selected examples are taken from Table 4.1 for D, the apparent lateral diffusion coefficient (A, this page) and f, the receptor mobile fraction (B, opposite page) for the temperatures indicated. Abbreviations: EGF, epidermal growth factor; NGF, nerve growth factor; V1, vasopressin type-1 (hepatic/smooth muscle) receptor; V2, vasopressin type-2 (renal) receptor; CCK_A, cholecystokinin type A receptor; PDGF, platelet-derived growth factor; LH, luteinizing hormone; Neu, p185^{neu} (Neu, gene product of the HER2 or c-ErbB-2 protooncogene).

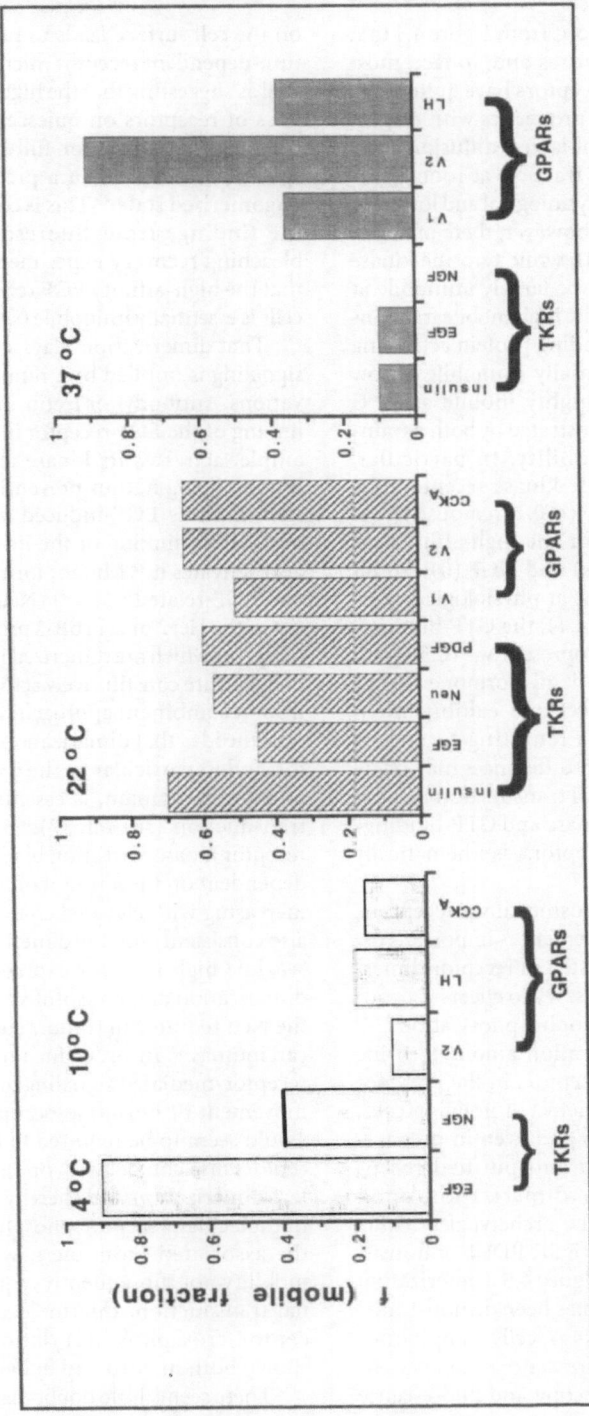

Fig. 4.4B. (see legend, opposite page)

is immediately obvious from Figure 4.4 that the two receptor classes and, in fact, most of the individual receptors have quite similar lateral mobility properties with respect to both the apparent lateral diffusion coefficient and mobile fraction at room temperature. At both physiological and low temperature (4-10°C), however, there are clear differences (Fig. 4.4); while tyrosine kinase receptors appear to be largely immobile at 37°C and exhibit quite high mobile fractions at 4°C, the GTP-binding protein activating receptors are essentially immobile at low temperature and highly mobile at 37°C (Fig. 4.4B), as demonstrated by both parameters of lateral mobility. In particular, whereas the tyrosine kinase receptors for insulin (< 0.1), EGF (< 0.1 in mouse fibroblasts and < 0.05 for the high-affinity receptor of A431 cells) and NGF (0) exhibit low mobile fractions at physiological temperature (see Table 4.1), the GTP-binding-protein coupled vasopressin V_1- (0.36) and V_2- (0.91) and luteinizing hormone (LH or lutropin -0.38) receptors exhibit much higher values. It is tempting to speculate[2,3,13,17,18] that such differences may relate to the differing signal transduction mechanisms of tyrosine kinase and GTP-binding-protein coupling receptors, as schematically shown in Figure 1.3.

In the case of tyrosine kinase receptors, receptor lateral movement is important for the initial signaling step of receptor dimerization which is necessary to effect intermolecular receptor autophosphorylation.[35-37] Receptor oligomerization among tyrosine kinase and other receptors in the presence of ligand has been detected in living cells, isolated membranes and even in preparations of solubilized and purified receptors.[6,37-44] Receptor dimerization upon ligand binding by the archetypal tyrosine kinase receptors for EGF, PDGF and insulin is illustrated in Figure 4.5. Dimerization of the EGF receptor has been demonstrated directly on single A431 cells using donor photobleaching fluorescence resonance energy transfer microscopy and fluorescence lifetime imaging microscopy.[40] The results indicate that EGF binding by the receptor

on the cell surface leads to rapid temperature-dependent receptor microclustering as well as suggesting that the high-affinity subclass of receptors on quiescent A431 cells, normally sufficient for full biological response, are present in a predimerized or oligomerized state.[40] This is consistent with the findings from fluorescence photobleaching recovery experiments indicating that the high-affinity EGF receptor of A431 cells is essentially immobile (see Table 4.1).[21]

That dimerization plays a direct role in signaling is implied by a number of observations. Antibody- or lectin-induced crosslinking of the EGF receptor in vitro, for example, activates its kinase activity, while immobilizing it, thus preventing dimerization, inhibits EGF-induced kinase activation.[39] Crosslinking of the insulin receptor also activates it.[42] Mutant forms of the EGF and EGF-related p185[neu] (Neu, gene product of the Her2 or c-ErbB-2 protooncogene) receptors which are dimerized in the absence of ligand are constitutively active.[45-47] Results from recombinant approaches further support the idea that dimerization of the receptor, and in particular of the receptor transmembrane domain, is essential for signal transduction (see ref. 37).[48] EGF-induced receptor kinase activation in vitro is directly dependent on the amount of EGF receptor, increasing with elevated concentration[39]— also consistent with the dimerization theory —while high receptor expression leads to dimerization and constitutive activation of the Neu receptor in tumors such as adenocarcinomas.[49] In order for tyrosine kinase receptor-mediated signaling to occur, rapid movement of hormone-occupied receptor would seem to be required to bring the receptors into contact with one another to effect dimerization and thereby facilitate intermolecular receptor phosphorylation of the associated monomers, with receptor mobility not subsequently required for signal transduction. Intermolecular EGF-receptor-crossphosphorylation has been shown both in vitro and in living cells.[50-52]

There seems little doubt that tyrosine kinase receptor dimerization or microaggregation effects receptor immobilization (see also

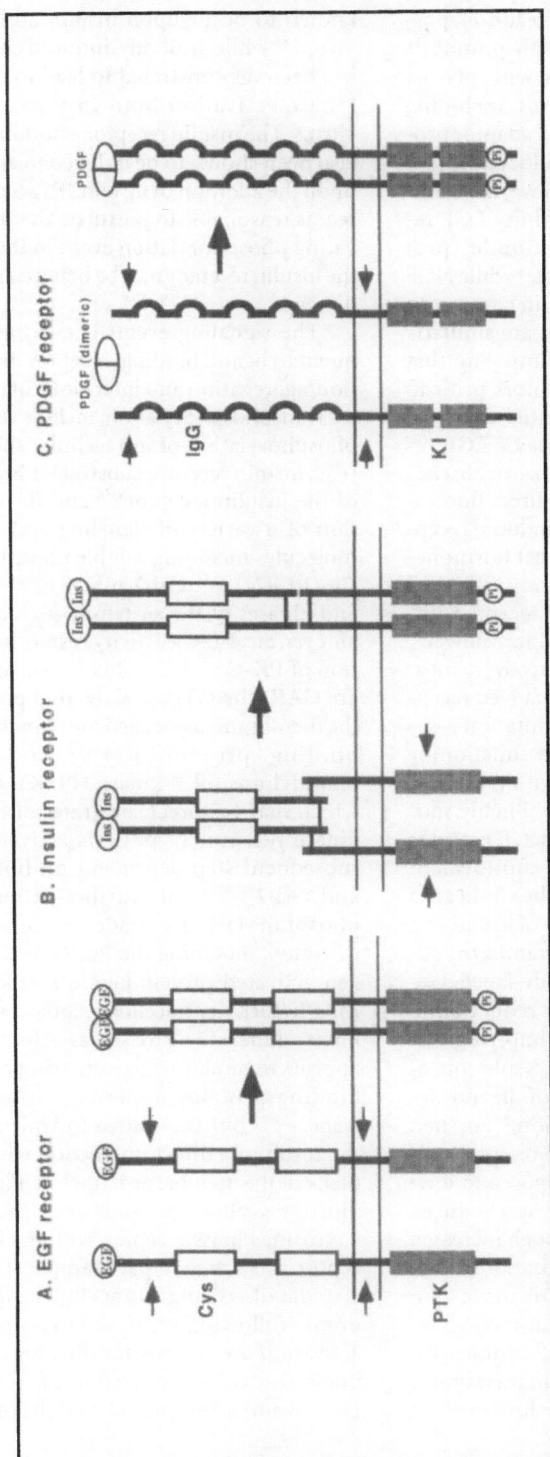

Fig. 4.5 Tyrosine kinase receptor-dimerization and ligand interactions. A. Dimerization of the epidermal growth factor (EGF) receptor upon EGF addition and induction of conformational changes. B. The insulin receptor is already dimerized before ligand addition. Insulin (Ins) binding induces conformational changes which alter disulfide-bond formation and stabilize the dimeric structure. Higher order insulin-receptor oligomerization (not shown) may also be important for signaling (see text). C. Dimerization of the platelet-derived growth factor (PDGF) receptor upon addition of the bivalent PDGF ligand. CSF-1 (the bivalent-ligand colony stimulating factor-1) and the CSF-1 receptor have a similar mechanism. Abbreviations: Cys, cysteine-rich domain; PTK, protein tyrosine kinase domain; IgG, immunoglobulin-like repeat sequences; KI, kinase insert domain.

chapter 3 section A; chapters 6 and 7), possibly through association with immobile structures, and in particular with plasma membrane adaptor complexes, including proteins such as α-adaptin and adaptor protein 2 (AP-2),[46] leading to endocytosis (see also ref. 53). As mentioned above, measurements show that the high-affinity EGF receptor of A431 cells is immobile (pre-aggregated) (D < 10^{-12} cm²/sec),[21] while NGF receptors on responsive cells, in contrast to those on nonresponsive cells, are similarly preclustered and immobile, implying that immobilization of NGF receptors prior to ligand binding may be essential to signal transduction.[23] Measurements of EGF receptor movement in the presence or absence of ligand using direct and indirect fluorescent probes and electric field-induced receptor asymmetry also indicate that hormone-occupied receptor is substantially less mobile than the unoccupied receptor, diffusing at a rate almost four times slower.[54] Fluorescence photobleaching recovery measurements have also shown that Neu receptor activation (induced by mutation—see Table 4.1—or by agonistically functioning antibodies) is concomitant with a marked reduction in mobile fraction.[46] Finally, mobility measurements of the PDGF receptor using fluorescently-labeled monovalent (Fab fragment) antibody probes indicate a mobile receptor in the absence of ligand (see Table 4.1 and legend).[5] Importantly, the addition of PDGF to Fab-antibody-labeled receptor results in a significant reduction in the receptor mobile fraction, implying that receptor immobilization occurs subsequent to, and probably as a result of, ligand-dependent receptor dimerization.[5] Further, direct measurements of PDGF receptor mobility suggest that high (as opposed to low) PDGF concentrations result in a reduced mobile fraction, probably through increased receptor aggregation/immobilization.[4]

As evident from Figure 4.5B, the insulin receptor theoretically does not require receptor lateral movement to effect dimerization since it is predimerized in the membrane. Receptor microclustering, however, is known to occur upon insulin addition in vivo,[37,44] while antibody-induced clustering has been demonstrated to lead to receptor kinase activation both in vivo[44] and in vitro.[42] The insulin receptor ectodomain has also been shown to be induced to aggregate upon the addition of insulin.[43] It accordingly seems reasonable to postulate that the activating phosphorylation event in the case of the insulin receptor may be between receptor dimers.

The signaling events occurring subsequent to ligand binding, receptor dimerization/aggregation and intermolecular receptor transphosphorylation, include the direct phosphorylation of intracellular substrates (e.g., insulin receptor substrate-1 in the case of the insulin receptor),[55] and the association of a variety of signaling and adaptor molecules including soluble phospholipase Cg (PL-Cg),[37,56] Grb2-mSOS1,[57,58] Shc,[57,58] and kinases of the src family such as p59^{fyn} and yes, etc. PL-Cg activity results in activation of PK-C and Ca²⁺ flux.[37] Also activated are GAP (the GTPase activating protein of the membrane-associated monomeric GTP-binding protein, p21^{ras}) and phosphatidylinositol 3-kinase (PI3K), both of which may be direct substrates of tyrosine kinase receptors.[37] p21^{ras} is activated in a subsequent step dependent on both GAP and Grb2,[37,57,58] and further intracellular phosphorylation cascades are ultimately activated, including the Raf/MAPK (mitogen-activated protein kinase)/MEK-1 (the MAPK/ERK - extracellular signal-regulated kinase-kinase-1) kinase cascade.[57,58] EGF appears to be able to activate trimeric GTP-binding proteins under certain circumstances[37,56] but the degree to which this is the result of a direct interaction within the plane of the membrane is unclear. Figure 4.6 illustrates cellular responses to activation of a tyrosine kinase receptor, with the EGF receptor as an archetypal example. The precise details of the intracellular signaling events following tyrosine kinase receptor activation are not within the scope of this book; readers are referred to the review by Ullrich and Schlessinger[37] and the other literature cited above (see also refs. 67 and 68).

In summary, the initial step of tyrosine kinase receptor activation involves a dimerization/aggregation event limiting lateral mobility, followed by the association with the activated receptor complex of a number of other cytoplasmic and membrane signaling molecules. Clearly, lateral movement of hormone-occupied receptor is only required initially to enable molecules to associate into aggregates, which then become largely immobilized. Subsequent to immobilization, all subsequent signaling events occur in the cytosolic phase and hence the relatively low tyrosine kinase receptor mobility measured at physiological and other temperatures (see Table 4.1) can be understood and rationalized in these terms.

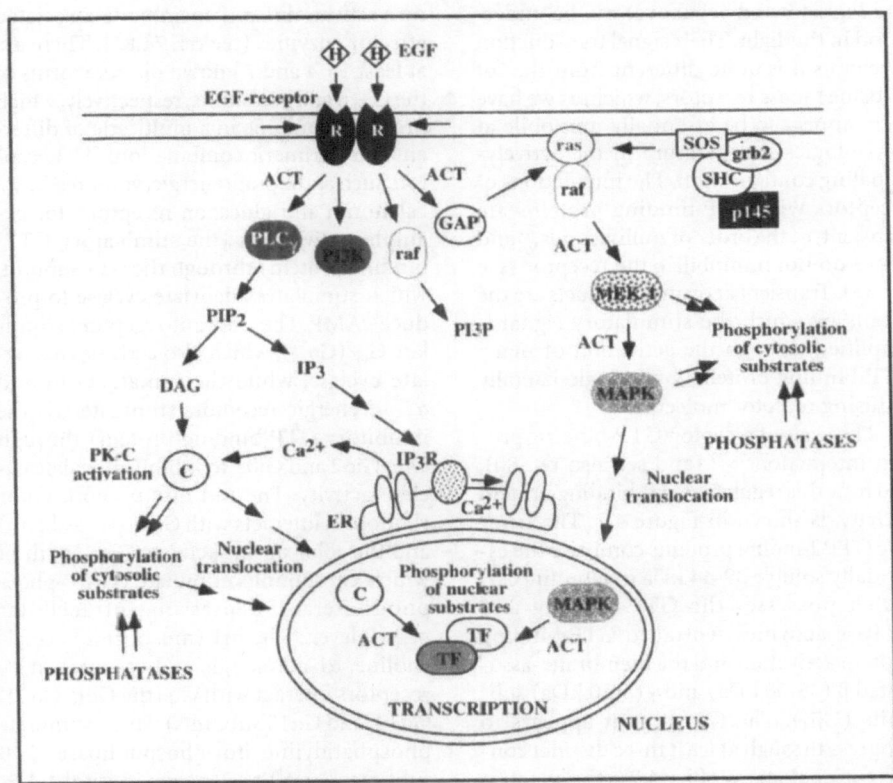

Fig. 4.6. The tyrosine kinase receptor-mediated signal transduction cascade, as typified by that for epidermal growth factor (EGF) and the EGF receptor (R). ACT indicates activating or agonistic action. Abbreviations: H, hormone (EGF); GAP, p21^{ras} (ras) GTPase activating protein; PLCγ, phospholipase Cγ; PI3K, phosphatidylinositol 3-kinase; Grb2, growth factor receptor binding protein 2; SOS, ras guanine nucleotide exchange factor; MAPK, mitogen activated protein kinase; MEK-1, MAPK/ERK (extracellular signal-regulated kinase) kinase-1; DAG, 1,2-diacyl glycerol; IP3, 1,4,5-inositol trisphosphate; PIP$_2$, phosphatidylinositol-4,5-bisphosphate; IP$_3$R, IP3 receptor; PK-C (C), Ca^{2+}, phospholipid-dependent protein kinase; ER, endoplasmic reticulum; TF, transcription factor. For distal signaling events of kinase (PK-C[59-61] and MAPK)[62-64] translocation to the nuclear envelope/nucleus and transcription factor activation, see refs. 37 and 65-68.

E. GTP-BINDING PROTEIN ACTIVATING RECEPTOR-MEDIATED SIGNAL TRANSDUCTION

In contrast to that by tyrosine kinase receptors, signal transduction by GTP-binding protein activating receptors requires interaction with other membrane proteins, namely with the trimeric GTP-binding proteins. The fact that receptors activating GTP-binding proteins appear to possess relatively high mobile fractions at 37°C (see Table 4.1 and Figs. 4.3 and 4.4 above) can be understood in this light. Their signal transduction mechanism is quite different from that of tyrosine kinase receptors, which, as we have seen, appear to be essentially immobile at physiological temperature in the actively-signaling configuration. The interactions of receptors with GTP-binding proteins are transient, of the order of milliseconds,[69] and hence do not immobilize the receptor (see ref. 53). Transient activating contacts are the means by which the stimulatory signal is amplified through the activation of many GTP-binding proteins by a single laterally diffusing receptor molecule.[70-73]

The cycle of receptor/GTP-binding protein interactions[74-79] (and see also ref. 80), and how this regulates GTP-binding protein activity, is shown in Figure 4.7. The trimeric GTP-binding proteins comprise the essentially soluble 39-44 kDa α-subunit (Gα) which possesses the GTP-binding and GTPase activities central to GTP-binding protein activation and the membrane-associated β (35-36 kDa) and γ (8-10 kDa) subunits (Gβγ). The Gα subunit appears to progress through at least three distinct conformational states (Fig. 4.7): the inactive GDP-bound state (Gα-GDP), the state in which the guanine nucleotide-binding site is empty (Gα-0) and the active GTP-bound state (Gα-GTP). Gα-GDP has a high affinity for the Gβγ subunits, with binding of Gβγ increasing its affinity for GDP. Ligand occupied/activated receptors (R*) bind to the heterotrimeric Gα-GDP-Gβγ complex to trigger release of GDP to yield the Gα-0-Gβγ conformation which R* binds tightly and

stabilizes. When GTP enters the empty nucleotide-binding site of Gα, the Gα-GTP assumes a new conformation which causes it to dissociate rapidly from both R* and Gβγ. Gα-GTP has a high affinity for binding to and activating appropriate effectors such as adenylate cyclase and phospholipase Cβ. Hydrolysis of the bound GTP terminates stimulation of the effector, returning Gα to its inactive GDP-bound state.[74]

Different receptors specifically activate different GTP-binding proteins which in turn activate different membrane-associated effector enzymes (see ref. 74,81). There are at least 16, 4 and 7 known discrete forms of the Gα, β and γ subunits, respectively, which probably combine in a multitude of different heterotrimeric combinations.[74,82] Receptors such as the β-adrenergic, vasopressin V_2, calcitonin and glucagon receptors, for example, activate Gs (the stimulatory GTP-binding protein) through the Gsα subunit, which stimulates adenylate cyclase to produce cAMP. The odorant receptors stimulate G_{olf} ($Gα_{olf}$), which also activates adenylate cyclase, while the somatostatin and α2-adrenergic receptors stimulate Gi (the inhibitory GTP-binding protein) through Giα, Giα2 and Giα3 to inhibit adenylate cyclase activity. The rod disc photoreceptor rhodopsin interacts with Gαt1 (transducin) and the color opsins activate Gαt2, both of which Gα subunits stimulate cGMP-phosphodiesterase, depressing intracellular cGMP levels. The m1 (muscarinic) acetylcholine, α1-adrenergic and vasopressin V_1 receptors interact with Gq (the Gαq, Gα11, Gα14, and Gα15 subunits) which stimulate phosphatidylinositol-phospholipase C (β subtypes) or A2 to generate inositol 1,4,5-trisphosphate and diacylglycerol. Other effector enzymes activated by GTP-binding proteins include channels for K^+ (opened by Gi) and Ca^{2+} (opened by Gsα and closed by Gαo and the m2 muscarinic acetylcholine receptor). All GTP-binding proteins modulate the levels of a secondary messenger molecule (e.g., cAMP, cGMP, Ca^{2+}, inositol 1,4,5-trisphosphate and diacylglycerol),[74,81] whose intracellular concentration regulates

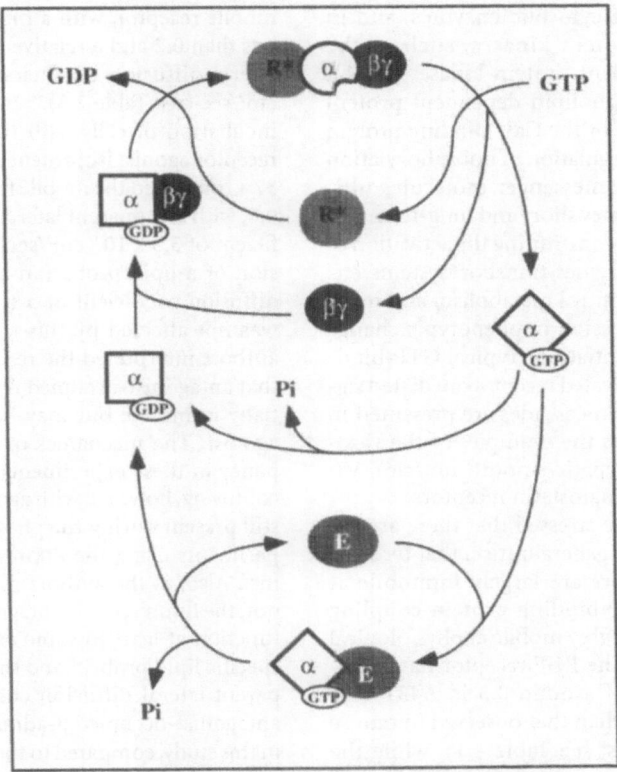

Fig. 4.7. The GTP-binding protein cycle. The Gα subunit passes through at least three distinct conformational states: the inactive GDP-bound state (Gα-GDP, shown as squares), the state in which the guanine nucleotide-binding site is empty (Gα-0, shown as circles) and the active GTP-bound state (Gα-GTP, shown as diamonds). Gα-GDP has a high affinity for the Gβγ subunits, with binding of Gβγ increasing its affinity for GDP. Ligand occupied/activated receptors (R*) bind to the heterotrimeric Gα-GDP-Gβγ complex to trigger release of GDP and generate Gα-0-Gβγ, which is bound tightly and stabilized by R*in the absence of added guanine nucleotides. When GTP enters the empty nucleotide-binding site of Gα, the Gα-GTP assumes a new conformation which causes it to dissociate rapidly from both R* and Gβγ. Gα-GTP has a high affinity for binding to and activating appropriate effectors (E → E*). Hydrolysis of the bound GTP terminates regulation of the effector, returning Gα to its inactive GDP-bound state.[74,80,81] Pi, inorganic phosphate.

the activity of cytosolic enzymes, and in particular protein kinases, such as the cAMP-dependent protein kinase (PK-A), the Ca^{2+},phospholipid-dependent protein kinase (PK-C) or the Ca^{2+}-binding protein calmodulin. Regulation of phosphorylation by the second messenger molecules ultimately modulates short and long-term cellular response, constituting the activation of preexisiting enzymes, transport systems, etc. in the cell to control metabolism and regulate gene expression in phenotypic change such as differentiation. Typical GTP-binding protein activated receptor-mediated signal transduction cascades are presented in Figure 4.8, with the examples of the vasopressin V_1- (hepatic/smooth muscle), V_2- (renal) and somatostatin receptors.

It should be stressed that there are exceptions to the generalization that tyrosine kinase receptors are largely immobile at 37°C and GTP-binding protein coupling receptors are highly mobile at physiological temperature. The EGF receptor has a mobile fraction of around 0.5 in A431 cells (much higher than that observed in mouse 3T3 fibroblasts; see Table 4.1), while the GTP-protein activating cholecystokinin type-A (CCK_A) receptor exhibits a very low mobile fraction at 37°C. With respect to the former, A431 cells have abnormally high receptor numbers—about 2×10^6/cell—and are killed by high ligand concentrations and, thus, arguably not very physiological (or at least significantly less so than 3T3 mouse fibroblasts).[87-89] With respect to the CCK_A receptor, it seems reasonable to suggest that the low mobile fraction results in part from immobilization prior to internalization, which occurs very rapidly in acinar pancreatic cells;[12] that receptor immobilization prior to internalization reduces the receptor mobile fraction has been demonstrated for the vasopressin type-2[16] and other receptors (see chapter 6, section C).

Mention should also be made of lateral mobility measurements of the Gs-activating β-adrenergic receptor using a fluorescently-labeled β-antagonist on Chang liver cells, interpreted as indicating a largely im-mobile receptor, with a mobile fraction of less than 0.2 and a relatively high apparent lateral diffusion coefficient of 1.4×10^{-9} cm^2/sec (see Table 2.3).[90] Intriguingly, pre-incubation of cells with the β-adrenergic receptor agonist isoprotenerol for 30 min at 37°C increased the mobile fraction to about 0.8, with an apparent lateral diffusion coefficient of 3.5×10^{-9} cm^2/sec.[90] Lateral diffusion of a lipid probe (an apparent lateral diffusion coefficient of 5-6×10^{-9} cm^2/sec) was not affected by this treatment.[90] The authors interpreted the results as meaning that antagonist-occupied receptor is essentially immobile but may be mobilized by agonist. The mechanics of receptor occupancy in these experiments are more than confusing, however, with agonist apparently still present during the photobleaching experiments using the fluorescent β-antagonist. Also, as the authors themselves point out, the fluorescent β-antagonist appears to function at least to some extent as a nonspecific lipid probe,[90] and the very high apparent lateral diffusion coefficient of the antagonist-occupied β-adrenergic receptor in this study, compared to those in Table 4.1, may be explicable in these terms. The results of this study should accordingly be treated with caution, and any conclusion that GTP-binding protein activating receptors binding nonpolypeptide hormones, such as the adrenergic receptors, may be fundamentally different from receptors binding polypeptide hormones based on this study alone seems preliminary. Rhodopsin, for example, is highly mobile in rod disc membranes, especially with respect to a high apparent lateral diffusion coefficient and mobile fraction of 0.4-0.6 (see Table 2.3). In the vasopressin V_2- and N-formyl-peptide receptor systems, antagonist-occupied receptors are as laterally mobile as agonist-occupied receptors and since they do not appear to be internalized seem to maintain a higher mobile fraction with time at physiological temperature (see chapter 6 section G). The idea that antagonist-occupied receptors are immobile (and hence inactive in signaling for that reason) clearly cannot be generalized.

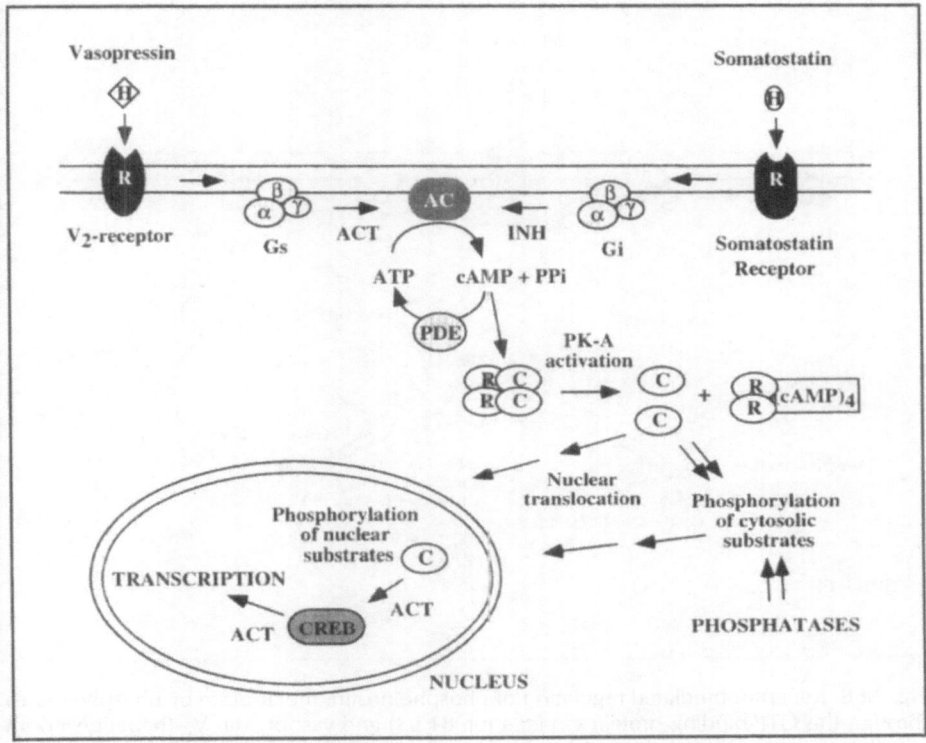

Fig. 4.8. GTP-binding protein activated receptor-mediated signal transduction cascades. ACT indicates activating or agonistic action, and INH inhibitory or antagonistic action. A. Receptor-mediated regulation of adenylate cyclase (AC) through the stimulatory (Gs) and inhibitory (Gi) GTP-binding proteins and the vasopressin V_2- (renal) and somatostatin receptors (R). For abbreviations, see legend to Fig. 4.8B on next page.

In this context, it should be remembered that the generalized comments with regard to the lateral mobility properties of the tyrosine kinase and GTP-binding protein activating receptor classes above are intended to be no less and no more than exactly that and in that respect are more than useful in considering the role of receptor lateral movement in membrane signal transduction. It is possible that particular receptors have their own unique lateral mobility properties (which may indeed be cell-type specific). Clearly, any broad conclusion with respect to a direct relationship between the receptor mobile fraction and signal transduction mechanism cannot be based on such observations alone but requires direct experimentation. The succeeding chapter will deal will the direct and indirect evidence for this in some detail.

F. STRUCTURAL CONSIDERATIONS

In comparing the lateral mobility properties of tyrosine kinase and GTP-binding protein coupling receptors, it should not be forgotten that there are quite profound structural differences between the two classes of receptors (see Fig. 4.9). De Haen[91] predicted over 20 years ago, prior to the discovery of GTP-binding proteins, that there should be a "basic structural similarity of all receptors for hormones activating cyclase." This has of course proven to be true, but the similarity extends much further, in fact, to all GTP-binding protein coupling

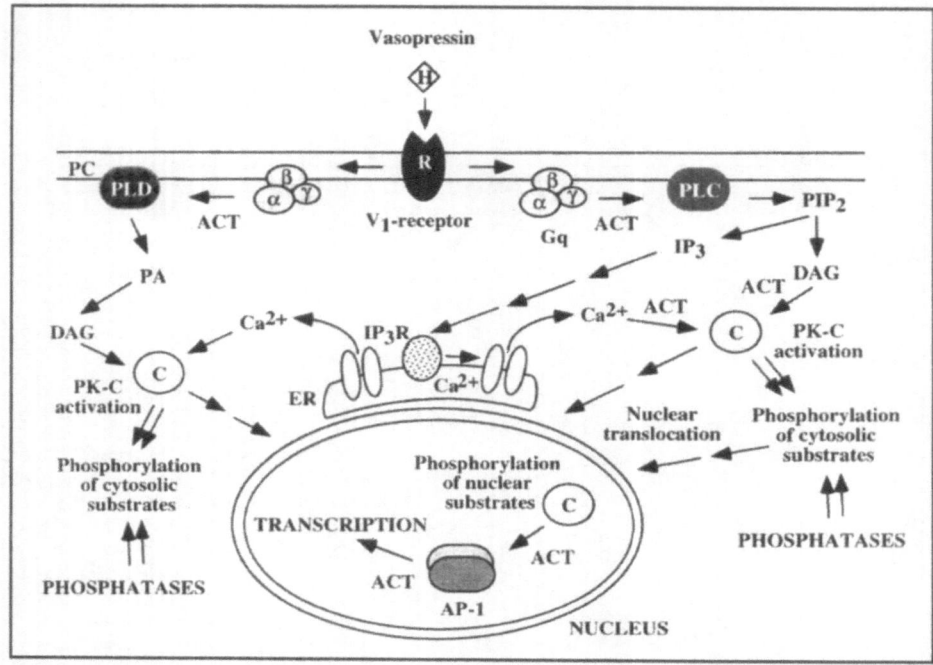

Fig. 4.8B. Receptor-mediated regulation of phosphoinositol metabolism by phospholipases through the GTP-binding protein Gq (see ref. 81,83) and vasopressin V$_1$- (hepatic/smooth muscle) receptor. Abbreviations: H, hormone; PK-A, cAMP-dependent protein kinase catalytic (C) and regulatory (R) subunits; PDE, phosphodiesterase; CREB, cAMP response element binding protein; PLC/D, phospholipase Cβ/D; PC, phosphatidyl choline; DAG, 1,2-diacyl glycerol; IP3, inositol 1,4,5-trisphosphate; PIP$_2$, phosphatidylinositol-4,5-bisphosphate; IP$_3$R, IP3 receptor; PK-C (C), Ca^{2+}, phospholipid dependent protein kinase; ER, endoplasmic reticulum; AP-1, activator protein-1 (heterodimer of the jun/fos transcription factors). Distal signaling events of kinase (PK-C[59-61] and PK-A C-subunit)[84-86] translocation to the nuclear envelope/nucleus and transcription factor activation are not the scope of this book (see however refs. 65,66).

(also known as "serpentine" because of their structure; see below) receptors characterized to date,[82,92,93] including rhodopsin. All have a common structure of seven transmembrane domains of alpha helical conformation (Fig. 4.9A) analogous to that of bacteriorhodopsin.[94] This holds true for the GTP-binding protein coupling α- and β-adrenergic, muscarinic, substance K, S12 serotonin, yeast pheromone,[95] vasopressin and oxytocin,[96] glucagon,[97] platelet activating factor, interleukin-8, Mip-1/Rantes, thrombin, formyl peptide and glycoprotein hormone (those for LH, follitropin or follicle

stimulating hormone, thyrotropin and thyroid stimulating hormone) receptors.[92,93]

Hydrophilicity plots of some GTP-binding protein activating receptors, compared to those of tyrosine kinase receptors, are depicted in Figure 4.9, while Figure 4.10 depicts the general structure of the GTP-binding protein coupling receptors in the membrane together with that of the GTP-binding protein trimer.[99] The critical receptor domains for GTP-binding protein activation are the second, third (two discrete regions thereof) and fourth cytoplasmic loops, corresponding to interaction regions for binding the Gα-subunit, while the cytoplasmic tail appears to interact with the Gβ-subunit (see Fig. 4.10 and refs. 82 and 99). This has been established by a variety of mutagenesis experiments as well as experiments using peptides encoding receptor and GTP-binding protein domain sequences (see also below).[82,99] The full complement of seven transmembrane helices appears to be essential for correct assembly of the receptor in the membrane and also hormone binding, as shown for the glucagon receptor.[97]

In contrast to the GTP-binding protein coupling receptors, receptors of the tyrosine kinase class have a single transmembrane region (see Figure 4.9A), with a large extracellular hormone binding domain and less extensive intracellular tyrosine kinase-containing domain. These include the receptors for:

i.   EGF, and the EGF-related Her2/Neu and Her3/c-erbB3 receptors and *Xmrk* (see ref. 37 and 68), with a large hormone-binding extracellular domain distinguished by two large cysteine-rich regions, an intracellular protein tyrosine kinase domain and a regulatory carboxy-terminal extension (see Fig. 4.5A);

ii. insulin, insulin-related growth factor (IGF-1) and IRR (see ref. 37 and 68), each of which have a dimeric structure with a monomer consisting of two polypeptides held together by disulphide bonds; one of which (the α-subunit) contains a large extracellular domain with a single cysteine-rich region and the other (the β-subunit), the transmembrane and intracellular protein tyrosine kinase domains (see Fig. 4.5B); and

iii. PDGF (A or B), colony stimulating factor and stem cell factor c-kit, each of which contain a large extracellular hormone-binding domain including five immunoglobulin (IgG) domain-like repeats and an intracellular protein kinase domain which has a "kinase insert region" that interrupts the kinase domain but is not involved in catalysis (see Fig. 4.5C).

Other tyrosine kinase receptor types include the receptors for NGF (TRKA) and hepatocyte growth factor (see ref. 68)—both of which resemble the EGF receptor in terms of the intracellular domain but have variant extracellular domains—while the acidic and basic fibroblast growth factor receptors (*flg* and *bek*, respectively), are structurally similar to PDGF receptors, including the kinase insert region, except that they contain three rather than five IgG-like repeat domains (see ref. 37 and 68).

Some of the regions within the tyrosine kinase receptor intracellular/protein tyrosine kinase domain which have been defined in terms of specific interaction properties are shown in Figure 4.11, with particular reference to the PDGF-β receptor. As can be seen, a variety of cytosolic factors recognize and bind to this domain upon hormone binding and phosphorylation and activation of the receptor tyrosine kinase activity.

Figure 4.9 enables the hydrophilicity plots of some of the tyrosine kinase receptors to be compared with those of GTP-binding protein activating receptors. When the structures depicted schematically in Figures 4.5 and 4.10 are compared, it is quite clear that the two receptor classes differ drastically in terms of the number of transmembrane regions and in the size and nature of the intracellular (cytoplasmic) domain: whereas tyrosine kinase receptors have a single c. 400-550 amino acid intracellular domain, the GTP-binding protein coupling receptors have three intracellular loops of between about 10 and 50 or so amino acids, each connecting the transmembrane domains, and an intracellular carboxy-terminus up to 90 or so amino acids in length.

Despite this large difference in structure between GTP-binding protein coupling and tyrosine kinase receptors, the apparent lateral diffusion coefficients are rather similar (see Table 4.1), although there is a trend of the tyrosine kinase receptors diffusing generally at a faster rate at 22°C and above (see Table 4.1; Fig. 4.4A); for example, the apparent lateral diffusion coefficients for the tyrosine kinase receptors at 22°C are around

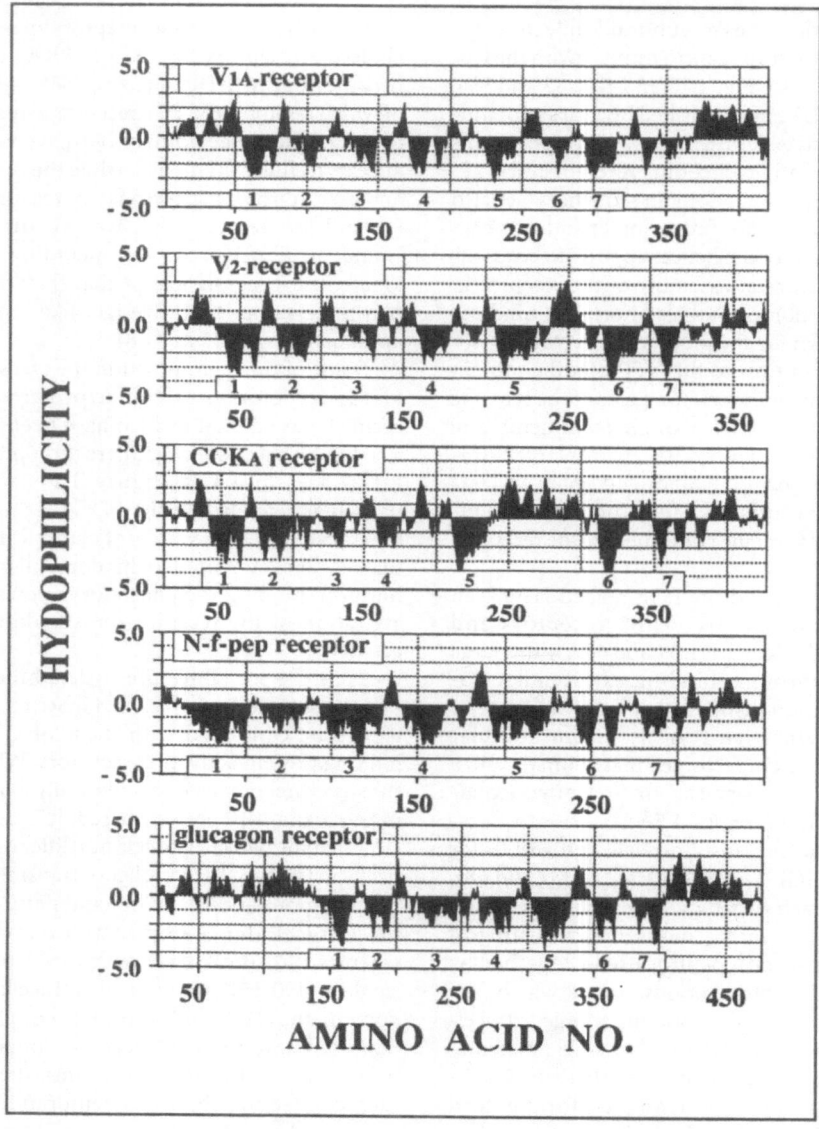

Fig. 4.9. Structural comparison of GTP-binding protein activating (A, this page) and tyrosine kinase (B, opposite page) receptors. Hydrophilicity plots using a window of seven amino acids with the Kyte and Doolittle scale[98] are shown for the sequences of the human receptors. The seven transmembrane helices of the GTP-binding protein activating receptors are numbered, while the single transmembrane domain (TM) of the tyrosine kinase receptors is indicated. The insulin receptor sequence includes both mature α and β receptor chains; the latter, which contains the transmembrane and tyrosine kinase domains, commences at amino acid 736. Abbreviations: cholecystokinin A (CCK$_A$), N-formyl peptide (N-f-pep), epidermal growth factor (EGF), platelet-derived growth factor (PDGF), nerve growth factor (NGF - TRKA).

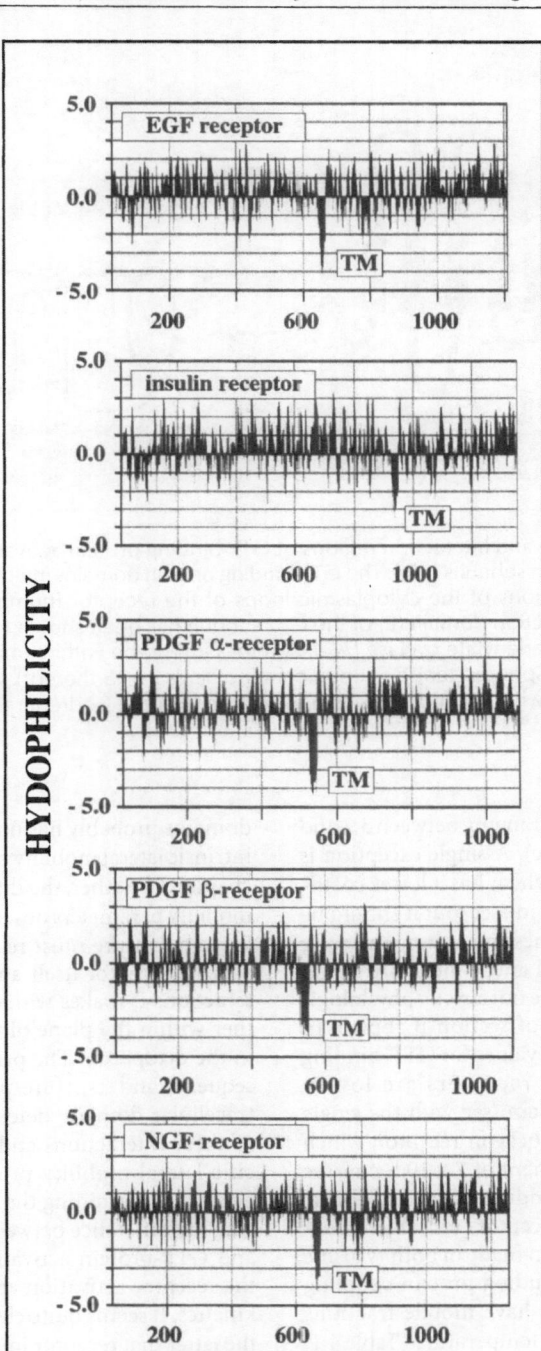

Fig. 4.9B.

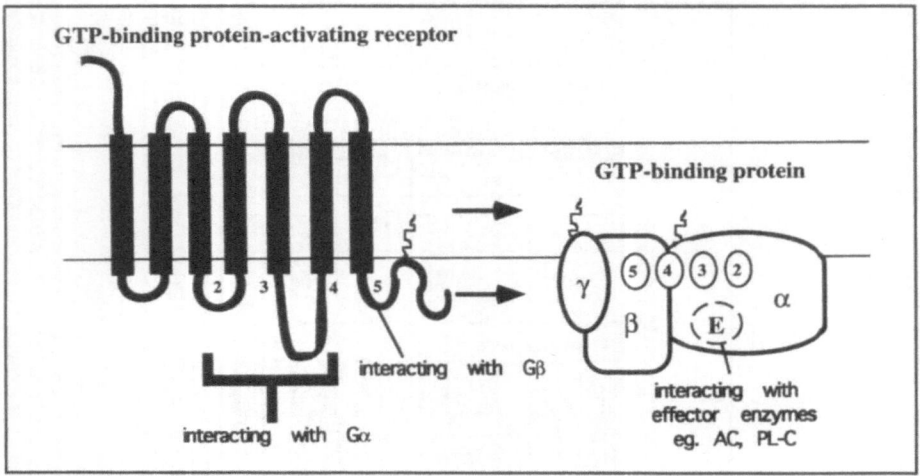

Fig. 4.10. Structure and interaction regions of GTP-binding protein activating receptors and GTP-binding protein subunits ($\alpha\beta\gamma$). The GTP-binding protein domains interacting with specific corresponding regions of the cytoplasmic loops of the receptor (numbered) are shown. The effector interaction domain (E) of the Gα subunit has been shown to be within amino acids 236-356 for adenylate cyclase (AC), and for Gαt1 to be within amino acids 293-314 for the cGMP phosphodiesterase γ-subunit (amino acids 24-45 thereof). G$\beta\gamma$ has also been shown to interact with phospholipase Cβ as well as with the β-adrenergic receptor kinase (both not shown).[74,82,99]

5-6 x 10^{-10} cm²/sec (ranging between 3.2 and 12.0 x 10^{-10} cm²/sec). A single exception is the Neu receptor, which has a lower coefficient of 2.3 x 10^{-10} cm²/sec, but it should be remembered that these measurements were made using labeled antibodies rather than ligand and hence are not strictly physiological (see beginning of section B above). In contrast to this, the values for GTP-binding protein coupling receptors are lower: around 2-3 x 10^{-10} cm²/sec with the single exception of the glucagon receptor, which has a higher coefficient of 7 x 10^{-10} cm²/sec (see Table 4.1). In addition, as noted above, almost all of the receptors for which measurements have been made in both tyrosine kinase and GTP-binding protein coupling receptor categories have mobile fractions around 0.6 at room temperature (Table 4.1; Fig. 4.4).

These results imply that the presence of either one or seven transmembrane helices or extensive intracellular and extracellular

domains probably has no marked effect on intrinsic lateral mobility properties (see also chapter 3). Rather, the differences in lateral mobility parameters observed at physiological temperature must relate directly to the specific receptor itself and the sorts of interactions it makes with other proteins either within the plane of the membrane or in the cytoplasm. The primary amino acid sequence and structure, particularly the intracellular domain, determines the nature of these interactions and thereby the specific lateral mobility properties. Also relevant in determining the latter and conferring the difference between tyrosine kinase and GTP-protein activating receptors, are the receptor activation and internalization kinetics. It seems quite clear with respect to the latter that receptor immobilization precedes internalization (see refs. 2, 3, 16, 46, 48, 101 and 102; see also ref. 53), so that receptor internalization kinetics are clearly also an important factor in determining lateral mobility parameters. This point, and its importance in the abrogation of response subsequent to hormonal stimulation, will be treated at some length in chapter 6, section E.

G. LATERAL MOBILITY MEASUREMENTS OF GTP-BINDING PROTEINS

Measurements of the lateral mobility of GTP-binding proteins have been restricted

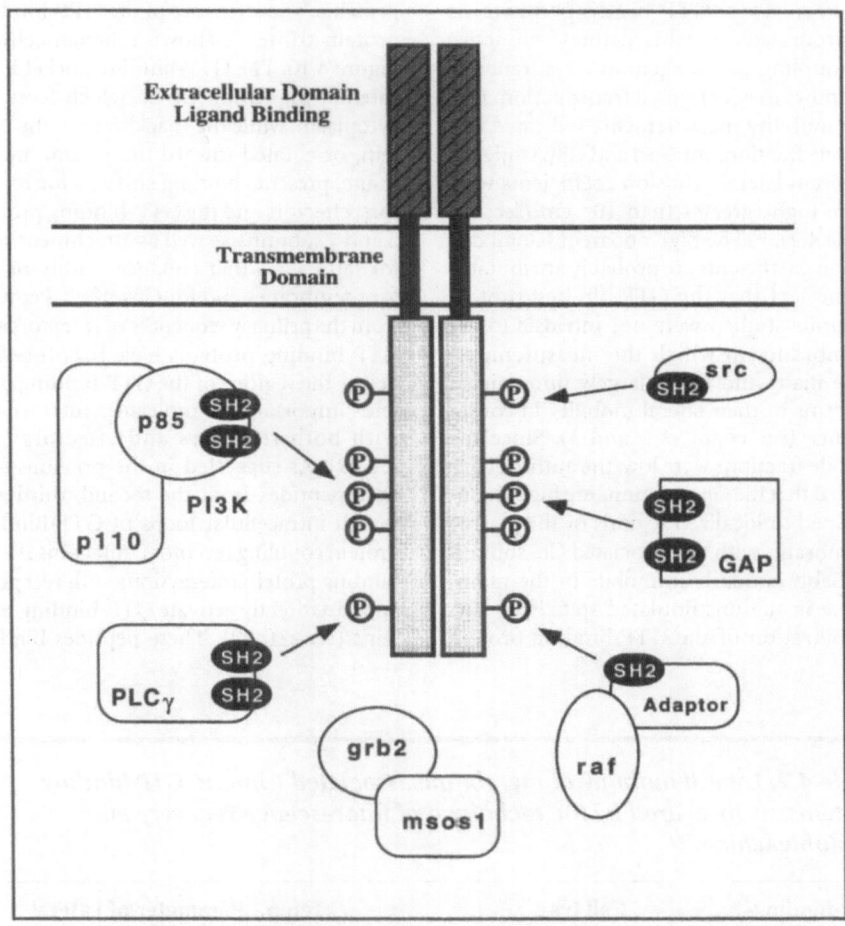

Fig. 4.11. Schematic representation of interactions with the intracellular domain of a tyrosine kinase receptor through phosphorylation sites and the src homology-2 (SH2) region.[68] In the case of the platelet-derived growth factor β-receptor,[67] the phosphoinositol 3-kinase (PI3K, comprising p85 and p110 subunits) recognizes tyrosines 740 and 751 (tyrosine 721 in the murine colony stimulating factor-1 receptor), and the GTPase activating protein (GAP) recognizes tyrosine 771, all within the kinase insert domain, while phospholipase C (PLC)-γ interacts with tyrosines 1009 and 1021 within the carboxy-terminus (tyrosines 992 and 766 in the epidermal growth factor and acidic fibroblast growth factor receptors, respectively) (see refs. 37, 67, 68 and 100).

to the study of Kwon et al[103] who used the approach of reconstituting fluorescently-labeled Gαo and Gβγ GTP-binding protein subunits into lipid vesicles which were then fused to living neuroblastoma cells, in order to obtain fluorescently-labeled GTP-binding protein subunits in biological membranes. The GTP-binding proteins incorporated using this method were active in coupling to endogenous α2-adrenergic receptors to effect signal transduction. Lateral mobility measurements indicated low mobile fractions for Gαo and Gβγ, while the apparent lateral diffusion coefficients were quite high (greater than 10^{-9} cm²/sec; see Table 4.2).[103] The high apparent lateral diffusion coefficients are probably attributable to the fact that the GTP-binding protein subunits studied were not intrinsic to the membranes in which the measurements were made and hence largely unrestricted in terms of their lateral mobility in consequence (see chapters 2 and 3). Since the mobile fractions were low, the authors concluded that the Gβγ-subunits are highly constrained to localized regions of the plasma membrane, with the associated Gα-subunit probably similarly immobile in the membrane in the unstimulated state.[103-105] Immobilization of the GTP-binding protein subunits would appear to be through linkage to the cytoskeleton (see also chapter 5, section F; chapter 8, section C).[103-105]

The relatively high apparent lateral diffusion coefficient is at least partly explicable in terms of the fact that GTP-binding proteins are essentially peripheral membrane proteins. The structure of the GTP-binding protein trimer is shown schematically in Figure 4.10. The GTP-binding pocket is located at the "front" of Gα which faces the cytoplasm, while the "back" face of the protein, orientated toward the plasma membrane, presents binding surfaces for receptors, effectors and the GTP-binding protein β and γ subunits as well as attachment sites for fatty acids that enhance avidity of Gα for membranes and for Gβγ.[74,99,106] Peptides from the primary sequences of receptors and GTP-binding proteins have been used to define the regions of the GTP-binding proteins important for molecular interactions with both receptors and effectors (see ref. 99). As suggested in the previous section, peptides from the second, third and fourth intracellular loops of GTP-binding protein coupling receptors antagonise GTP-binding protein interactions with receptors and can directly activate GTP-binding proteins (see ref. 99). These peptides bind to

---

Table 4.2. Lateral mobility of membrane-associated trimeric GTP-binding proteins, as measured by the technique of fluorescence recovery after photobleaching.[103]

GTP-binding-protein subunit[#]	Cell type	Temp.	Parameter of Lateral Mobility*	
			D (10^{-10}cm²/sec)	f
bovine brain Goα	NG-108-15 neuroblastoma x glioma cells	22°C	40	0.34
bovine brain Goβγ			20	0.16

[#]Fluorescently-labeled Gαo and Gβγ subunits were incorporated into lipid vesicles which were then fused to NG-108-15 cells using polyethylene glycol.[103]
*Values are for the apparent lateral diffusion coefficient (D) and mobile fraction (f).

sites that include the amino and carboxy terminal regions of the Gα subunit and an as yet to be identified region of the Gβ subunit.

One can imagine that lateral movement of hormone-occupied receptor brings the GTP-binding protein interaction regions of the receptor into contact with the appropriate regions of the Gα subunit in the GTP-binding protein trimer, as shown in Figure 4.10, resulting in activation of the latter. There is a body of evidence indicating an active signaling role for the Gβγ subunits in interacting with various effector enzymes either independently of the Gα subunit or synergistically or antagonistically to Gα-stimulated effects,[99,104,107] which will be dealt with in more detail in chapter 5 (section E). The fact that GTP-binding proteins may be largely immobile as indicated by the results in Table 4.2 implies that lateral movement of GTP-binding protein activating receptors may be an integral part of signal transduction in terms of bringing receptors and GTP-binding proteins into contact with one another (see chapter 5).

H. LATERAL MOBILITY OF CYTOKINE RECEPTORS

Cytokine receptors play an essential role in immune responses, including regulating the differentiation and proliferation of hemapoietic cell lineages. Many cytokine receptors comprise multiple subunits, e.g., those for IL-3, IL-5 and GM-CSF (granulocyte macrophage colony-stimulating factor), which consist of α and β subunits (of which the latter is shared between the three receptors) and the IL-2 and IL-4 receptors which comprise α, β and γ subunits. Lateral movement is postulated to be required for complexation of the respective subunits of the receptor upon ligand binding (see chapter 1 section C; Fig. 1.3). As already mentioned at the beginning of section C of this chapter, measurement of lateral mobility of cytokine receptor proteins is restricted to the IL-2 receptor α-subunit.[29] Results (Table 4.1) indicate an apparent lateral diffusion coefficient and mobile fraction at 30°C not radically different from those of receptors in the

tyrosine kinase or GTP-binding protein activating class, but it should be remembered that measurements were performed using labeled monoclonal antibody rather labeled IL-2 and hence are not strictly physiological (see beginning of section B above).

The mobile fraction of 0.37 for the IL-2 receptor α-subunit is not very high, possibly indicative of precomplexation of receptor subunits. Although there is evidence that the IL-3 α/β receptor subunits may not be crosslinked in the absence of IL-3,[108] there does indeed appear to be some evidence for preformed α/β subunits in the absence of ligand.[109] This implies that the IL-2 receptor lateral mobility measurements referred to may represent, at least in part, the α/β subunit complex, which may be immobilized to some extent. No information is available as to whether the IL-2 receptor γ-subunit may be precomplexed with the α/β subunits.

In broad structural terms, cytokine receptors resemble those of the tyrosine kinase receptor class in possessing a single transmembrane domain and large extracellular domains involved in ligand binding.[30,110] Some subunits of oligomeric cytokine receptors contain SH2 (src homology-2 domain) sequence motifs. Although cytokine receptor subunits lack intrinsic tyrosine kinase activity, their signal transduction pathway is comparable to that of tyrosine kinase receptors (and different to that of GTP-binding protein activating receptors) in that receptor subunit complexation is the only signaling event occurring within the plane of the membrane; subsequent signaling events occur in the cytosol and involve association of a variety of soluble signaling components,[30,110-112] including cytosolic tyrosine kinases of the Janus class (JAKs) as well as p59[fyn]/p56[lyn],[30] which trigger intracellular phosphorylation cascades. Signals are ultimately communicated to the nucleus through tyrosine phosphorylation-induced nuclear translocation (see refs. 113 and 114) of the normally cytoplasmic transcription factors STAT (signal transducers and activators of transcription).[115-117]

The limited nature of the primary data clearly hampers any meaningful conclusions being drawn with respect to the role of cytokine receptor subunit mobility in signaling. It seems reasonable to speculate, however, that since cytokine receptors functionally and structurally resemble tyrosine kinase receptors and have analogous signal transduction pathways, lateral mobility of these receptors/their receptor subunits is likely to play a similar role in signal transduction to that of mobility in the case of tyrosine kinase receptors.

I. SUMMARY AND IMPLICATIONS FOR SIGNAL TRANSDUCTION

It is evident from this chapter that it is both possible to perform measurements of lateral mobility plasma membrane integral receptors and that the results are informative with respect to the basic differences between the two classes of receptor for which most measurements have been made. It is clear that plasma membrane polypeptide hormone receptors are intrinsically mobile in the lipid bilayer as demonstrated most conclusively by measurements at room temperature, where the rates of receptor internalization are much slower than at 37°C, and corollaries of signal transduction interactions, such as the association with other proteins both within the plane of the membrane and in the cytoplasm, do not have such an effect on receptor lateral mobility.

It can reasonably be concluded from direct measurements of the lateral mobility of plasma membrane polypeptide hormone receptors that tyrosine kinase receptors and GTP-binding protein activating receptors are essentially different and that these differences probably relate to their distinct signal transduction mechanisms. Tyrosine kinase receptors require receptor dimerization and the subsequent association of a variety of cytosolic signaling molecules in order to effect signal transduction, and the fact that they are largely immobile at physiological temperature is consistent with this. Rapid movement of hormone-occupied receptor

is only required to effect dimerization, activated receptors being aggregated and essentially immobile, and this is reflected in the low mobile fraction at 37°C.

In contrast, GTP-binding protein activating receptors must interact with membrane associated GTP-binding proteins in order to effect signal transduction and hence are necessarily mobile within the plane of the membrane. Consistent with this, they exhibit a relatively high receptor mobile fraction at physiological temperature. Because GTP-binding proteins appear to be largely immobile,[103-105] lateral movement of GTP-binding protein activating receptors is essential in bringing receptors and GTP-binding proteins into contact with one another and hence for signal transduction. The indirect and direct evidence validating this conclusion will be discussed in detail in the next and subsequent chapters.

REFERENCES

1. Fahrenholz F, Jans DA, Peters R. Lateral mobility of the V_1- and V_2-receptors in plasma membranes: a role in signal transduction and receptor down-regulation. Colloques INSERM: Vasopressin 1991; 208:49-56.

2. Jans DA. The mobile receptor hypothesis revisited: a mechanistic role for hormone receptor lateral mobility in signal transduction. Biochim Biophys Acta 1992; 1113:271-276.

3. Jans DA, Pavo I. A mechanistic role for polypeptide hormone receptor lateral mobility in signal transduction. Amino Acids 1995; 9:93-109.

4. Ljungquist P, Wasteson A, Magnusson K-E. Lateral diffusion of plasma membrane receptors labelled with either platelet-derived growth factor (PDGF) or wheat germ agglutinin (WGA) in human leukocytes and fibroblasts. Bioscience Reports 1989; 9:63-73.

5. Ljungquist-Hoeddelius P, Lirvall M, Wasteson A et al. Lateral diffusion of PDGF-β receptor in human fibroblasts. Bioscience Reports 1991; 11(1):43-52.

6. Levi A, Schechter Y, Neufeld EJ et al. Mobility, clustering and transport of

nerve growth factor in embryonal sensory cells and in a sympathetic neuronal cell line. Proc Natl Acad Sci USA 1980; 77:3469-3473.

7. Schlessinger J, Schechter Y, Cuatrecasas P et al. Quantitative determination of the lateral diffusion coefficients of the hormone-receptor complexes of insulin and epidermal growth factor on the plasma membrane of cultured fibroblasts. Proc Natl Acad Sci USA 1978; 75:5353-5357.

8. Zidovetzki R, Yarden Y, Schlessinger, J et al. Rotational diffusion of epidermal growth factor complexed to surface receptors reflects rapid microaggregation and endocytosis of occupied receptors. Proc Natl Acad Sci USA 1981; 78:6981-6985.

9. Hillman GM, Schlessinger J. The lateral diffusion of epidermal growth factor complexed to its surface receptors does not account for the thermal sensitivity of patch formation and endocytosis. Biochemistry 1982; 21:1667-1672.

10. Goncalves E, Yamada K, Thatte HS et al. Optimizing transmembrane domain helicity accelerates insulin receptor internalization and lateral mobility. Proc Natl Acad Sci USA 1993; 90:5762-5766.

11. Gilboa L, Ben-Levy R, Yarden Y et al. Roles for a cytoplasmic tyrosine and tyrosine kinase activity in the interactions of Neu receptors with coated pits. J Biol Chem 1995; 270:7061-7067.

12. Roettger BF, Rentsch RU, Hadac EM et al. Insulation of a G protein-coupled receptor on the plasmalemmal surface of the pancreatic acinar cell. J Cell Biol 1995; 130:579-590.

13. Jans DA, Peters R, Zsigo J et al. The adenylate cyclase-coupled vasopressin V_2-receptor is highly laterally mobile in membranes of LLC-PK$_1$ renal epithelial cells at physiological temperature. EMBO J 1989; 8(9):2431-2438.

14. Jans DA, Peters R, Fahrenholz F. Lateral mobility of the phospholipase-C-activating vasopressin V_1-type receptor in A7r5 smooth muscle cells:a comparison with the adenylate cyclase-coupled V_2-receptor. EMBO J 1990; 9(9):2693-2699.

15. Jans DA, Peters R, Jans P et al. Ammonium chloride affects receptor number and lateral mobility of the vasopressin V_2-type receptor in the plasma membrane of LLC-PK$_1$ renal epithelial cells:role of the cytoskeleton. Exper Cell Res 1990; 191:121-128.

16. Jans DA, Peters R, Fahrenholz F. An inverse relationship between receptor internalization and the fraction of laterally mobile receptors for the vasopressin renal-type V_2-receptor; an active role for receptor immobilization in down-regulation? FEBS Lett 1990; 274:223-226.

17. Jans DA, Peters R, Jans P et al. Vasopressin V_2-receptor mobile fraction and ligand-dependent adenylate cyclase-activity are directly correlated in LLC-PK$_1$ renal epithelial cells. J Cell Biol 1991; 114(1):53-60.

18. Pavo I, Jans DA, Peters R et al. A vasopressin antagonist that binds to the V_2-receptor of LLC-PK$_1$ renal epithelial cells is highly laterally mobile but does not effect ligand-induced receptor immobilization. Biochim Biophys Acta 1994; 1223:240-246.

19. Schlessinger J, Axelrod D, Koppel DE et al. Lateral transport of a lipid probe and labeled proteins on a cell membrane. Science 1977; 195:307-309.

20. Livneh E, Benveniste M, Prywes R et al. Large deletions in the cytoplasmic kinase domian of the epidermal growth factor receptor do not affect its lateral mobility. J Cell Biol 1986; 103:327-331.

21. Rees AR, Gregoriou M, Johnson P et al. High affinity epidermal growth factor receptors on the surface of A-431 cells have restricted lateral diffusion. EMBO J 1984; 3:1843-1847.

22. Roess DA, Rahman NA, Kenny N. Molecular dynamics of luteinizing hormone receptors on rat luteal cells. Biochim Biophys Acta 1992; 1137:309-316.

23. Venkatakrishnan G, McKinnon CA, Pilapil CG et al. Nerve growth factor receptors are preaggregated and immobile on responsive cells. Biochemistry 1991; 30(11):2748-2753.

24. Niswender GD, Roess DA, Sawyer HR et al. Differences in the lateral mobility of receptors for luteinizing hormone (LH) in the luteal plasma membrane when occupied by ovine LH versus human

chorionic gonadotropin. Endocrinol 1985; 116:164-169.

25. Roess DA, Niswender GD, Barisas BG. Cytocholasins and colchicine increase the lateral mobility of human chorionic gonadotropin-occupied luteinizing hormone receptors on ovine luteal cells. Endocrinol 1988; 122:261-269.

26. Philpott CJ, Rahman NA, Kenny N et al. Rotational dynamics of luteinizing hormone receptors and MHC class I antigens on murine Leydig cells. Biochim Biophys Acta 1995; 1235(1):62-68.

27. Johansson B, Wymann MP, Holmgren-Peterson K et al. N-formyl peptide receptors in human neutrophils display distinct membrane distribution and lateral mobility when labeled with agonist and antagonist. J Cell Biol 1993; 121:1281-1289.

28. Gupte SS. Localization and diffusion of glucagon receptor in rat hepatocytes. Receptor 1994; 4(3):175-190.

29. Edidin M, Aszalos A, Damjanovich S et al. Lateral diffusion measurements give evidence for association of the Tac peptide of the IL-2 receptor with the T27 peptide in the plasma membrane of HUT-102-B2 T cells. J Immunol 1988; 141:1206-1210.

30. Taniguchi T, Minami Y. The IL-2/IL-2 receptor system: a current overview. Cell 1993; 73:5-8.

31. Jans DA, Resink TJ, Wilson E-L et al. Isolation of a mutant LLC-PK₁ cell line defective in hormonal responsiveness: a pleiotropic lesion affecting receptor function. Eur J Biochem 1986; 160:407-412.

32. Jans DA, Resink TJ, Hemmings BA. Complementation between LLC-PK₁ mutants affected in polypeptide hormone receptor function. Eur J Biochem 1987; 162:571-576.

33. Luzius H, Jans DA, Jans P et al. Isolation and genetic characterization of a renal epithelial cell mutant defective in vasopressin V₂-type receptor binding and function. Exper Cell Res 1991; 195:478-484.

34. Helmreich EJM, Elson EL. Protein and lipid mobility. Adv in Cyclic Nucleotide and Prot Phosphor Res 1984; 18:1-62.

35. Schlessinger J. Signal transduction by allosteric receptor oligomerization. Trends Biochem Sci 1988; 13:443-447.

36. Schlessinger J. The epidermal growth factor receptor as a multifunctional allosteric protein. Biochemistry 1989; 27:3119-3123.

37. Ullrich A, Schlessinger J. Signal transduction by receptors with tyrosine kinase activity. Cell 1990; 61:203-212.

38. Yarden Y, Schlessinger J. Epidermal growth factor induces rapid, reversible aggregation of the purified epidermal growth factor receptor. Biochem 1987; 26:1443-1451.

39. Yarden Y, Schlessinger J. Self-phosphorylation of epidermal growth factor: evidence for a model of intermolecular allosteric activation. Biochem 1987; 26:1434-1442.

40. Gadella TW Jr, Jovin TM. Oligomerization of epidermal growth factor receptors on A431 cells studied by time-resolved fluorescence imaging microscopy. A stereochemical model for tyrosine kinase receptor activation. J Cell Biol 1995; 129(6):1543-1558.

41. Cochet C, Kashles O, Chambaz EM et al. Demonstration of epidermal growth factor-induced receptor dimerization in living cells using a chemical covalent cross-linking agent. J Biol Chem 1988; 263:3290-3295.

42. Heffetz D, Zick Y. Receptor aggregation is necessary for activation of the soluble insulin receptor kinase. J Biol Chem 1986; 261:889-894.

43. Johnson JD, Wong H-L, Rutter WJ. Properties of the insulin receptor ectodomain. Proc Natl Acad Sci USA 1988; 85:7516-7520.

44. Kahn CR, Baird KL, Jarrett DB et al. Direct demonstration that receptor crosslinking or aggregation is important in insulin action. Proc Natl Acad Sci USA 1978; 75(9):4209-4213.

45. Sorokin A. Activation of the EGF receptor by insertional mutations in its juxtamembrane regions. Oncogene 1995; 11(8):1531-1540.

46. Gilboa L, Ben Levy R, Yarden Y et al. Roles for a cytoplasmic tyrosine and tyrosine kinase activity in the interactions

of Neu receptors with coated pits. J Biol Chem 1995; 270(13):7061-7067.

47. Weiner DB, Lui J, Cohen JA et al. A point mutation in the neu oncogene mimics ligand induction of receptor aggregation. Nature 1989; 339:230-231.

48. Yamada K, Goncalves E, Carpentier JL et al. Transmembrane domain inversion blocks ER release and insulin receptor signaling. Biochemistry 1995; 34(3):946-954.

49. Kraus MH, Popescu NC, Amsbaugh SC et al. Overexpression of the EGF receptor-related proto-oncogene erbB-2 in human mammary tumor cell lines by different molecular mechanisms. EMBO J 1987; 6(3):605-610.

50. Honegger AM, Kris RM, Ullrich A et al. Evidence that autophosphorylation of solubilized EGF-receptors is mediated by intermolecular cross phosphorylation. Proc Natl Acad Sci USA 1989; 86:925-929.

51. Honegger AM, Schmidt A, Ullrich A et al. Evidence for EGF-induced autophosphorylation of the EGF-receptor in living cells. Mol Cell Biol 1990; 10(8):4035-4044.

52. Ballotti R, Lammers R, Scimeca I-C et al. Intermolecular transphosphorylation between insulin receptors and EGF-insulin receptor chimerae. EMBO J 1989; 8:3303-3309.

53. Fire E, Zwart DE, Roth MG et al. Evidence from lateral mobility studies for dynamic interactions of a mutant influenza hemagglutinin with coated pits. J Cell Biol 1991; 115:1585-1594.

54. Giugni TD, Braslau DL, Haigler HT. Electric field-induced redistribution and postfield relaxation of epidermal growth factor receptors on A431 cells. J Cell Biol 1987; 104(5):1291-1297.

55. Heller-Harrison RA, Morin M, Czech MP. Insulin regulation of membrane-associated insulin receptor substrate 1. J Biol Chem 1995; 270(41):24442-24450.

56. Noh DY, Shin SH, Rhee SG. Phosphoinositide-specific phospholipase C and mitogenic signaling. Biochim Biophys Acta 1995; 1242(2):99-113.

57. Rozakis F, Adcock M, van der Geer P et al. MAP kinase phosphorylation of mSos1 promotes dissociation of mSos1-

Shc and mSos1-EGF receptor complexes. Oncogene 1995; 11(7):1417-1426.

58. Langlois WJ, Sasaoka T, Saltiel AR et al. Negative feedback regulation and desensitization of insulin- and epidermal growth factor-stimulated p21ras activation. J Biol Chem 1995; 270(43):25320-25323.

59. Eldar H, Ben-Chaim J, Livneh E. Deletions in the regulatory or kinase domains of protein kinase C-alpha cause association with the cell nucleus. Exper Cell Res 1992; 202:259-266.

60. Leach KL, Powers EA, Ruff VA et al. Type 3 protein kinase C localization to the nuclear envelope of phorbol ester-treated NIH 3T3 cells. J Cell Biol 1992; 109:685-695.

61. Leach KL, Ruff VA, Jarpe MB et al. α thrombin stimulates nuclear diglyceride levels and differential nuclear localization of protein kinase C isozymes in IIC9 cells. J Biol Chem 1992; 267:21816-21822.

62. Chen R-H, Sarnecki C, Blenis J. Nuclear localization and regulation of erk- and rsk-encoded protein kinases. Mol Cell Biol 1992; 12:915-927.

63. Ettehadieh E, Sanghera JS, Pelech SL et al. Tyrosyl phosphorylation and activation of MAP kinases by p56^{lck}. Science 1992; 255:853-855.

64. Gronowski AM, Rotwein P. Rapid changes in nuclear protein tyrosine phosphorylation after growth hormone treatment in vivo. Identification of phosphorylated mitogen-activated protein kinase and STAT91. J Biol Chem 1994; 269:7874-7878.

65. Jans DA. Nuclear signaling pathways for polypeptide ligands and their membrane-receptors? FASEB J 1994; 8:841-847.

66. Jans DA. Regulation of protein transport to the nucleus by phosphorylation. Biochem J 1995; 311:705-716.

67. Heldin C-H. Structural and functional studies on platelet-derived growth factor. EMBO J 1992; 11:4251-4259.

68. Panaotou G, Waterfield MD. The assembly of signalling complexes by receptor tyrosine kinases. Bioessays 1993; 15(3):171-177.

69. Neubig RR, Sklar LA. Subsecond modulation of formyl peptide-linked guanine

nucleotide-binding proteins by guanosine 5'-O-(3-thio)triphosphate in permeabilized neutrophils. Mol Pharmacol 1993; 43(5):734-740.

70. Brandt DR, Ross EM. Catecholamine-stimulated GTPase cycle; multiple sites of regulation by β-adrenergic receptor and Mg²⁺ studied in reconstituted receptor-Gₛ vesicles. J Biol Chem 1986; 261:1656-1664.

71. Orly J, Schramm M. Fatty acids as modulators of membrane functions; catecholamine-activated adenylate cyclase of the turkey erythrocyte. Proc Natl Acad Sci USA 1975; 72:3433-3437.

72. Ransnas LA, Insel PA. Subunit dissociation is the mechanism for hormonal activation of the Gs protein in native membranes. J Biol Chem 1988; 263(33):17239-17242.

73. Alousi AA, Jasper JR, Insel PA et al. Stoichiometry of receptor-Gs-adenylate cyclase interactions. FASEB J 1991; 5:2300-2303.

74. Conklin BR, Bourne HR. Structural elements of Gα subunits that interact with Gβγ, receptors, and effectors. Cell 1993; 73:631-641.

75. Bourne HR, Sanders DA, McCormick F. The GTPase superfamily: a conserved switch for diverse cell functions. Nature 1990; 348:125-132.

76. Bourne HR, Sanders DA, McCormick F. The GTPase superfamily: conserved structure and molecular mechanism. Nature 1991; 349:117-127.

77. Birnbaumer L. G proteins in signal transduction. Annu Rev Pharmacol Toxicol 1990; 30:675-705.

78. Freissmuth M, Casey PJ, Gilman AG. G proteins control diverse pathways of transmembrane signalling. FASEB J 1989; 3:2125-2131.

79. Ross EM. Signal sorting and amplification through G protein-coupled receptors. Neuron 1989; 3:141-152.

80. Lamb TD. Stochastic simulation of activation in the G-protein cascade of photoinduction. Biophys J 1994; 67:1439-1454.

81. Johnson GL, Dhanasekaran N. The G-protein family and their interactions with receptors. Endocrine Reviews 1992; 10(3):317-331.

82. Strader CD, Fong TM, Tota MR et al. Structure and function of G protein-coupled receptors. Ann Rev Biochem 1994; 63:101-132.

83. Thibonnier M. Signal transduction of V₁-vascular vasopressin receptors. Regulatory Peptides 1992; 38:1-11.

84. Meinkoth JL, Taylor SS, Feramisco JR. Dynamics of the distribution of cyclic AMP-dependent protein kinase in living cells. Proc Natl Acad Sci USA 1990; 87:9595-9599.

85. Nigg EA, Hilz H, Eppenberger HM et al. Rapid and reversible translocation of the catalytic subunit of cAMP-dependent protein kinase type II from the Golgi complex to the nucleus. EMBO J 1985; 4:2801-2807.

86. Pearson D, Nigg EA, Nagamine Y et al. Mechanisms of cAMP-mediated gene induction; examination of renal epithelial cell mutants affected in the catalytic subunit of the cAMP-dependent protein kinase. Exper Cell Res 1991; 192:315-318.

87. Gill GN, Lazar CS. Increased phosphotyrosine content and inhibition of proliferation in EGF-treated A431 cells. Nature 1981; 293:305-307.

88. Gregoriou M, Rees AR. Properties of a monoclonal antibody to epidermal growth factor receptor with implications for the mechanism of action of EGF. EMBO J 1984; 3:929-937.

89. Barnes DW. Epidermal growth factor inhibits growth of A431 human epidermoid carcinoma in serum-free cell culture. J Cell Biol 1982; 93(1):1-4.

90. Henis YI, Hekman M, Elson EL et al. Lateral diffusion of β-receptors in membranes of cultured liver cells. Proc Natl Acad Sci USA 1982; 79:2907-2911.

91. De Haen C. The non-stoichiometric floating receptor model for hormone sensitive adenylyl cyclase. J Theor Biol 1976; 58:383-400.

92. Gerard C, Gerard NP. C5A anaphylatoxin and its seven transmembrane-segment receptor. Annu Rev Immunol 1994; 12:775-808.

93. Saless R, Remy JJ, Levin JM et al. Towards understanding the glycoprotein hormone receptors. Biochimie 1991; 73(1):109-120.

94. Henderson R, Baldwin JM, Ceska TA. Model for the structure of bacteriorhodopsin based on high-resolution elec-

tron cryo-microscopy. J Mol Biol 1990; 213:899-929.

95. Hausdorff WP, Caron MG, Lefkowitz RJ. Turning off the signal: desensitisation of the β-adrenergic receptor function. FASEB J 1990; 4:2881-2889.

96. Bichet DG. Molecular and cellular biology of vasopressin and oxytocin receptors and action in the kidney. Curr Opin Hephrol Hypertens 1994; 3(1):46-53.

97. Unson CG, Cypess AM, Kim HN et al. Characterization of deletion and truncation mutants of the rat glucagon receptor. Seven transmembrane segments are necessary for receptor transport to the plasma membrane and glucagon binding. J Biol Chem 1995; 270:27720-27727.

98. Kyte J, Doolittle RF. A simple method for displaying the hydropathic character of a protein. J Mol Biol 1982; 157(1):105-132.

99. Taylor JM, Neubig RR. Peptides as probes for G protein signal transduction. Cell Signal 1994; 6(8):841-849.

100. Kazlauskas A, Cooper JA. Autophosphorylation of the PDGF receptor in the kinase insert region regulates interactions with cell proteins. Cell 1989; 58:1121-1132.

101. Schwartz AL, Fridovich SE, Lodish HF. Kinetics of internalization and recycling of the asialoglycoprotein receptor in a hepatoma cell line. J Biol Chem 1982; 257(8):4230-4237.

102. Schlessinger J, Schechter Y, Willingham MC et al. Direct visualization of binding, aggregation, and internalization of insulin and epidermal growth factor on living fibroblastic cells. Proc Natl Acad Sci USA 1978; 75:2659-2663.

103. Kwon G, Axelrod D, Neubig RR. Lateral mobility of tetramethylrhodamine (TMR) labelled G protein α and βγ subunits in NG108-15 cells. Cellular Signalling 1994; 6(6):663-679.

104. Neubig RR. Membrane organization in G-protein mechanisms. FASEB J 1994; 8:939-945.

105. Graeser D, Neubig RR. Compartmentation of receptors and guanine nucleotide-binding proteins in NG108-15 cells: lack of cross-talk in agonist binding among the alpha 2-adrenergic, muscarinic, and opiate receptors. Mol Pharmacol 1993; 43(3):434-443.

106. Neer EJ, Smith TF. G protein heterodimers: new structures propel new questions. Cell 1996; 84:175-178.

107. Clapham DE, Neer EJ. New roles for G-protein βγ-dimers in transmembrane signalling. Nature 1993; 365:403-406.

108. Stomski FC, Sun Q, Bagley CJ et al. Human interleukin-3 (IL-3) induces disulfide-linked IL-3 receptor alpha- and beta-chain heterodimerization, which is required for receptor activation but not high-affinity binding. Mol Cell Biol 1996; 16(6):3035-3046.

109. Roessler E, Grant A, Ju G et al. Cooperative interactions between the interleukin 2 receptor alpha and beta chains alter the interleukin-2-binding affinity of the receptor subunits. Proc Natl Acad Sci USA 1994; 91(8):3344-3347.

110. Taga T, Kishimoto T. Cytokine receptors and signal transduction. FASEB J 1992; 6:3387-3396.

111. Fung MR, Scearce RM, Hoffman JA et al. A tyrosine-kinase physically associates with the beta-subunit of the human IL-2 receptor. J Immunol 1991; 147(4):1253-1260.

112. Sato S, Katagiri T, Takaki S et al. IL-5 receptor-mediated tyrosine phosphorylation of SH2/SH3-containing proteins and activation of Bruton's tyrosine and Janus 2 kinases. J Exp Med 1994; 180:2101-2111.

113. Jans DA. Regulation of protein transport to the nucleus by phosphorylation. Biochem J 1995; 311:705-716.

114. Jans DA, Huebner. Regulation of protein transport to the nucleus - the central role of phosphorylation. Physiol Rev 1996; 76:651-685.

115. Schindler C, Shuai K, Prezioso VR et al. Interferon-dependent tyrosine phosphorylation of a latent cytoplasmic transcription factor. Science 1992; 257:809-813.

116. Shuai K, Schindler C, Prezioso VR et al. Activation of transcription by IFN-gamma: tyrosine phosphorylation of a 91-kD DNA binding protein. Science 1992; 258:1808-1812.

117. Shuai K, Ziemiecki A, Wilks AF et al. Polypeptide signalling to the nucleus through tyrosine phosphorylation of Jak and Stat proteins. Nature 1993; 366:580-583.

CHAPTER 5

EVIDENCE FOR THE ROLE OF MEMBRANE RECEPTOR LATERAL MOVEMENT IN GTP-BINDING PROTEIN-MEDIATED SIGNAL TRANSDUCTION

A. INTRODUCTION

As we saw in the previous chapter, direct measurements indicate that plasma membrane polypeptide hormone receptors are mobile within the plane of the lipid bilayer. Tyrosine kinase and GTP-binding protein activating receptors appear to be essentially different in terms of their lateral mobility properties, and it seems plausible to relate these differences to their distinct signal transduction mechanisms.[1-8] As discussed in chapter 4, lateral movement of hormone-occupied tyrosine kinase receptors is required to effect receptor dimerization which triggers the association of a variety of cytosolic signaling molecules in order to effect signal transduction.[9-13] Movement of the hormone-occupied receptor is not required subsequent to dimerization, and consistent with this, activated receptors appear to be aggregated and essentially immobile at physiological temperature (see section E, chapter 4).

In contrast to tyrosine kinase receptors, GTP-binding protein activating receptors must by definition interact with membrane-associated GTP-binding proteins in order to effect signal transduction and thus need to be mobile within the plane of the membrane. Consistent with this, they exhibit a relatively high receptor mobile fraction at physiological temperature (see chapter 4, Fig. 4.4B). Hormone-occupied GTP-binding protein activating receptors are not immobilized by their interaction with GTP-binding proteins and hence are able to activate multiple GTP-binding proteins and thus amplify the hormonal signal at the level of the membrane. Since GTP-binding proteins are largely immobile (see chapter 4, section G), lateral movement of hormone-occupied GTP-binding protein activating receptors is essential to bring receptors and GTP-binding proteins into contact with one another and hence for signal transduction. The indirect and direct evidence validating this conclusion will be discussed in detail in this chapter. In particular, the question of whether lateral movement of hormone-occupied GTP-binding protein activating receptors is central to signal transduction at the level of the membrane will be addressed.

The Mobile Receptor Hypothesis: The Role of Membrane Receptor Lateral Movement in Signal Transduction, edited by David A. Jans. © 1997 R.G. Landes Company.

B. KINETIC CONSIDERATIONS IN GTP-BINDING PROTEIN-MEDIATED RECEPTOR-EFFECTOR SYSTEMS

Detailed kinetic analyses using the β-adrenergic receptor/adenylate cyclase system in membranes isolated from turkey erythrocytes indicate that hormone-dependent adenylate cyclase activation is a diffusion-controlled process. Experimental approaches have included varying:

1. the amounts of the various signaling components,
2. their state of activation using agonists stimulating the particular components (e.g., pertussis or cholera toxins which activate the GTP-binding protein subunits Giα and Gsα subunits, respectively, and forskolin, a diterpene from the Indian plant *Coleus forskolii*, which binds and activates adenylate cyclase directly), and
3. their selective inhibition to differing degrees using affinity labels in the case of the β-adrenergic receptor, or *p*-hydroxymercuri-benzoate in the case of adenylate cyclase.[14-20]

Results indicate that hormone-dependent adenylate cyclase activation is independent of the concentrations of adenylate cyclase or stimulatory GTP-binding protein (Gs) as well as of GDP release from the Gsα-subunit.[14-20] Interaction between activated Gs and the adenylate cyclase catalytic subunit also does not appear to be the rate-limiting step.[16] Rather, the rate determining step in hormone-mediated adenylate cyclase activation is that of the interaction between the hormone-occupied receptor and Gs components.[14-20]

As already seen in chapter 1 (section C), the sequence of events in hormone-dependent GTP-binding protein-mediated effector enzyme activation is as follows:

$$H + R \overset{k_1}{\Longleftrightarrow} HR' \quad [1.1] \text{ (extracellular)}$$

$$HR' + G \overset{k_2}{\Longleftrightarrow} HR - G'' \overset{k_3}{\longrightarrow} HR' + G'$$
$$[1.5] \text{ (intramembrane)}$$

$$G' + E \overset{k_4}{\Longleftrightarrow} E - G' \overset{k_5}{\Longleftrightarrow} G' + E'$$
$$[1.6] \text{ (cytosolic)},$$

where k_1 to k_5 are the rate constants for the respective reactions; H, hormone; R, receptor; G, GTP-binding protein; ' denotes activation; '' denotes a transient (millisecond) complex; and E, effector enzyme (e.g., adenylate cyclase). Activation in the case of the GTP-binding protein is its conversion to the activated (GTP-bound) state. Cycles of [1.5] amplify the signal at the level of the membrane.

The kinetic analyses discussed above[14-20] imply that [1.5] is the rate-determining step in hormone-mediated adenylate cyclase activation. In terms of the Mobile Receptor Hypothesis, the fact that interaction between hormone-occupied receptor and Gs is rate-determining is completely consistent with the notion that lateral movement of the hormone-receptor complex within the plane of the plasma membrane lipid bilayer brings it into transient, activating contacts with the membrane-associated trimeric GTP-binding protein complex.[21] That membrane diffusion events are rate limiting can be understood in terms of the fact that GTP-binding protein activation appears to lead to release of the Gα-subunit into the aqueous phase and that interaction between activated GTP-binding proteins and effector enzymes ([1.6] above and see also section E below) accordingly occurs in the cytosolic phase.[1,22] Since the rate of lateral diffusion in cytosol is much greater than that within the plane of the membrane (compare Tables 2.1 and 2.3), the kinetics of the interactive events within the membrane ([1.5]: rate constant k_2 and k_3 above) are much slower than those within either the extracellular (rate constant k_1) or cytosolic (rate constants k_4 and k_5) aqueous phases. Because the hormone-receptor-GTP-binding protein complex is only transient and thus dissociates rapidly (i.e., $k_2 >> k_3$), diffusion-controlled interaction between the receptor and GTP-binding protein membrane-components is the rate-limiting step of effector enzyme activation. It should be noted that the postulated existence of pre-coupled GTP-binding proteins and effector enzymes (e.g., refs. 16, 23) is also kinetically consistent with the interaction between receptor and GTP-binding proteins being rate-limiting.

Some authors have interpreted particular kinetic studies as being evidence for the existence of precoupled receptor/GTP-binding protein complexes (e.g., ref. 24), even though, as already mentioned in chapter 1, the general consensus appears to be that a number of different receptors can interact with the same GTP-binding protein within the same cell and hence are unlikely to be precoupled (see ref. 25).[24,26-29] The results of Neubig et al[24] suggest that in the α2-adrenergic receptor/adenylate cyclase system of human platelets about a third of receptors are precoupled to Gi in the absence of agonist, a third may be unable to couple to Gi and the remaining third interacts with Gi upon hormone addition through collision coupling in the plane of the membrane. The established low-affinity nature of the interaction of receptors and GTP-binding proteins in the absence of hormone in most systems implies that the existence of precoupled receptor-GTP-binding proteins is simply the result of the fact that there is a given probability of collision and a weak interaction between receptor and GTP-binding protein components all the time (see also chapter 6 section G). Compartmentalization at the level of the membrane (see section F below) through preformed low-affinity complexes of this nature may enable particular vectorial interactions (e.g., association of a particular hormone-receptor complex with a specific GTP-binding protein).[25-29]

In summary, kinetic evidence from a number of systems and laboratories strongly implies that the rate limiting step of effector activation is that of receptor-hormone complex interaction with the GTP-binding protein, which occurs within the plane of the membrane. Findings are clearly consistent with the idea that signal transduction at the level of the membrane is diffusion-controlled.[1,2,16,17,20]

C. INDIRECT EVIDENCE FOR A ROLE OF RECEPTOR LATERAL MOVEMENT IN GTP-BINDING PROTEIN-MEDIATED SIGNAL TRANSDUCTION

As hinted at in chapter 3 (section B), a number of studies have taken the approach of examining signal transduction kinetics under conditions in which protein lateral mobility is perturbed through modulating membrane fluidity (see ref. 23). This has been variously achieved by looking at cells whose membrane lipid composition has been modified through dietary, nutritional, aging or enzymatic means.[17,19,23,30-42] Highly fluid mouse LM fibroblast cell membranes enriched with linoleate and depleted in oleate exhibited about 4-fold higher prostaglandin E_1-stimulated adenylate cyclase activity compared to normal.[33,34] In contrast, basal and fluoride (GTP-binding protein activating)-stimulated activities were unaffected,[33,34] indicating that the increased receptor-mediated adenylate cyclase activation was not due to structural effects of the altered membrane fluidity on the GTP-binding protein trimer or adenylate cyclase. In similar fashion, liver plasma membranes isolated from rats fed a diet lacking essential fatty acids but supplemented with unsaturated fatty acids, and hence of increased fluidity, exhibited elevated hormone-dependent adenylate cyclase activity.[20] Orly and Schramm[19] and Hanski et al[17] among others (see ref. 23) introduced *cis*-vaccenic acid into turkey erythrocyte membranes to increase membrane fluidity and tested adenylate cyclase activation in response to epinephrine in the presence of the nonhydrolyzable GTP analog Gpp(NH)p (5'-guanylylimidodiphosphate). The rate constant of adenylate cyclase activation was increased maximally up to nearly 20-fold. Results could not be attributed to effects on either the number or binding affinity of the β-adrenergic receptor, leading the authors to conclude that adenylate cyclase activation by the β-receptor is a diffusion-controlled process.[17] That membrane lipid fluidity can be a critical parameter in determining membrane signaling kinetics is also implied by experiments in which the cyclic macrolide filipin, which forms complexes with membrane cholesterol and thus perturbs membrane integrity, decreases catecholamine-stimulated adenylate cyclase activity.[36] Filipin does not affect receptor binding activity,[37] and hence its effects on receptor-dependent adenylate cyclase activation are not attributable to structural effects on the receptor itself.

Changes in membrane fluidity due to aging or to delipidation treatments can also have significant effects on signal transduction, as has been shown for several GTP-binding protein coupling hormone receptors.[30-32,38-40] Moscona-Amir et al[30,31] compared 5 and 14 day old rat heart myocytes for their signal transduction properties with respect to muscarinic M2-receptor-mediated inhibition of isoprotenerol-stimulated cAMP accumulation. Aging of the myocyte cultures was accompanied by a marked reduction in the inhibition of cAMP accumulation, through less efficient coupling of activation of the M2 receptor upon hormone binding to stimulation of activity of the inhibitory Gi GTP-binding protein.[31] Treatment of aging cultures with phosphatidylcholine liposomes under conditions that yielded a lipid composition resembling that of five day old cultures restored the muscarinic receptor-mediated effect on cAMP accumulation,[31] implying that the effects on coupling at the level of the membrane stemmed directly from the changes in lipid fluidity.[30]

Zakharova et al[32] measured the lateral mobility of membrane integral proteins in reticulocyte plasma membranes which were treated to reduce the "fluid lipid" fraction through phospholipase A_2 hydrolysis, and found a clear effect in terms of a reduction in both the apparent lateral diffusion coefficient and mobile fraction (see Fig. 3.3). Concomitant with this was a clear diminution in isoprotenerol-stimulated adenylate cyclase activity, demonstrating that protein lateral movement, modulated in turn by

lipid fluidity, is a critical factor in signal transduction in this system.[32,38-40] Zakharova et al[32] also directly perturbed plasma membrane protein mobility using polylysines which effect protein aggregation and immobilization (reduction of both the apparent lateral diffusion coefficient and mobile fraction) and found that this also greatly impaired hormone-mediated adenylate cyclase activation. Neither phospholipase A_2 nor polylysine treatment affected the β-adrenergic receptor number or binding affinity.[32,38-40] Phospholipase treatment was also shown not to affect GTP-binding protein or adenylate cyclase activities,[33-35] while the fact that polylysines of larger molecular weight at constant residue concentration had a greater effect in terms of reducing both membrane protein mobility and hormone-stimulated adenylate cyclase activity implied that the effects on signal transduction were not due to direct effects of polylysine on GTP-binding proteins or adenylate cyclase, but to increased protein aggregation and immobilization.[27,35] The results of these experiments are summarized in Figure 5.1, the conclusion being that reduction of membrane protein mobility, and in particular of the mobile fraction, concomitantly reduces hormone-dependent adenylate cyclase activation. The data is plotted reciprocally with respect to the general membrane protein mobile fraction in Figure 5.2 to demonstrate the direct dependence of hormone-dependent adenylate cyclase activation on the membrane protein mobile fraction. The clear implication is that membrane protein lateral movement is required for signal transduction.

In a comparable approach to the use of polylysine by Zakharova et al.,[32] Atlas et al[48] employed cationized ferritin to inhibit the lateral mobility of membrane proteins in isolated turkey erythrocyte membranes, as determined by electron microscopic demonstration of large-scale inhibition of membrane protein cold-induced clustering or "thermotropic separation", where, under non cross linked conditions, membrane proteins can be induced to aggregate by low

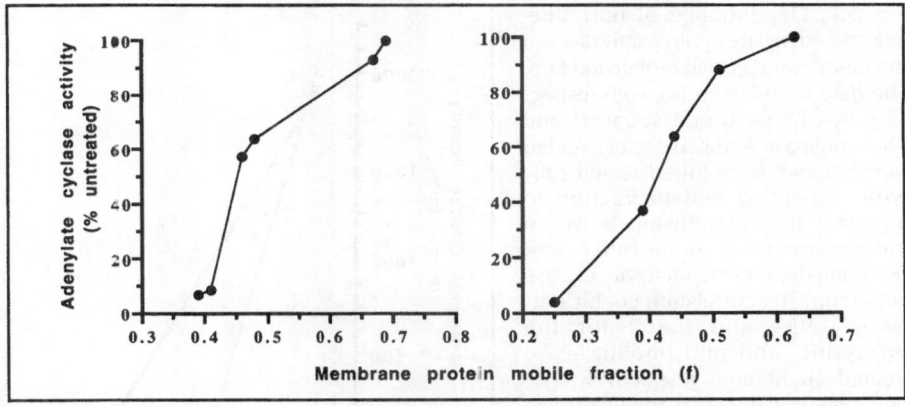

Fig. 5.1. Dependence of signal transduction at the level of the membrane on membrane protein lateral mobility. The mobile fraction in isolated rabbit reticulocyte plasma membranes was modified either by treatment with poly-L-lysine to induce protein aggregation (left panel) or modulation of the membrane fluid lipid fraction using phospholipase A_2 treatment (right panel; see also Fig. 3.3); isoprotenerol-stimulated adenylate cyclase activity is expressed as a percentage of that of untreated membranes.[32] It should be remembered that general membrane mobility, rather than the lateral mobility of specific signaling components, was measured.

temperature.[49] A concomitant reduction in the activation of adenylate cyclase by epinephrine-occupied β-adrenergic receptor was observed, which was not attributable to direct effects of cationized ferritin on either the β-receptor or adenylate cyclase moiety.[48] Cationized ferritin-induced inhibition of the hormone-dependent cyclase activity was concluded to result from the inhibition of the lateral mobility of the β-receptor and a concomitant decrease in the bimolecular rate of interaction between the receptor and adenylate cyclase[48] (note that this study was performed prior to the discovery of GTP-binding proteins).

As already alluded to in chapter 3, the physiological relevance of many studies comparable to those referred to here is questionable (see ref. 14), since perturbation of membrane structure and hence membrane protein function is a likely consequence of the enzymatic and other treatments. Although this clearly has been demonstrated not to be the basis of the effects on signal transduction in the examples discussed

above, there are instances in which receptor number and affinity and basal effector enzyme activities are indeed affected by the treatments modulating membrane lipid composition (see refs. 23, 31 and 41), meaning that effects cannot be ascribed to changes in membrane protein lateral mobility. Further, in most cases where membrane viscosity/fluidity has actually been measured in such experimental systems, e.g., through fluorescence polarization measurements[23,42] or fluorescence photobleaching recovery measurements of lipid probes,[14,43] there appears to be doubt as to whether the treatments used really do affect membrane fluidity and/or membrane component lateral movement (see ref. 23). Clearly, however, these reservations do not apply to the studies of Zakharova et al,[32] where the fluid lipid fraction was quantified using a spin probe and ESR (electron spin resonance) spectroscopy, and of Hanski et al,[17] where fluorescence polarization measurements were used to quantify and confirm effects on membrane fluidity. Of the studies mentioned in

Fig. 5.2. Dependence of hormone-induced adenylate cyclase activation on the membrane protein mobile fraction. The data from Figure 5.1 with respect to poly-L-lysine (open squares) and phospholipase A_2 (filled circles) treated membranes[32] is replotted reciprocally with respect to mobile fraction to highlight the relationship between membrane protein mobility and hormone-dependent adenylate cyclase activation. The correlation coefficients were 0.954 and 0.962 for the polylysine- and phospholipase A_2-treated membranes respectively. The plot was not linear at lower mobile fractions (not shown), indicating a threshold effect (see also refs. 28, 44-47).[32,38-40] Again, it should be remembered that general membrane protein mobility, rather than the lateral mobility of specific signaling components, was measured.

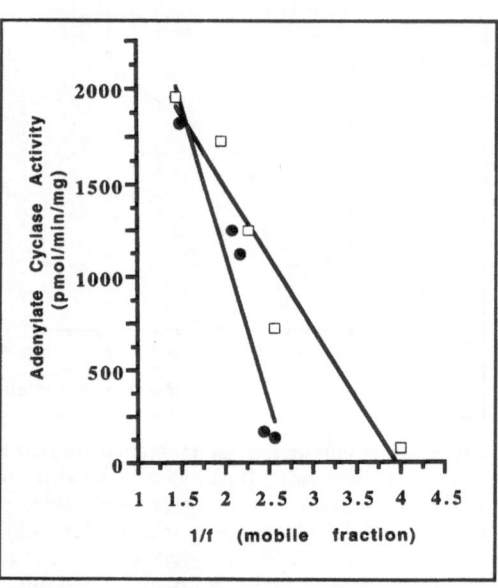

this section, that of Zakharova et al[32] is also exceptional in that general protein lateral mobility was directly measured and shown to be affected by the treatments used, rather than simply being inferred to be influenced. Studies including parallel direct measurements of signal transduction kinetics and of the lateral mobility of specific membrane signaling components (i.e., receptors and/or GTP-binding proteins as opposed to general membrane protein lateral mobility as in the study of Zakharova et al)[32] appear to be restricted to the vasopressin V_2-receptor/Gs/adenylate cyclase system (see next section).[1,2,5,7,8]

D. DIRECT EVIDENCE FOR A ROLE OF RECEPTOR LATERAL MOVEMENT IN GTP-BINDING PROTEIN-MEDIATED SIGNAL TRANSDUCTION

As indicated in chapter 4, direct fluorescence photobleaching recovery measurements indicate that, in contrast to tyrosine kinase receptors, GTP-binding protein activating receptors display significant lateral mobility at 37°C (see Table 4.1; Figs. 4.3 and

4.4.).[3,4,8,50-52] As noted, the Gs- and adenylate cyclase-activating renal vasopressin V_2-type receptor mobile fraction is highest at physiological temperature and lowest at 4°C (see Fig. 4.3), which implies a mechanistic role for receptor lateral mobility in signal transduction.[3] While other nonreceptor proteins are known to exhibit such properties with respect to the mobile fraction (e.g., refs. 53 and 54), this may be peculiar to the V_2-receptor in terms of receptors. Alternatively, this may be largely attributable to the fact that most other GTP-binding protein activating receptors (e.g., the CCK_A receptor) are more rapidly immobilized prior to internalization than the V_2-receptor, so that lateral mobility measurements cannot be made sufficiently quickly in order to enable demonstration of a high mobile fraction at 37°C (see next chapter, section E and refs. 1, 2, 6 and 8).[55-58]

Interestingly and importantly, the V_2-receptor mobile fraction can be reversibly reduced in cells of the LLC-PK$_1$ renal epithelial line using pretreatments of either low temperature (see Fig. 5.3)[5] or the weak base NH_4Cl (see Fig. 8.2).[7] The apparent lateral

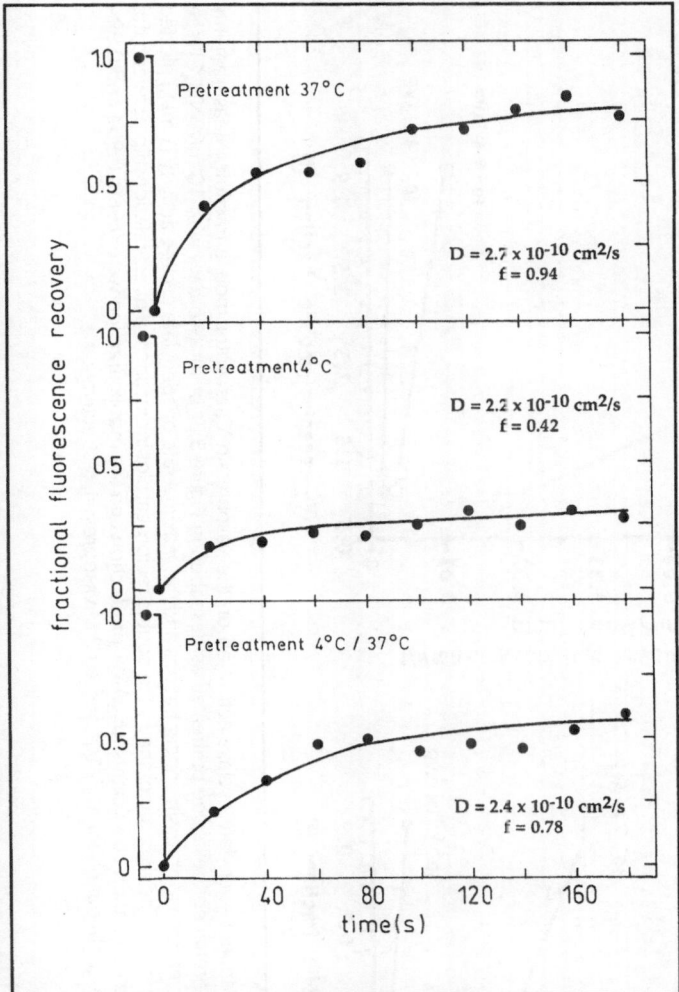

Fig. 5.3. Reversible reduction of vasopressin V_2-receptor lateral mobility through temperature pretreatment. Cells of the LLC-PK$_1$ renal epi-thelial line were pretreated for one hour either at 37 (top panel) or 4°C (middle panel), or for one hour at 4°C followed by a further hour at 37°C (bottom panel), prior to measurement of V_2-receptor apparent lateral diffusion coefficient (D) or mobile fraction (f) using fluorescence photo-bleaching recovery.[7] Measure-ments were performed as described in the legend to Fig. 4.2.

diffusion coefficient does not appear to be affected by the treatments (Figs. 5.3,8.2),[5,7] both of which have been shown to result in the immobilization of other membrane proteins (e.g., ref. 59). Modulation of the V_2-receptor mobile fraction using such pretreatments has enabled the role of receptor lateral mobility in signal transduction to be tested directly; cells can be pretreated to reduce the V_2-receptor mobile fraction and then stimulated with hormone or the receptor-independent adenylate cyclase activator forskolin.[5,7] Adenylate cyclase activation can

then be measured to assess the effect of reduced V_2-receptor mobile fraction on signal transduction kinetics.[7] Responses to vasopressin were found to be markedly reduced in terms of the maximal rates of cAMP production in cells pretreated to reduce the V_2-receptor mobile fraction (Fig. 5.4).[5,7] Elevated cAMP production as the result of hormonal (or other) stimulation leads to activation of transcription and production of the urokinase (urokinase-type plasminogen activator) protease as a longer-term response; hormone-stimulated

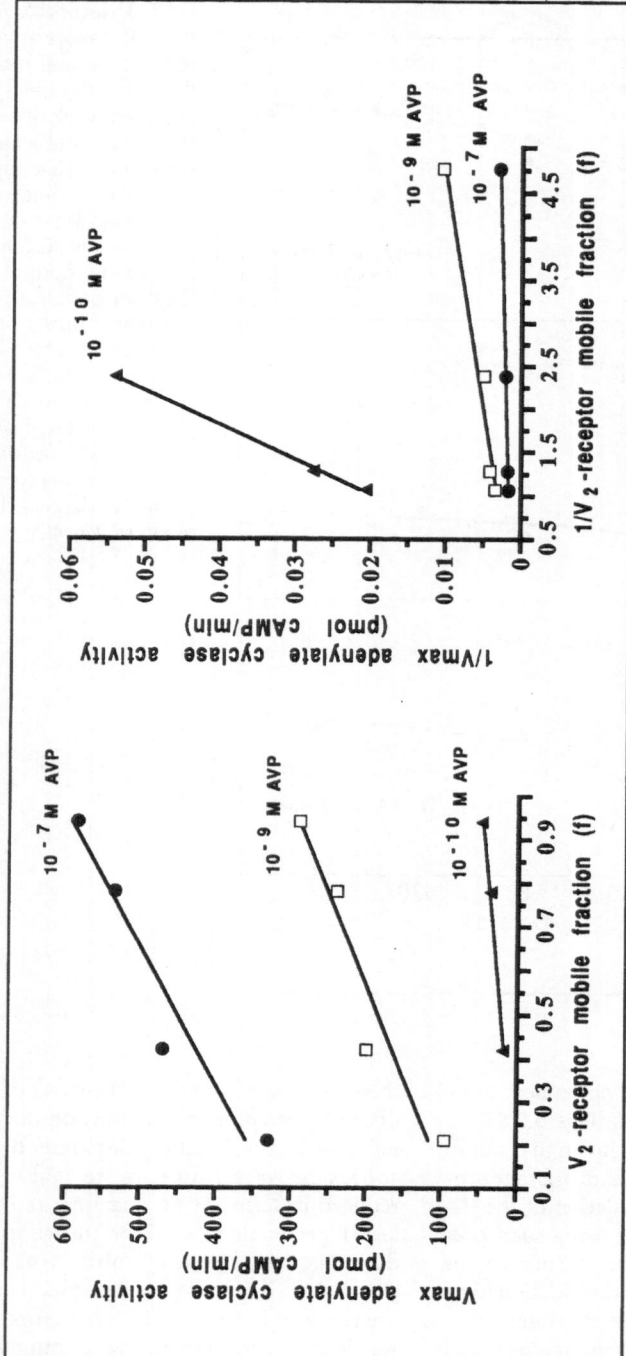

Fig. 5.4. Dependence of hormone-dependent adenylate cyclase activation on the vasopressin V_2-receptor mobile fraction. Measurements were performed on cells of the LLC-PK$_1$ renal epithelial line pretreated as described in Figure 5.3$\frac{2}{3}$ or for two days with 10 mM NH$_4$Cl (see Fig. 8.2).[5] Cells were stimulated with increasing vasopressin concentrations and the maximal rate of adenylate cyclase activity (Vmax) plotted with respect to the mobile fraction (left panel). The right panel shows the double reciprocal plot indicating the direct relationship between the receptor mobile fraction and Vmax. The correlation coefficients for cAMP production and the mobile fraction were 0.960, 0.954 and 0.994 (left panel) and 0.996, 0.985, and 0.999 (right panel) for 10^{-7}, 10^{-9} and 10^{-10} M vasopressin, respectively.

urokinase production is also impaired in cells pretreated to reduce the V_2-receptor mobile fraction.[7] In contrast to hormone-dependent adenylate cyclase activation, stimulation of adenylate cyclase activity by forskolin is completely normal,[5] implying that receptor-independent responses are unaffected by the pretreatments and indicating that the effects on hormone-dependent adenylate cyclase activation cannot be attributed to direct effects of the pretreatments on adenylate cyclase.[5]

The maximal rate of vasopressin-stimulated cAMP production correlates directly with the magnitude of the receptor mobile fraction (see Fig. 5.4),[7] suggesting a direct role for V_2-receptor lateral mobility in hormone-mediated adenylate cyclase activation.[3,5,7] That low V_2-receptor mobile fraction results in a more pronounced reduction of maximal ligand-stimulated adenylate cyclase activation at vasopressin concentrations below the receptor dissociation constant (K_D of 1.9×10^{-9} M)[7,8,60-62] is consistent with the receptor mobile fraction being particularly crucial under physiological conditions of low ligand concentrations and receptor occupancy (see Fig. 5.4).[7] The experimental data for cAMP production and V_2-receptor lateral mobility (Fig. 5.4 left panel)[7] is plotted in reciprocal form in the right panel of Figure 5.4 to reveal the direct relationship between adenylate cyclase activation and the V_2-receptor mobile fraction. The results imply that only mobile receptors participate in signal transduction, with receptor lateral diffusion being rate limiting in terms of adenylate cyclase activation.

The biochemical basis of the reversible V_2-receptor immobilization effected by low temperature treatment is not completely understood at this stage but may relate to the cold-induced aggregation of membrane proteins alluded to above (section C).[21] That cold treatment can disproportionately affect hormone-dependent adenylate cyclase activation has been demonstrated in the turkey erythrocyte membrane (see refs. 21 and 23); interestingly, activation can be restored through the introduction of *cis*-vaccenic

acid to increase lipid fluidity.[63] Interestingly, low temperature treatment has been shown to induce changes in the actin cytoskeleton of LLC-PK$_1$ cells, as does the NH_4Cl treatment which also affects the V_2-receptor mobile fraction (see chapter 8, section C; Figs. 8.1 and 8.2).[5,7] This implies that the cytoskeleton may directly modulate V_2-receptor mobility as has been shown for other membrane proteins (see chapter 3; Table 3.4; refs. 52 and 64-68)[59] and thereby play a central regulatory role in signal transduction. This will be discussed in some detail in section C of chapter 8 (see also chapter 6, section E).

It should be stressed that the conclusions here with respect to a mechanistic role for receptor lateral movement in membrane signal transduction are based on results from in vivo experiments and hence of considerable physiological relevance.[3,5,7,8] The measurement of lateral mobility of a specific receptor under conditions in which the receptor mobile fraction can be modulated, and the effect of this on the kinetics of adenylate cyclase activation enable conclusions to be drawn with some confidence. In a theoretical treatment of hormone-dependent adenylate cyclase activation kinetics, Gupte[69] drew similar conclusions with respect to the role of receptor lateral movement in signal transduction. Using the results of direct measurements of the lateral mobility parameters of the glucagon receptor (Table 4.1), he concluded that the hormone-induced activation kinetics were completely consistent with the diffusion of individual components constituting the rate-limiting step in the signal transduction process.[69] Theoretical studies of GTP-binding protein-coupled systems (e.g., ref. 70), according to collision coupling models, clearly predict a reduction in the rate of effector activation effected by a decrease in the number of hormone-occupied receptors.[14,25] The results shown here for the dependence of the maximal rate of adenylate cyclase activation on the V_2-receptor mobile fraction correlates with this prediction; effector activation corresponds exactly to the fraction of mobile receptors. That is, immobile receptors can effectively be seen as non-hormone-occupied receptors; only mobile receptors participate in signal transduction.

E. Gα SIGNALING IN THE CYTOSOLIC PHASE

As mentioned briefly above, the fact that interaction of activated receptor and GTP-binding protein, rather than that of the activated GTP-binding protein and the effector enzyme, is rate-limiting in hormone-dependent effector activation is understandable in diffusion-controlled terms if one considers that the subsequent interaction of activated GTP-binding proteins with effector enzymes occurs in the cytosol rather than within the plane of the membrane.[71-80] The Gα subunit is normally anchored to the membrane by the Gβγ subunits,[73,74,81] but, upon receptor- or toxin-mediated activation, it redistributes to the cytosolic fraction. This idea is supported by a number of in vivo studies using a variety of receptor/GTP-binding protein systems and agonists.[71-80] Examples include dissociation from the membrane of:

1. Gsα induced by the guanine nucleotides GTP and the nonhydrolyzable analog guanosine 5'-O-thiotriphosphate (GTPγS) but not by GDP, as well as the β-adrenergic receptor agonist isoprotenerol in S49 mouse lymphoma cells,[71,72]
2. Gsα effected by treatment of rat liver membranes with cholera toxin,[74]
3. Gsα induced by stimulation of the prostacyclin receptor in mouse mastocytoma cells,[79]
4. Goα stimulated by nonhydrolyzable guanine nucleotides in NG108-15 neuroblastoma cells,[78]
5. Giα induced by treatment with Bordetella pertussis toxin (islet-activating protein–IAP) in rat liver membranes,[79]
6. Giα effected by α2-receptor-stimulation in human platelets,[77] and
7. Gi2α induced by IAP in neutrophils.[73]

Interestingly, although Giα subunits do

dissociate from the membrane,[74,77] the extent of redistribution into the cytosol appears to be quantitatively markedly less than that of other GTP-binding proteins examined (see ref. 74).

Once liberated into the cytosol, the Gα subunit diffuses rapidly along the cytoplasmic side of the membrane to contact and activate membrane-associated effector enzymes, such as adenylate cyclase and phospholipase Cβ. As already noted, because protein lateral diffusion is much faster in the aqueous than in the membrane phase (see chapter 2),[82] the rate-limiting step of adenylate cyclase or phospholipase Cβ activation is that of the diffusion-driven collisionary contacts between the hormone-receptor and GTP-binding protein complexes within the lipid bilayer which induce liberation of Gα into the aqueous phase.[1,7,22]

The more traditional view of GTP-binding protein-mediated signaling subsequent to GTP-binding protein activation has been that the Gα-subunit has the primary and central role in signaling in terms of interacting with effector enzymes, but mention should be made of the fact that there is a body of evidence indicating an active signaling role for the Gβγ subunit dimer.[83-92] A passive signaling role was originally assumed based on observations that excess Gβγ subunits reversed stimulation of adenylate cyclase by GTPγS-bound Gsα,[83-85] presumed to be through complexing Gsα into the inactive GTP-binding protein trimer (refer to Fig. 4.7 for the GTP-binding protein cycle).[83,86] A more detailed analysis was provided by molecular cloning, which revealed that there are at least four distinct adenylate cyclase forms: type I is activated by GTPγS-bound Gsα and inhibited by Gβγ, type III is unaffected by Gβγ and types II and IV are actually activated by Gβγ although only in the presence of Gsα.[87,88] Other Gβγ isoforms including transducin Gβγ (Gtβγ) could also activate adenylate cyclase type II to differing extents.[87] The first indication that Gβγ could directly interact with and activate effectors came from the observation that bovine brain Gβγ could activate cardiac K⁺-

channel normally regulated by the muscarinic acetylcholine receptor.[89-91] Giα can also independently regulate the cardiac K^+-channel activity.[89,90,92] Further examples of effectors regulated by Gβγ include phospholipase A_2, which can be activated by Gtβγ,[93] and phospholipase Cβ, which can be regulated by both Gqα and Gqβγ upon M1 muscarinic receptor activation.[94] Genetic evidence for a role of Gβγ in transducing signals from a cell surface-receptor comes from the yeast mating pheromone pathway, where null mutations in Gβ (the STE4 gene) and Gγ (STE18) block mating and pheromone responses, while Gα (SCG1 or GPA1) disruption leads to a constitutive response.[95,96]

It should be pointed out that the concentrations of Gβγ for biological effects are far in excess of those for activated Gα subunit,[83] so that the in vivo significance of Gβγ signaling may be minimal in most systems. One possible exception is phospholipase A_2 activation in retinal membranes, where the rod outer segment Gtβγ concentration of 500 μM easily exceeds the 0.6 μM required for effector activation;[93] however, the extent of Gt activation under physiological conditions and whether this is of the order of 0.1% of total Gtβγ protein has not been formally demonstrated. Since the Gβγ subunit dimer is more tightly membrane-associated than the Gα subunit as demonstrated for bovine brain Go using fluorescently-labeled subunits and resonance energy transfer (and see also refs. 73 and 74),[81] and the presence of Gβγ in the cytosolic phase upon GTP-binding protein activation has not been demonstrated,[72-74,77] it seems reasonable to postulate that Gβγ action in vivo may predominantly take place within the plane of the membrane. Experiments in which purified Gβγ subunits are added to isolated membranes to demonstrate effects on membrane effectors[87-90,92] can accordingly be questioned in terms of their physiological relevance from this point of view. Since Gβγ subunit signaling, in contrast to that of Gα, is probably restricted to the membrane, it can be assumed that signal transduction

mediated by Gα is more rapid than that mediated by Gβγ and hence kinetically much more significant.

An additional important role of Gβγ subunits appears to be interaction with receptor kinases and the mediation thereby of receptor phosphorylation at the membrane, as part of the desensitization of response subsequent to hormonal stimulation.[83,96-102] Hormone treatment in a number of receptor/effector systems has been shown to result in phosphorylation of the receptor through specific kinases, such as the muscarinic acetylcholine,[97-100] β-adrenergic[98,102] and rhodopsin[97,101] receptor kinases. This has been shown to be able to be mediated in most cases by the Gβγ subunits, which induce receptor phosphorylation by up to 10-fold (see ref. 83), apparently through providing membrane attachment sites to anchor the kinase in the vicinity of the receptor.[101,102] The region of the β-adrenergic receptor kinase responsible for interaction with Gβγ has been mapped to the C-terminal 130 amino acids of the kinase.[98,100] Unlike the β-adrenergic receptor kinase, the rhodopsin kinase does not directly associate with the Gβγ subunits, presumably because it lacks a region homologous to the C-terminus of the β-adrenergic receptor kinase.[98,100] A number of the antagonistic effects of the Gβγ subunits may be explained in part in terms of kinase-mediated effects on receptor activity. Gβγ subunit-induced phosphorylation of receptors by other types of kinase such as the Ca^{2+}-phospholipid-dependent kinase (PK-C) may also play a role in downregulation.[96] The mechanistic role of receptor phosphorylation in abrogation of the stimulatory signal subsequent to hormonal stimulation will be addressed in chapter 6 (section F). Since the kinetics of signaling mediated by the Gβγ subunit are probably quite slow (see previous paragraph), it may well be that the primary role of Gβγ in signal transduction in many systems is in downregulation through controlling receptor phosphorylation.

F. STOICHIOMETRIC CONSIDERATIONS AND TRIMERIC GTP-BINDING PROTEIN IMMOBILITY

It should not be overlooked that there is generally a large excess of GTP-binding proteins relative to the number of receptors in biological membranes.[1,2,27,28,103] In the adenylate cyclase system, this excess of Gs proteins to receptors is over 100- to 1000-fold for the β-adrenergic receptor of human neutrophils,[27] the β-adrenergic receptor of S49 mouse lymphoma cells[103] and the V_2-receptor of LLC-PK$_1$ renal epithelial cells.[7] The ratio of α2-adrenergic receptors to Gi molecules in membranes of human platelets is similarly high (1:20-100),[28] while the ratio of N-formyl-peptide receptors to G_{i2} molecules is about 1:200 in human neutrophils (50,000 receptors/cell).[27] One exception to this excess of GTP-binding proteins is the toad rod disk membrane,[104] where there are 10 times more receptor (the photoreceptor rhodopsin) molecules than Gt,[104-106] which itself is at a ratio of 27:1 with respect to the effector enzyme cGMP phosphodiesterase.[104,107] Perhaps significantly in this context, rhodopsin has a high apparent lateral diffusion coefficient of 3-4 x 10^{-9} cm^2/sec (see Table 2.3), which may be necessary to signaling in such a situation where GTP-binding proteins are at a premium relative to receptor.

The large excess of GTP-binding proteins with respect to both receptor and effector enzymes in the adenylate cyclase system, taken into consideration with the fact that GTP-binding proteins appear to be largely immobile within the plane of the membrane, makes completely understandable the fact that in kinetic terms, Gs and the adenylate cyclase catalytic subunit can be regarded as being a single unit (see refs. 15,25,70,108 and 109 for kinetic analyses in the A2 adenosine/ adenylate cyclase system), i.e. the interaction step between Gs and adenylate cyclase is not the rate-limiting event in signal transduction (see also chapter 1, section B). The fact that there is a large excess of GTP-binding proteins relative to the number of effectors in all systems examined (e.g. refs. 103 and 105) further ensures that the crucial step is that of receptor-mediated GTP-binding protein activation.

The vast excess of GTP-binding proteins with respect to receptors puts into question the exact basis of the need for mobile receptors for signal transduction. If GTP-binding proteins were as mobile as receptors within the lipid bilayer, one could assume that receptor mobility would be largely superfluous since GTP-binding proteins are in such excess. As seen in chapter 4 (section G), however, direct measurements of GTP-binding protein lateral mobility using fluorescence photobleaching recovery,[110] indicate that GTP-binding proteins are largely immobile. Conclusions based on this study alone should be treated with some caution, especially as the measurements were made using GTP-binding proteins reconstituted into lipid vesicles subsequently fused to neuroblastoma cells, rather than on GTP-binding proteins in situ in living cells (although it is probably reasonable to speculate that the measurements, if anything, are likely to overestimate, rather than underestimate, GTP-binding protein lateral mobility parameters since the reconstituted protein presumably lacks restrictive intrinsic cytoskeletal linkages; see chapter 3). However, this study is consistent with a number of other studies that indicate that the accessibility of GTP-binding proteins is severely restricted at the level of the membrane,[26,28,111] and it seems reasonable to assert that the apparent lack of or restriction to mobility on the part of GTP-binding proteins explains the need for mobile receptors in signal transduction.

The mechanistic basis of immobility of GTP-binding protein appears to be through linkage to the cytoskeleton. There is a body of evidence (see ref. 111) for direct association of GTP-binding protein subunits with both microtubules and actin:

i. Gsα and G_{i2}α subunits appear to be linked to the cytoskeleton in neutrophils in the absence of stimulation (but released upon hormonal stimulation).[112]

ii. The microtubule inhibitors colchicine/vinblastin increase non-hydrolyzable GTP analog-stimulated adenylate cyclase activity (see ref. 111)[113] in S49 lymphoma cells, implying GTP-binding protein association with microtubules.

iii. Guanine nucleotide transfer from tubulin to Gi has been demonstrated in neuronal cells,[114] as has direct binding of ^{125}I-labeled tubulin by Gsα, G$_i$α and other Gα subunits.[115,116]

iv. Gi proteins have been found to colocalize in actin/myosin/fodrin-rich structures in a mouse T cell lymphoma line.[117]

v. The Gβγ subunit complex in S49 lymphoma cells has been demonstrated in actin-rich Triton X-100 insoluble pools.[118]

Significantly, the membrane-linked cytoskeletal proteins spectrin/fodrin (spectrin β$_{II}$) and dynamin (the microtubule motor protein), among other proteins, have a "pleckstrin homology domain", which was originally characterized in pleckstrin, a major PK-C phosphorylation substrate of human platelets. The domain appears to be a sequence motif mediating specific protein-protein interactions in analogous fashion to SH2 or SH3 domains (*src* homology domains 2 or 3; see Fig. 4.11). Significantly, pleckstrin homology domains have been implicated in binding GTP-binding proteins and the Gβγ subunits in particular (see refs. 111, 119 and 120), and it appears not inconceivable that the basis of GTP-binding protein immobility is via linkage to the cytoskeleton through pleckstrin homology domains in spectrin/fodrin and other cytoskeletal components. Interestingly, the region of the β-adrenergic receptor kinase necessary for Gβγ binding, absent from the rhodopsin kinase, also contains a pleckstrin homology domain (see above).[111,121,122] The apparent immobility of GTP-binding proteins in turn may not only be central to the requirement for receptor lateral movement within the plane of the membrane in GTP-binding protein-mediated signal transduc-

tion, but also have a role in mediating receptor phosphorylation and thereby the downregulation of response subsequent to hormonal stimulation. The central role of the cytoskeleton in these processes will be dealt with in a little more detail in chapter 8 (section C).

G. AMPLIFICATION IN GTP-BINDING PROTEIN-MEDIATED RECEPTOR-EFFECTOR SYSTEMS THROUGH RECEPTOR LATERAL MOVEMENT

Unlike tyrosine kinase receptors where hormonal binding leads to receptor dimerization intermolecular phosphorylation, association of cytosolic signaling components, and subsequent rapid immobilization (see chapter 4, section D, Figs. 4.5 and 4.6), GTP-binding protein activating receptors are not immobilized upon activation. Receptor contacts with GTP-binding proteins are transient, on the order of milliseconds; e.g. the rate constant of association of the neutrophil N-formyl-peptide-receptor-ligand complex and GTP-binding protein trimer is 2 x 10[7]/M/sec, with a half time of redissociation of 140 milliseconds.[21] Clearly, GTP-binding protein activating receptors are not immobilized upon interaction with and activation of GTP-binding proteins. This means that GTP-binding protein activating receptors can activate more than one GTP-binding protein, this representing one level at which the stimulatory signal represented by hormone binding is amplified.

Signal amplification is the means by which low receptor occupancy can lead to maximal biological response and is a well-documented characteristic of GTP-protein activating receptor systems.[19,20,23,24,27,29,71,103,105,123-128] Amplification occurs at multiple levels of the signal transduction cascade through the fact that:

i. as stated, one receptor molecule can activate multiple GTP-binding proteins,

ii. one adenylate cyclase molecule produces many molecules of cAMP, and

iii. one activated cAMP-dependent protein kinase molecule phosphorylates multiple substrate proteins, etc.

Analyses in the S49 mouse lymphoma cell system indicate that each β-adrenergic receptor is able to activate greater than or equal to 100 molecules of Gs in native membranes.[71] The precise degree of coupling for a particular receptor in a specific cell type can, of course, vary according to many factors, including the number of receptors and/or GTP-binding proteins, the down-regulation kinetics, etc. In human neutrophils, for example, massive amplification occurs at the level of Gs activation in response to hormone binding to the β-adrenergic receptor of which there are 500/cell.[27] Hormone binding by the β-adrenergic receptor results in the turnover of 100 molecules of GTP/min/ligand-occupied receptor through activation of the GTPase activity of Gsα, ultimately resulting in the production of up to 10,000 molecules of cAMP/hormone-bound receptor.[27]

The substantial amplification effect at the level of the membrane permitting the activation of multiple GTP-binding proteins by a single hormone-occupied receptor can be understood quite simply in terms of receptor lateral movement within the plane of the membrane. Lateral diffusion of the hormone-receptor complex brings it into transient, activating contacts with membrane-associated GTP-binding protein trimers; these contacts do not result in immobilization of the receptor, so that successive GTP-binding protein trimers can be contacted and activated to bring about amplification of the stimulatory signal (i.e., through repeated cycles of [1.5] above). It is noteworthy in this context perhaps, that hepatic membranes have been demonstrated to contain GTP-binding protein trimer aggregates,[129-132] which dissociate upon glucagon addition,[131,132] implying that collisionary contacts of receptor with GTP-binding protein multimers could potentially activate multiple GTP-binding proteins with each contact in such membrane systems, thus releasing multiple Gsα molecules into the cytosol.

H. SUMMARY

While there remain a number of questions, and one could obviously decry the lack of more extensive experimental substantiation, there appears to be a significant body of evidence supporting the basic tenet of the Mobile Receptor Hypothesis or Collision Coupling Theory with respect to GTP-binding protein activating receptors—that receptor lateral movement appears to be an important parameter in signal transduction. With particular reference to experiments pertaining to the β-adrenergic- and vasopressin V_2-receptors and the adenylate cyclase system, it seems reasonable to postulate the following:

1. The agonistic (stimulatory) signal is directly dependent on the receptor mobile fraction; only mobile receptors appear to participate in signal transduction.[7] This is particularly important at low receptor occupancy; that is at physiologically low hormone concentrations (see Fig. 5.4).[5,7]

2. Receptor lateral movement is the rate-limiting step of signal transduction (see also ref. 22). The mechanistic basis of this is that the Gα subunit appears to dissociate from the membrane upon agonistic stimulation;[71-80] since diffusion in the cytosolic phase is much faster than that within the plane of the membrane,[23,82] lateral diffusion events in the lipid bilayer primarily determine the rate of signal transduction.

3. Receptor lateral movement is probably the mechanistic basis of signal amplification. Through continued diffusion within the plane of the membrane, one receptor is able to activate multiple GTP-binding protein complexes through transient activating contacts with successive GTP-binding proteins.

4. Factors such as membrane fluidity or protein aggregation which perturb membrane protein lateral motion, impinge on GTP-binding protein activating receptor-dependent signal

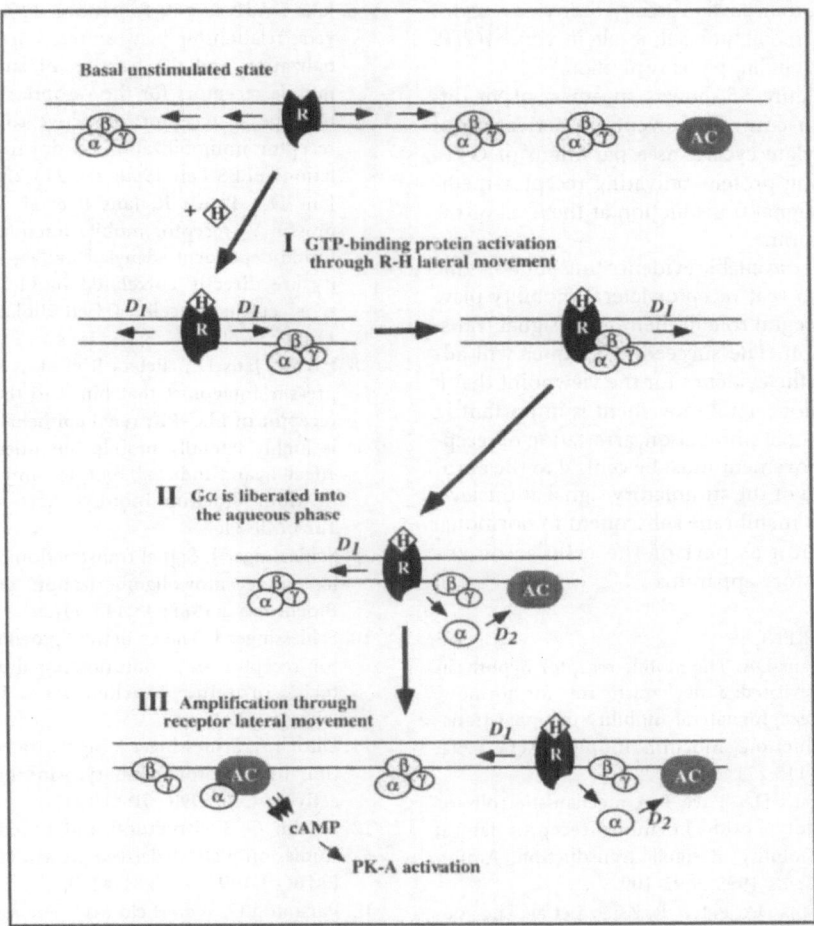

Fig. 5.5. Schematic representation of the diffusion-controlled events in activation of adenylate cyclase as a paradigm of GTP-binding protein activating receptor-mediated signal transduction at the level of the membrane. I. Upon agonist addition, hormone ligand (H) binds specifically to receptor (R) and lateral diffusion of R-H results in collisionary activation of the GTP-binding protein trimeric complex ($\alpha\beta\gamma$). II. Activated $Gs\alpha$ (GTP bound) is released into the aqueous phase.[71-80] Rapid local diffusion in the cytosol of $Gs\alpha$ along the inner membrane surface leads to interaction with and activation of the AC catalytic subunit (AC) to stimulate cAMP production. III. Signal amplification is effected by continued diffusion of R-H to activate multiple GTP-binding protein complexes[19,71,72] AC activation leads to activation of the cAMP-dependent protein kinase (PK-A). Since the rate of receptor lateral movement (D_1) is much slower than $Gs\alpha$ movement in the aqueous phase (D_2), the former is rate-limiting. The concentration of H determines the number of receptors that are occupied by hormone according to the dissociation constant (K_D - see chapter 2, Fig. 2.5B) and thereby the absolute number of mobile hormone-occupied receptors. This in turn determines the rate of AC activation (see Fig. 5.4). No attempt has been made to include a signaling role on the part of $G\beta\gamma$ in this figure (see section E).[83,98-102]

transduction because decreased receptor mobility directly results in reduced GTP-binding protein activation.[32,38-41]

Figure 5.5 shows a summary of the diffusion-controlled events in activation of adenylate cyclase as a paradigm of GTP-binding protein activating receptor-mediated signal transduction at the level of the membrane.

The available evidence thus supports the notion that receptor lateral mobility plays an integral role in membrane signal transduction. The succeeding chapter will address the evidence for the viewpoint that if receptor lateral movement is important in hormonal stimulation, arrestation of receptor movement must be central to the abrogation of the stimulatory signal at the level of the membrane subsequent to hormonal addition as part of the cellular down-regulatory apparatus.

REFERENCES

1. Jans DA. The mobile receptor hypothesis revisited: a mechanistic role for hormone receptor lateral mobility in signal transduction. Biochim Biophys Acta 1992; 1113:271-276.

2. Jans DA, Pavo I. A mechanistic role for polypeptide hormone receptor lateral mobility in signal transduction. Amino Acids 1995; 9:93-109.

3. Jans DA, Peters R, Zsigo J et al. The adenylate cyclase-coupled vasopressin V_2-receptor is highly laterally mobile in membranes of LLC-PK$_1$ renal epithelial cells at physiological temperature. EMBO. J 1989; 8(9):2431-2438.

4. Jans DA, Peters R, Fahrenholz F. Lateral mobility of the phospholipase-C-activating vasopressin V_1-type receptor in A7r5 smooth muscle cells: a comparison with the adenylate cyclase-coupled V_2-receptor. EMBO J 1990; 9(9):2693-2699.

5. Jans DA, Peters R, Jans P et al. Ammonium chloride affects receptor number and lateral mobility of the vasopressin V_2-type receptor in the plasma membrane of LLC-PK$_1$ renal epithelial cells: role of the cytoskeleton. Exper Cell Res 1990; 191:121-128.

6. Jans DA, Peters R, Fahrenholz F. An inverse relationship between receptor internalization and the fraction of laterally mobile receptors for the vasopressin renal-type V_2-receptor; an active role for receptor immobilization in down-regulation? FEBS Lett 1990; 274:223-226.

7. Jans DA, Peters R, Jans P et al. Vasopressin V_2-receptor mobile fraction and ligand-dependent adenylate cyclase-activity are directly correlated in LLC-PK$_1$ renal epithelial cells. J Cell Biol 1991; 114(1):53-60.

8. Pavo I, Jans DA, Peters R et al. A vasopressin antagonist that binds to the V_2-receptor of LLC-PK$_1$ renal epithelial cells is highly laterally mobile but does not effect ligand-induced receptor immobilization. Biochim Biophys Acta 1994; 1223:240-246.

9. Schlessinger J. Signal transduction by allosteric receptor oligomerization. Trends Biochem Sci 1988; 13:443-447.

10. Schlessinger J. The epidermal growth factor receptor as a multifunctional allosteric protein. Biochemistry 1989; 27:3119-3123.

11. Ullrich A, Schlessinger J. Signal transduction by receptors with tyrosine kinase activity. Cell 1990; 61:203-212.

12. Heldin C-H. Structural and functional studies on platelet-derived growth factor. EMBO J 1992; 11:4251-4259.

13. Panaotou G, Waterfield MD. The assembly of signalling complexes by receptor tyrosine kinases. Bioessays 1993; 15(3):171-177.

14. Tolkovsky AM, Levitzki A. Mode of coupling between the β-adrenergic receptor and adenylate cyclase in turkey erythrocytes. Biochemistry 1978; 17:3795-3810.

15. Tolkovsky AM, Levitzki A. Coupling of a single adenylate cyclase to two receptors: adenosine and catecholamine. Biochemistry 1978; 17:3811-3817.

16. Tolkovsky A, Braun S, Levitzki A. Kinetics of interaction between β-receptors, GTP-proteins and the catalytic subunit of turkey erythrocyte adenylate cyclase. Proc Natl Acad Sci USA 1982; 79:213-217.

17. Hanski E, Rimon G, Levitzki A. Adenylate cyclase activation by the β-adrenergic receptor as a diffusion controlled process. Biochemistry 1979; 18:846-853.

18. Rimon G, Hanski E, Levitzki A. Temperature dependence of beta receptor, adenosine receptor, and sodium fluoride stimulated adenylate cyclase from turkey erythrocytes. Biochemistry 1980; 19:4451-4460.

19. Orly J, Schramm M. Fatty acids as modulators of membrane functions; catecholamine-activated adenylate cyclase of the turkey erythrocyte. Proc Natl Acad Sci USA 1975; 72:3433-3437.

20. Bergman RN, Hechter H. Neurophyseal hormone-responsive renal adenylate cyclase IV. A random-hit matrix model for coupling in a hormone-sensitive adenylate cyclase system. J Biol Chem 1978; 253:3238-3250.

21. Neubig RR, Sklar LA. Subsecond modulation of formyl peptide-linked guanine nucleotide-binding proteins by guanosine 5'-O-(3-thio)triphosphate in permeabilized neutrophils. Mol Pharmacol 1993; 43(5):734-740.

22. Chabre M. The G protein connection: is it in the membrane or the cytoplasm. Trends in Biochem Sci 1987; 12:213-215.

23. Helmreich EJM, Elson EL. Protein and lipid mobility. Adv in Cyclic Nucleotide and Prot Phosphor Res 1984; 18:1-62.

24. Neubig RR, Gantzos RD, Thomsen WJ. Mechanism of agonist and antagonist binding to alpha 2 adrenergic receptors: evidence for a precoupled receptor-guanine nucleotide protein complex. Biochemistry 1988; 27(7):2374-2384.

25. Gross W, Lohse MJ. Mechanism of activation of A_2 adenosine receptors II. A restricted collision-coupling model of receptor-effector interaction. Mol Pharmacol 1991; 39:524-530.

26. Graeser D, Neubig RR. Compartmentation of receptors and guanine nucleotide-binding proteins in NG108-15 cells: lack of cross-talk in agonist binding among the alpha 2-adrenergic, muscarinic, and opiate receptors. Mol Pharmacol 1993; 43(3):434-443.

27. Mueller H, Weingarten R, Ransnas LA et al. Differential amplification of antagonistic receptor pathways in neutrophils. J Biol Chem 1991; 266(20):12939-12943.

28. Neubig RR, Gantzos RD, Brasier RS. Agonist and antagonist binding to alpha 2-adrenergic receptors in purified membranes from human platelets. Implications of receptor-inhibitory nucleotide-binding protein stoichiometry. Mol Pharmacol 1985; 28(5):475-486.

29. Omann GM, Harter JM, Hassan N et al. A threshold level of coupled G-proteins is required to transduce neutrophil responses. J Immunol 1992; 149(6):2172-2178.

30. Moscona-Amir E, Henis YI, Yechiel E et al. Role of lipids in age-related changes in the properties of muscarinic receptors in cultured rat heart myocytes. Biochemistry 1986; 25(24):8118-8124.

31. Moscona-Amir E, Henis YI, Sololovsky M. Aging of rat heart myocytes disrupts muscarinic receptor coupling that leads to inhibition of cAMP accumulation and alters the pathway of muscarinic-stimulated phosphoinositide hydrolysis. Biochemistry 1989; 28(17):7130-7137.

32. Zakharova OM, Rosenkranz AA, Sobolev AS. Modification of fluid lipid and mobile protein fractions of reticulocyte plasma membranes affects agonist-stimulated adenylate cyclase. Application of the percolation theory. Biochim Biophys Acta 1995; 1236:177-184.

33. Engelhard VH, Esko JD, Storm DR et al. Modification of adenylate cyclase activity in LM cells by manipulation of the membrane phospholipid composition in vivo. Proc Natl Acad Sci USA 1976; 73:4482-4486.

34. Engelhard VH, Glaser M and Storm DR. Effect of membrane compositional changes on adenylate cyclase activity in LM cells. Biochemistry 1978; 17:3191-3200.

35. Brivio-Haugland RP, Louis SL, Musch K et al. Liver plasma membranes from essential farry acid deficient rats: isolation, fatty acid composition and activities of 5'-nucleotidase, ATPase and adenylate cyclase. Biochim Biophys Acta 1976; 433:150-163.

36. Puchwein G, Pfeuffer T, Helmreich EJM. Uncoupling of catecholamine activation of pigeon erythrocyte membrane adeylate cyclase by filipin. J Biol Chem 1974; 249:3232-3240.

37. Limbird LE, Gill DM, Lefkowitz RJ. Agonist-promoted coupling of the β-adrenergic receptor with the guanine nucleotide regulatory protein of the adenylate cyclase system. Proc Natl Acad Sci USA 1980; 77:775-779.

38. Sobolev AS, Rosenkranz AA, Kazarov AR. Interaction of proteins of the adenylate cyclase complex: area-limited mobility of movement along the whole membrane? Analysis with the application of the percolation theory. Biosc Rep 1984; 4:897-902.

39. Sobolev AS, Kazarov AR, Rosenkranz AA. Application of percolation theory principles to the analysis of interaction of adenylate cyclase complex proteins in cell membranes. Mol Cell Biochem 1988; 81(1):19-28.

40. Kazarov AR, Rosenkranz AA, Sobolev AS. Membrane regulation of hormonal activation of adenylate cyclase in rat reticulocyte. Studia Biophysica 1989; 134:67-71.

41. Bakardjieva A, Gulla HJ, Helmreich EJM. Modulation of the β-receptor adenylate cyclase interactions in cultured Chang liver cells by phospholipid enrichment. Biochemistry 1979; 18:3016-3023.

42. Shinitzky M, Barenholz Y. Fluidity parameters of lipid regions determined by fluorescence polarization. Biochim Biophys Acta 1978; 515:367-394.

43. Henis YI, Rimon G, Felder S. Lateral mobility of phospholipids in turkey erythrocytes: implications for adenylate cyclase activation. J Biol Chem 1982; 257:1407-1411.

44. Saxton MJ. The membrane skeleton of erythrocytes. A percolation model. Biophys J 1990; 57:1167-1177.

45. Saxton MJ. The spectrin network as a barrier to lateral diffusion in erythrocytes: a percolation analysis. Biophys J 1989; 55:21-28.

46. Saxton MJ. Lateral diffusion in an archipelago: distance dependence of the diffusion coefficient. Biophys J 1989; 56:615-622.

47. Saxton MJ. Single-particle tracking: effects of corrals. Biophys J 1995; 69(2):389-398.

48. Atlas D, Volsky DJ, Levitzki A. Lateral mobility of beta-receptors involved in adenylate cyclase activation. Biochim Biophys Acta 1980; 597(1):64-69.

49. Volsky DJ, Loyter A. Inhibition of membrane fusion by suppression of lateral movement of membrane proteins. Biochim Biophys Acta 1978; 514:213-224.

50. Roess DA, Rahman NA, Kenny N. Molecular dynamics of luteinizing hormone receptors on rat luteal cells. Biochim Biophys Acta 1992; 1137:309-316.

51. Niswender GD, Roess DA, Sawyer HR et al. Differences in the lateral mobility of receptors for luteinizing hormone (LH) in the luteal plasma membrane when occupied by ovine LH versus human chorionic gonadotropin. Endocrinol 1985; 116:164-169.

52. Roess DA, Niswender GD, Barisas BG. Cytocholasins and colchicine increase the lateral mobility of human chorionic gonadotropin-occupied luteinizing hormone receptors on ovine luteal cells. Endocrinol 1988; 122:261-269.

53. Jacobson K, O'Dell D, August JT. Lateral diffusion of an 80,000-dalton glycoprotein in the plasma membrane of murine fibroblasts: relationships to cell structure and function. J Cell Biol 1984; 99(5):1624-1633.

54. Duband J-L, Nuckolls GH, Ishihara A et al. Fibronectin receptor exhibits high lateral mobility in embryonic locomoting cells but is immobile in focal contacts and fibrillar streaks in stationary cells. J Cell Biol 1988; 107:1385-1396.

55. Roettger BF, Rentsch RU, Hadac EM et al. Insulation of a G protein-coupled receptor on the plasmalemmal surface of the pancreatic acinar cell. J Cell Biol 1995; 130:579-590.

56. Johansson B, Wymann MP, Holmgren-Peterson K et al. N-formyl peptide receptors in human neutrophils display distinct membrane distribution and lateral mobility when labeled with agonist and antagonist. J Cell Biol 1993; 121:1281-1289.

57. Zidovetzki R, Yarden Y, Schlessinger, J et al. Rotational diffusion of epidermal growth factor complexed to its surface receptor the rapid microaggregation and endocytosis of occupied receptors. Proc Natl Acad Sci USA 1981; 78:6981-6985.

58. Hillman GM, Schlessinger J. The lateral diffusion of epidermal growth factor complexed to its surface receptors does not account for the thermal sensitivity of patch formation and endocytosis. Biochemistry 1982; 21:1667-1672.

59. Fire E, Zwart DE, Roth MG et al. Evidence from lateral mobility studies for dynamic interactions of a mutant influenza hemagglutinin with coated pits. J Cell Biol 1991; 115:1585-1594.

60. Jans DA, Resink TJ, Wilson E-L et al. Isolation of a mutant LLC-PK$_1$ cell line defective in hormonal responsiveness: a pleiotropic lesion affecting receptor function. Eur J Biochem 1986; 160:407-412.

61. Jans DA, Resink TJ, Hemmings BA. Complementation between LLC-PK$_1$ mutants affected in polypeptide hormone receptor function. Eur J Biochem 1987; 162:571-576.

62. Luzius H, Jans DA, Jans P et al. Isolation and genetic characterization of a renal epithelial cell mutant defective in vasopressin V$_2$-type receptor binding and function. Exper Cell Res 1991; 195:478-484.

63. Briggs MM, Lefkowitz RJ. Parallel modulation of catecholamine activation of adenylate cyclase and formation of the high-affinity agonist receptor complex in turkey erythrocyte membranes by temperature and *cis*-vaccenic acid. Biochemistry 1980; 19:4461-4466.

64. Golan DE, Veatch W. Lateral mobility of band 3 in the human erythrocyte membrane studied by fluorescence photobleaching recovery: evidence for control by cytoskeletal interactions. Proc Natl Acad Sci USA 1980; 77:2537-2541.

65. Tsuji A, Ohnishi S. Restriction of the lateral motion of Band 3 in the erythrocyte membrane by the cytoskeletal network: dependence on spectrin association state. Biochemistry 1986; 25:6133-6139.

66. Jacobson K, O'Dell D, August JT. Lateral diffusion of an 80,000-dalton glycoprotein in the plasma membrane of murine fibroblasts: relationships to cell structure and function. J Cell Biol 1984; 99(5):1624-1633.

67. Paller MS. Lateral mobility of Na,K-ATPase and membrane lipids in renal cells. Importance of cytoskeletal integrity. J Membr Biol 1994; 142(1):127-135.

68. Smith PR, Stoner JC, Viggiano SC et al. Effects of vasopressin and aldosterone on the lateral mobility of eptihelial Na$^+$ channels in A6 epithelial cells. J Memb Biol 1995; 147(2):195-205.

69. Gupte SS. Localization and diffusion of glucagon receptor in rat hepatocytes. Receptor 1994; 4(3):175-190.

70. Braun S, Levitzki A. The attenuation of epinephrine-dependent adenylate cyclase by adenosine and the characteristics of the adenosine stimulatory and inhibitory sites. Mol Pharmacol 1979; 16:737-748.

71. Ransnäs LA, Insel PA. Subunit dissociation is the mechanism for hormonal activation of the G$_s$ protein in native membranes. J Biol Chem 1988; 263:17239-17242.

72. Ransnäs LA, Svoboda P. Jasper JR et al. Stimulation of β-receptors of S49 lymphoma cells redistributes the subunit of the stimulatory G protein between cytosol and membranes. Proc Natl Acad Sci USA 1989; 86:7900-7903.

73. Bokoch GM, Bickford K, Bohl BP. Subcellular localization and quantitation of the major neutrophil pertussis toxin substrate, Gn. J Cell Biol 1988; 106(6):1927-1936.

74. Lynch CJ, Morbach L, Blackmore PF et al. α-subunits of N$_s$ are released from the plasma membrane following cholera toxin activation. FEBS Lett. 1986; 200:333-336.

75. Deckmyn H, Tu SM, Majerus PW. Guanine nucleotides stimulate soluble phosphoinositide-specific phospholipase C in the absence of membranes. J Biol Chem 1986; 261(35):16553-16558.

76. Molina Y, Vedia L, Lapetina EG. Iloprost-induced translocation of a 23-kDa protein that is recognized by a Gsα antiserum. Proc Natl Acad Sci USA 1989; 86(3):868-870.

77. Zamorski MA, Ferraro JC, Neubig RR. Subcellular distribution of α2-adrenergic receptors, pertussis-toxin substrate and adenylate cyclase in human platelets. Biochem J 1990; 265(3):755-762.

78. McArdle H, Mullaney I, Magel A et al. GTP analogues cause release of the α subunit of the GTP binding protein, G$_o$, from the plasma membrane of NG108-15 cells. Biochem Biophys Res Commun 1988; 152:243-251.

79. Negishi M, Hashimoto H, Ichikawa A. Translocation of α subunits of stimula-

tory guanine nucleotide-binding proteins through stimulation of the prostacyclin receptor in mouse mastocytoma cells. J Biol Chem 1992; 267:2364-2369.

80. Stryer L, Bourne HR. G proteins: a family of signal transducers. Annu Rev Cell Biol 1986; 2:391-419.

81. Remmers AE, Neubig RR. Resonance energy transfer between guanine nucleotide binding protein subunits and membrane lipids. Biochemistry 1993; 32(9):2409-2414.

82. Peters R. Fluorescence microphotolysis to measure nucleocytoplasmic transport and intracellular mobility. Biochim Biophys Acta 1986; 864:305-359.

83. Clapham DE, Neer EJ. New roles for G-protein βγ-dimers in transmembrane signalling. Nature 1993; 365:403-406.

84. Bourne HR, Sanders DA, McCormick F. The GTPase superfamily: conserved structure and molecular mechanism. Nature 1991; 349:117-127.

85. Spiegel AM, Shenker A, Weinstein LS. Receptor-effector coupling by G proteins: implications for normal and abnormal signal transduction. Endocrine Rev 1992; 13:536-565.

86. Conklin BR, Bourne HR. Structural elements of Gα subunits that interact with Gβγ, receptors, and effectors. Cell 1993; 73:631-641.

87. Tang WJ, Gilman AG. Type-specific regulation of adenylyl cyclase by G protein βγ subunits. Science 1991; 254(5037):1500-1503.

88. Warner DR, Basi NS, Rebois RV. Cell-free synthesis of functional type IV adenylyl cyclase. Anal Biochem 1995; 232(1):31-36.

89. Logothetis DE, Kurachi Y, Galper J et al. The βγ subunits of GTP-binding proteins activate the muscarinic K+ channel in heart. Nature 1987; 325:321-326.

90. Codina J, Yatani A, Grenet D et al. The α-subunit of the GTP binding protein Gk opens atrial potassium channels. Science 1987; 236:536-538.

91. Krapivinsky G, Krapivinsky L, Wickman K et al. Gβγ binds directly to the G protein-gated K+ channel, IKACh. J Biol Chem 1995; 270(49):29059-29062.

92. Schreibmayer W, Dessauer CW, Vorobiov D et al. Inhibition of an inwardly rectifying K+ channel by G-pro-

tein α-subunits. Nature 1996; 380(6575):624-627.

93. Jelsema CL, Axelrod J. Stimulation of phospholipase A2 activity in bovine rod outer segments by the βγ subunits of transducin and its inhibition by the α subunit. Proc Natl Acad Sci USA 1987; 84(11):3623-3627.

94. Katz A, Wu D, Simon MI. Subunits βγ of heterotrimeric G protein activate β2 isoform of phospholipase C. Nature 1992; 360:686-689.

95. Leberer E, Dignard D, Hougan L et al. Dominant-negative mutants of a yeast G-protein β subunit identify two functional regions involved in pheromone signalling. EMBO J 1992; 11(13):4805-4813.

96. Leberer E, Dignard D, Harcus D et al. The protein kinase homologue Ste20p is required to link the yeast pheromone response G-protein βγ subunits to downstream signalling components. EMBO J 1992; 11(13):4815-4824.

97. Haga K, Haga T. Activation by G protein βγ subunits of agonist- or light-dependent phosphorylation of muscarinic acetylcholine receptors and rhodopsin. J Biol Chem 1992; 267(4):2222-2227.

98. Kameyama K, Haga K, Haga T et al. Activation by G protein βγ subunits of β-adrenergic and muscarinic receptor kinase. J Biol Chem 1993; 268(11):7753-7758.

99. Haga K, Kameyama K, Haga T. Synergistic activation of a G protein-coupled receptor kinase by G protein βγ subunits and mastoparan or related peptides. J Biol Chem 1994; 269(17):12594-12599.

100. Haga T, Haga K, Kameyama K et al. Phosphorylation of muscarinic receptors: regulation by G proteins. Life Sci 1993; 52(5-6):421-428.

101. Inglese J, Koch WJ, Caron MG. Isoprenylation in regulation of signal transduction by G-protein-coupled receptor kinases. Nature 1992; 359(6391):147-150.

102. Pitcher JA, Inglese J, Higgins JB et al. Role of βγ subunits of G proteins in targeting the β-adrenergic receptor kinase to membrane-bound receptors. Science 1992; 257(5074):1264-1267.

103. Alousi AA, Jasper JR, Insel PA et al. Stoichiometry of receptor-Gs-adenylate cyclase interactions. FASEB J 1991; 5:2300-2303.

104. Dratz EA, Miljanich GP, Nemes PP et al. The structure of rhodopsin and its disposition in the rod outer segment disk membrane. Photochem Photobiol 1979; 29(4):661-670.

105. Lamb TD. Stochastic simulation of activation in the G-protein cascade of phototransduction. Biophys J 1994; 67:1439-1454.

106. Hamm HE, Bownds MD. Protein complement of rod outer segments of frog retina. Biochemistry 1986; 25(16):4512-4523.

107. Dumke CL, Arshavsky VY, Calvert PD et al. Rod outer segment structure influences the apparent kinetic parameters of cyclic GMP phosphodiesterase. J Gen Physiol 1994; 103(6):1071-1098.

108. Braun S, Levitzki A. Adenosine receptor permanently coupled to turkey erythrocyte adenylate cyclase. Biochemistry 1979; 18:2134-2138.

109. Lohse MJ, Klotz KN, Schwabe U. Mechanism of A2 adenosine receptor activation. I. Blockade of A2 adenosine receptors by photoaffinity labeling. Mol Pharmacol 1991; 39:517-523.

110. Kwon G, Axelrod D, Neubig RR. Lateral mobility of tetramethylrhodamine (TMR) labelled G protein alpha and beta gamma subunits in NG 108-15 cells. Cell Signal 1994; 6(6):663-679.

111. Neubig RR. Membrane organization in G-protein mechanisms. FASEB J 1994; 8:939-946.

112. Sarndahl E, Bokoch GM, Stendahl O et al. Stimulus-induced dissociation of alpha subunits of heterotrimeric GTP-binding proteins from the cytoskeleton of human neutrophils. Proc Natl Acad Sci USA 1993; 90:6552-6556.

113. Leiber D, Jasper JR, Alousi AA et al. Alteration in Gs-mediated signal transduction in S49 lymphoma cells treated with inhibitors of microtubules. J Biol Chem 1993; 268:3833-3837.

114. Rasenick MM, Wang N. Exchange of guanine nucleotides between tubulin and GTP-binding proteins that regulate adenylate cyclase: cytoskeletal modification of neuronal signal transduction. J Neurochem 1988; 51:300-311.

115. Rasenick MM, Wang N, Yan K. Specific associations between tubulin and G proteins: participation of cytoskeletal elements in cellular signal transduction. Adv Second Messenger Phosphoprot Res 1990; 24:381-386.

116. Wang N, Rasenick MM. Tubulin-G protein interactions involve microtubule polymerization domains. Biochemistry 1991; 30:10957-10965.

117. Bourguignon LYW, Walker G, Huang HS. Interactions between a lymphoma membrane-associated guanosine 5'-triphosphate-binding protein and the cytoskeleton during receptor patching and capping. J Immunol 1990; 144:2242-2252.

118. Carlson KE, Woolkalis MJ, Newhouse MG et al. Fractionation of the beta subunit common to guanine nucleotide-binding regulatory proteins with the cytoskeleton. Mol Pharmacol 1986; 30:463-468.

119. Macias MJ, Musacchio A, Ponstingl H et al. Structure of the pleckstrin homology domain from beta-spectrin. Nature 1994; 369:675-677.

120. Lombardo CR, Weed SA, Kennedy SP et al. Beta II-spectrin(fodrin) and beta I epsilon 2-spectrin (muscle) contain NH2- and COOH-terminal membrane association domains (MAD1 and MAD2). J Biol Chem 1994; 269(46):29212-29219.

121. Musacchio A, Gibson T, Rice P et al. The PH domain: a common piece in the structural patchwork of signalling proteins. Trends Biochem Sci 1993; 18(9):343-348.

122. Shaw G. Identification of novel pleckstrin homology (PH) domains provides a hypothesis for PH domain function. Biochem Biophys Res Commun 1993; 195(2):1145-1151.

123. Brandt DR, Ross EM. Catecholamine-stimulated GTPase cycle; multiple sites of regulation by β-adrenergic receptor and Mg^{2+} studied in reconstituted receptor-G$_s$ vesicles. J Biol Chem 1986; 261:1656-1664.

124. Ciccarelli-E, Svoboda-M, De-Neef-P et al. Pharmacological properties of two recombinant splice variants of the PACAP type I receptor, transfected and stably expressed in CHO cells. Eur J Pharmacol 1995; 288(3):259-267.

125. Baylor D. How photons start vision. Proc Natl Acad Sci USA 1996; 93(2):560-565.

126. Nanoff C, Mitterauer T, Roka F et al. Species differences in A1 adenosine receptor/G protein coupling: identification of a membrane protein that stabilizes the association of the receptor/G protein complex. Mol Pharmacol 1995; 48(5):806-817.

127. Marcil J, Anand-Srivastava MB. Defective ANF-R2/ANP-C receptor-mediated signalling in hypertension. Mol Cell Biochem 1995; 149-150:223-231.

128. Carroll RC, Morielli AD, Peralta EG. Coincidence detection at the level of phospholipase C activation mediated by the m4 muscarinic acetylcholine receptor. Curr Biol 1995; 5(5):536-544.

129. Schlegel W, Kempner ES, Rodbell M. Activation of adenylate cyclase in hepatic membranes involves interactions of the catalytic unit with multimeric complexes of regulatory proteins. J Biol Chem 1979; 254(12):5168-5176.

130. Rodbell M. The role of GTP-binding proteins in signal transduction: from the sublimely simple to the conceptually complex. Curr Top Cell Regul 1992; 32:1-47.

131. Nakamura S, Rodbell M. Glucagon induces disaggregation of polymer-like structures of the α-subunit of the stimulatory G protein in liver membranes. Proc Natl Acad Sci USA 1991; 88(16):7150-7154.

132. Nakamura S, Rodbell M. Octyl glucoside extracts GTP-binding regulatory proteins from rat brain "synaptoneurosomes" as large, polydisperse structures devoid of beta gamma complexes and sensitive to disaggregation by guanine nucleotides. Proc Natl Acad Sci USA 1990; 87(16): 6413-6417.

EVIDENCE FOR THE ROLE OF RECEPTOR IMMOBILIZATION IN DESENSITIZATION SUBSEQUENT TO HORMONAL STIMULATION

A. INTRODUCTION

We saw in the previous chapter that a body of indirect and direct experimental evidence supports the notion that receptor lateral mobility plays an integral role at the level of the membrane in transducing the stimulatory signal represented by hormone binding to receptor. This chapter intends to discuss the evidence for the assertion that if receptor lateral movement is important in hormonal stimulation, as it appears to be, arrestation of receptor movement must be central to the abrogation of the stimulatory signal subsequent to hormonal addition, as part of the cellular downregulatory apparatus.[1,2] Particularly in the case of GTP-binding protein activating receptors where only mobile receptors appear to participate in signal transduction,[1-5] it seems reasonable to suggest that desensitization of response subsequent to hormone addition involves the abrogation of receptor movement as an initial step. The evidence for this will be examined in some detail below, the conclusion being that agonistic stimulation triggers receptor immobilization prior to internalization. Receptor immobilization does not appear to exclusively play a role in downregulation of response subsequent to stimulation, however, but is also central to eliciting the stimulatory signal in several receptor systems, including those of tyrosine kinase receptors (as already mentioned in chapter 4) and receptors mediating cell-cell interaction or cell adhesion to the substratum, and these will be dealt with in chapter 7.

B. RECEPTOR INTERNALIZATION IN DESENSITIZATION OF RESPONSE

Desensitization or downregulation is the process by which the stimulatory signal elicited by hormonal treatment is switched off. Just as stimulation and amplification occurs at a number of levels in all signal transduction cascades especially in terms of cytosolic responses (see chapter 5 section G), desensitization similarly takes place at multiple levels (see refs. 6 and 7). Whereas stimulation of the vasopressin renal V_2-receptor for example triggers GTP-binding protein activation, cAMP production by adenylate cyclase, phosphorylation of specific protein substrates by the activated cAMP-dependent protein kinase (PK-A) and cAMP-dependent gene induction mediated by transcription factors, such as the cAMP-response element binding factor CREB (see chapter 4; Fig. 4.8A),[8] it also elicits a multitude of desensitization phenomena, including GTP hydrolysis to inactivate Gs, cAMP breakdown by phosphodiesterase, cAMP efflux, degradation of the PK-A catalytic subunit[7,9-11] and dephosphorylation of phosphorylated proteins, etc.

At the level of the membrane, receptor endocytosis or internalization subsequent to ligand binding is a central part of desensitization in most signal transduction systems (see refs. 2,4,12-17). Occurring prior to endocytosis, receptor phosphorylation (see section F below), either by specific receptor kinases[18-22] (as alluded to in chapter 5, section D) or other kinases,[20,23-26] is a further key event. Endocytosis of hormone-occupied, activated receptors performs at least two functions in the desensitization process:

i. Activated receptors are removed from the membrane so that the stimulatory signal represented by the hormone-receptor complex is abrogated.

ii. Receptors are removed from the membrane to prevent subsequent binding to and stimulation by hormone.

During endocytosis, polypeptide hormone receptors are delivered to intracellular membrane-enveloped compartments, including the endosomes, and subsequently either to the lysosomes as part of the degradative pathway in which receptors and ligands are proteolysed[16,17,27,28] or back to the plasma membrane as part of the recycling pathway.[4,13,15,16,29-32] Receptor internalization is an energy- and temperature-dependent process involving the actin filaments and microtubules of the cytoskeleton[33] as well as the Ca^{2+}-binding protein calmodulin (the δ-subunit of phosphorylase kinase) and calcium itself (see ref. 15).[34] Membrane-associated coated pit structures, the site of endocytosis of many receptors (see refs. 33-35),[13,16,36-38] are invaginations into the plasma membrane, which are "coated" with a flexible scaffold structure of a polyhedral arrangement of the membrane-associated clathrin protein. They are able to pinch off inside the cell during endocytosis and can ultimately fuse with degradative membrane vesicles such as lysosomes.

C. RECEPTOR IMMOBILIZATION PRIOR TO INTERNALIZATION

A number of studies have shown that membrane protein reduced mobility or immobilization is integrally associated with endocytosis.[1,2,33,34,39,40] This has been demonstrated, for example, by a point mutant of the influenza hemagglutinin molecule which, unlike wild type, is able to be internalized through coated pits and concomitantly reduced in terms of the apparent lateral diffusion coefficient.[35] In similar fashion, mutations to the C1 complement receptor CR1 cytoplasmic domain, which confer association with coated pits, decrease CR1 lateral mobility and increase its internalization.[40] Receptor aggregation is known to precede internalization for a variety of plasma membrane integral receptors, including those for insulin, epidermal growth factor (EGF), vasopressin, nerve growth factor (NGF), asialoglycoprotein, cholecystokinin, luteinizing hormone (LH) and Fc receptors.[2,39-52] Based on the strong correlation between receptor aggregation and immobilization (see chapter 3),[40,52-54] it seems reasonable to postulate that receptor immobilization precedes internalization.

That this appears to be the case has been shown for the vasopressin V$_2$-receptor of renal epithelial cells, where the relatively slow kinetics of receptor internalization permit such an analysis.[39] As illustrated in Figure 6.1, cells show diffuse labeling with a fluorescently-labeled vasopressin agonist at 4°C (panel F) at which temperature endocytosis does not occur, but aggregated staining after 30 min at 37°C (panel B; see ref. 39) does occur. Fluorescence photobleaching recovery measurements of V$_2$-receptor lateral movement with time at system 37°C, illustrated in Figure 6.2, imply that

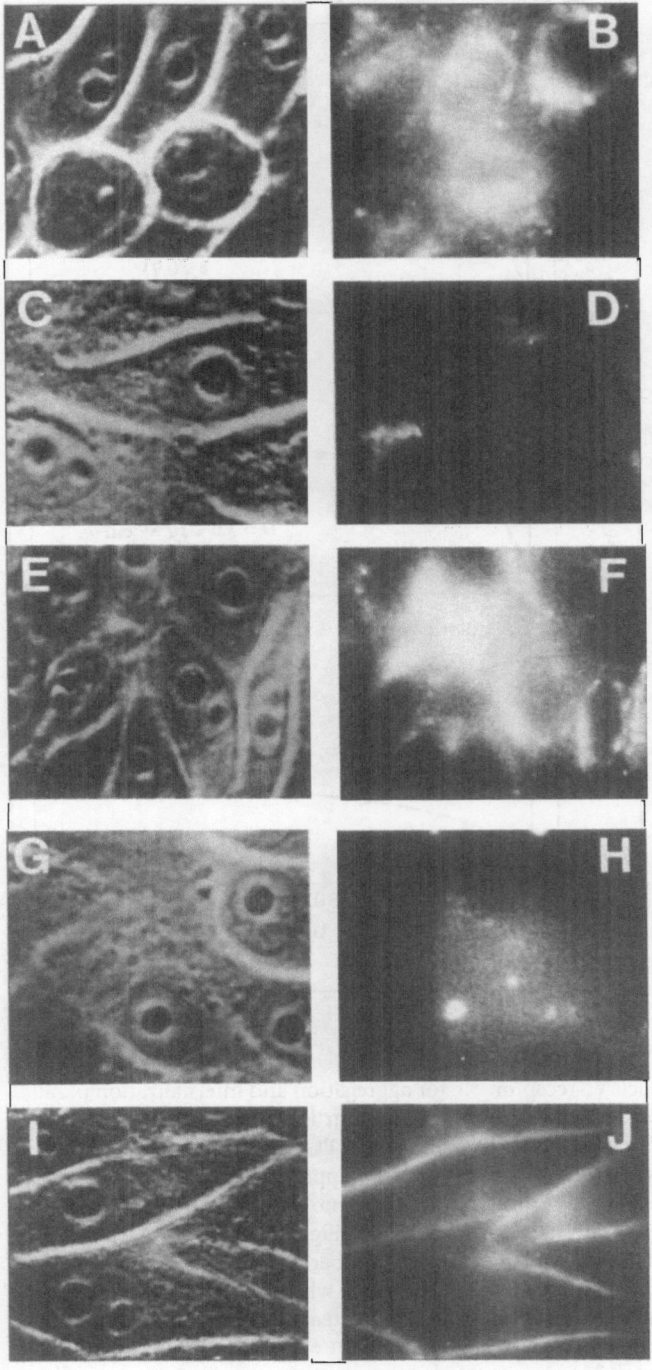

Fig. 6.1. Temperature-dependent vasopressin V_2-receptor aggregation in renal epithelial cells. Cells of the LLC-PK$_1$ cell line were incubated with a specific fluorescently-labeled vasopressin agonist (TR-LVP -deamino[Lys[8](tetramethyl-rhodamyl-aminothio-carbonyl)] vasopressin) at the temperatures indicated, in the absence (panels B and F) and presence (panels D and H) of an excess of unlabeled vasopressin (nonspecific binding). Phase contrast pictures are presented on the left. Diffuse labeling is evident for the cells incubated at 4°C (panel F), at which temperature endocytosis does not occur, but highly fluorescent aggregates are evident after 30 min at 37°C (panel B).[39] For comparison, cells labeled with the fluorescent lipid probe DiOC$_{14}$(3) (3,3'-ditetradecyloxacarbo-cyanine iodide) are shown (bottom panels) exhibiting much more homogeneous plasma membrane staining (panel J).

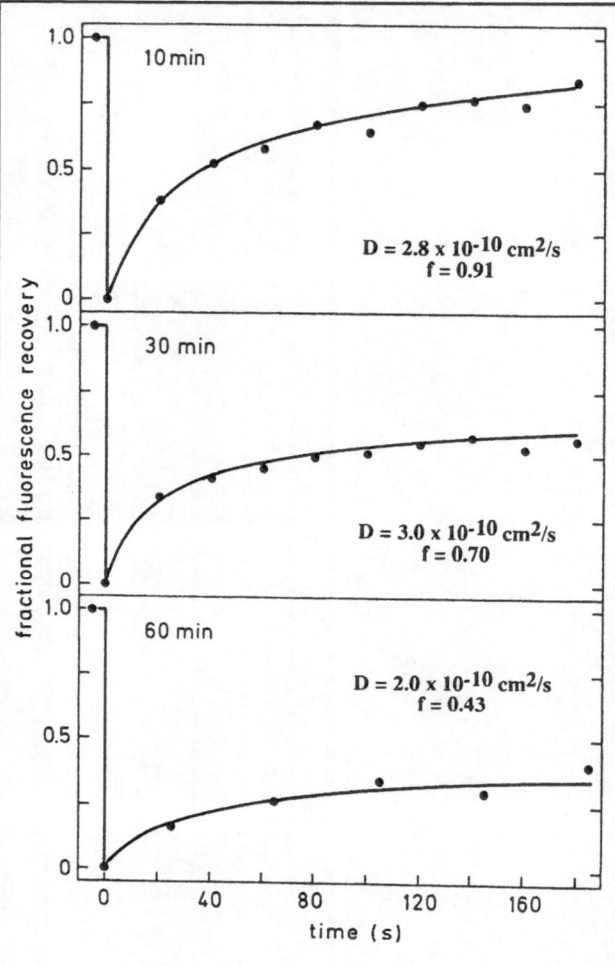

Fig. 6.2. Decline of the vasopressin type-2 receptor mobile fraction with time at 37°C. Measurements were performed by the fluorescence photobleaching recovery technique in renal epithelial cells (the LLC-PK$_1$ line) on cells at various times after hormone addition, essentially as outlined in the legend to Figure 4.2. Values for the apparent lateral diffusion coefficient (D) and mobile fraction (f) are shown.[39]

internalization directly parallels V_2-receptor immobilization through reduction of the receptor mobile fraction.[2,39] The V_2-receptor mobile fraction decreases with time at 37°C as the receptor is immobilized prior to/concomitant with receptor internalization.[39] The apparent lateral diffusion coefficient in contrast is essentially unaffected,[39] implying that receptor immobilization occurs through irreversible receptor binding to immobile structures (see ref. 35).

Similar qualitative results have been obtained for the NGF receptor, where recep-

tor aggregation and internalization parallel a reduction in receptor lateral mobility,[41] and for the EGF and insulin receptors, where both the apparent lateral diffusion coefficient and mobile fraction fall with time at 37°C (see Fig. 6.3).[55] Table 6.1 shows data for a number of polypeptide hormone receptors for which the receptor mobile fraction and extent of internalization are known, while Figure 6.4 shows a plot of the relationship between the two parameters, indicating a linear relationship. Similar results with respect to the direct relationship between

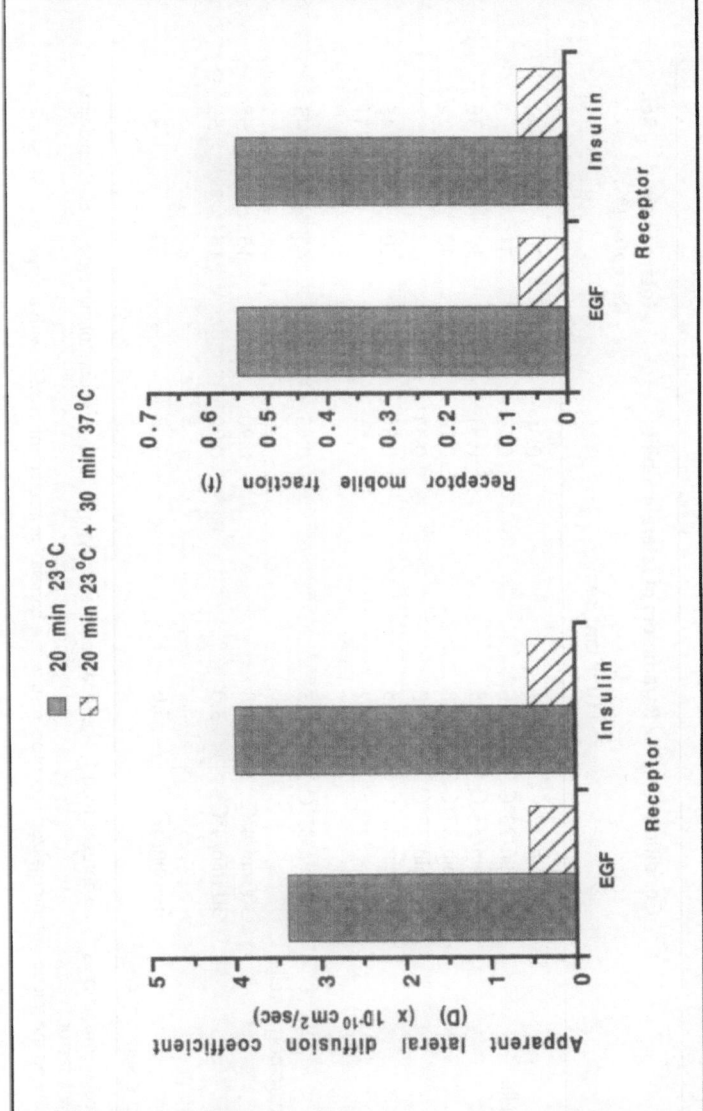

Fig. 6.3. Decline of tyrosine kinase receptor lateral mobility with time at 37°C. Values for the apparent lateral diffusion coefficient (D) and mobile fraction (f) are shown for mouse 3T3 fibroblasts.[55] The metabolic inhibitor sodium azide prevents this process (not shown) through inhibition of endocytosis.[55]

Table 6.1. Relationship between lateral diffusion properties of plasma membrane receptors and receptor internalization[#]

Receptor (cell line)	Condition	Parameters of lateral mobility*		Internalized Receptor (%)	Ref.
		D (10^{-10} cm^2/sec)	f		
V$_1$-receptor (A7r5 rat smooth muscle cells)	10 min/37°C	5.1	0.36	75	44
	10 min/23°C	3.6	0.50	51	
V$_2$-receptor (LLC-PK$_1$ porcine renal epithelial cells)	10 min/37°C	2.5	0.91	10	39
	30 min/37°C	2.8	0.70	40	
	60 min/37°C	2.0	0.43	64	
	10 min/23°C	1.5	0.65	20	45
CCK$_A$-receptor (pancreatic acinar cells)	5 min/24°C	2.2	0.51	65	31,47
EGF-receptor (A431 human epidermoid carcinoma cells)	10 min/37°C	8.5	0.45	45	43
(mouse 3T3 fibroblasts)	30 min/4°C	2.8	0.90	15	55
insulin receptor (mouse 3T3 fibroblasts)	30 min/4°C	3.0	0.90	15	55
NGF receptor (chicken embryonal immature sensory cells)	60 min/37°C[+]	5.6	0.22	70	41

Abbreviations: V$_1$, hepatic/vascular smooth muscle vasopressin type 1; V$_2$, renal vasopressin type 2; CCK$_A$, cholecystokinin type A; EGF, epidermal growth factor; NGF, nerve growth factor.

[#]Analysis of the pooled data of f and the fraction of internalized receptors results in a correlation coefficient of 0.95, with a slope of - 0.98 (see Fig. 6.4; see also Refs. 1,2,39).

*D, apparent lateral diffusion coefficient; f, mobile fraction.

[+]Measurements of mobility made at room temperature.[41]

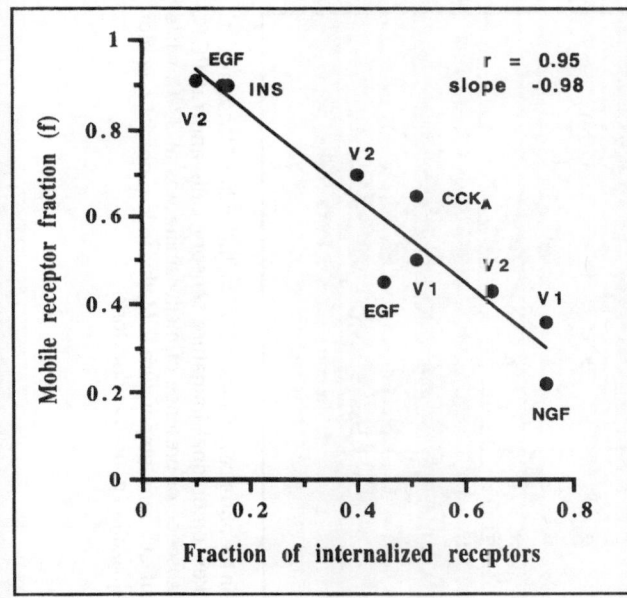

Fig. 6.4. The relationship of the polypeptide hormone receptor mobile fraction to receptor internalization. Pooled data indicate an inverse linear relationship (the correlation coefficient - r - is shown). Receptor data is from Table 6.1 (see also refs. 1,2 and39), with receptor abbreviations as follows: V2, vasopressin V_2; V1, vasopressin V_1; EGF, epidermal growth factor; INS, insulin; CCK_A, cholecystokinin type A and NGF, nerve growth factor.

receptor immobilization efficiency and internalization have been obtained for the high-affinity IgE Fc receptor of basophils (see chapter 7; Fig. 7.3).[52] The clear implication is that receptor immobilization precedes and is a prerequisite for endocytosis in many receptor systems.[1,2,39]

In a number of studies particularly with respect to tyrosine kinase receptors, it has proved necessary to use either low temperatures or inhibitors to block receptor internalization (e.g., sodium azide or 2-deoxyglucose)[42,55] or aggregation (methylamine or bacitracin),[27,41,56] in order to be able to make a reasonable estimate of receptor lateral mobility. In the case of the α_2-macroglobulin receptor, values for the apparent lateral diffusion coefficient (D) and mobile fraction (f) are 7.2 x 10^{-10} cm^2/sec and 0.65 at 4°C but much lower after 30 min at 23°C (5.5 x 10^{-10} cm^2/sec and 0.37 for D and f, respectively).[27] That these lower values are due to receptor aggregation is demonstrated by the fact that treatment with methylamine or bacitracin reverses this effect (e.g., values in the presence of bacitracin of 9.6 x 10^{-10} cm^2/sec and 0.56 for D and f, respectively).[27]

Comparable experiments for the NGF receptor[41] are shown in Figure 6.5, where methylamine similarly prevents receptor clustering, yielding higher D and f values. The lateral mobility of the cholecystokinin type A (CCK_A) receptor cannot be measured at 37°C, apparently as a result of the rapid aggregation/immobilization/internalization kinetics.[47] That immobilized receptors, however, are still at the plasma surface (and not already internalized) has been demonstrated using colloidal gold and electron microscopy,[47] further implying that receptor immobilization precedes endocytosis.

D. RECEPTOR MOVEMENT REQUIRED FOR INTERNALIZATION

That receptor lateral movement within the plane of the membrane plays a role in receptor desensitization has been shown in several studies.[36,37,57-60] In the case of receptors internalized through coated pits or other structures in particular locations at the membrane surface, lateral movement of the receptor-hormone complex has been shown to be required to bring the receptor into

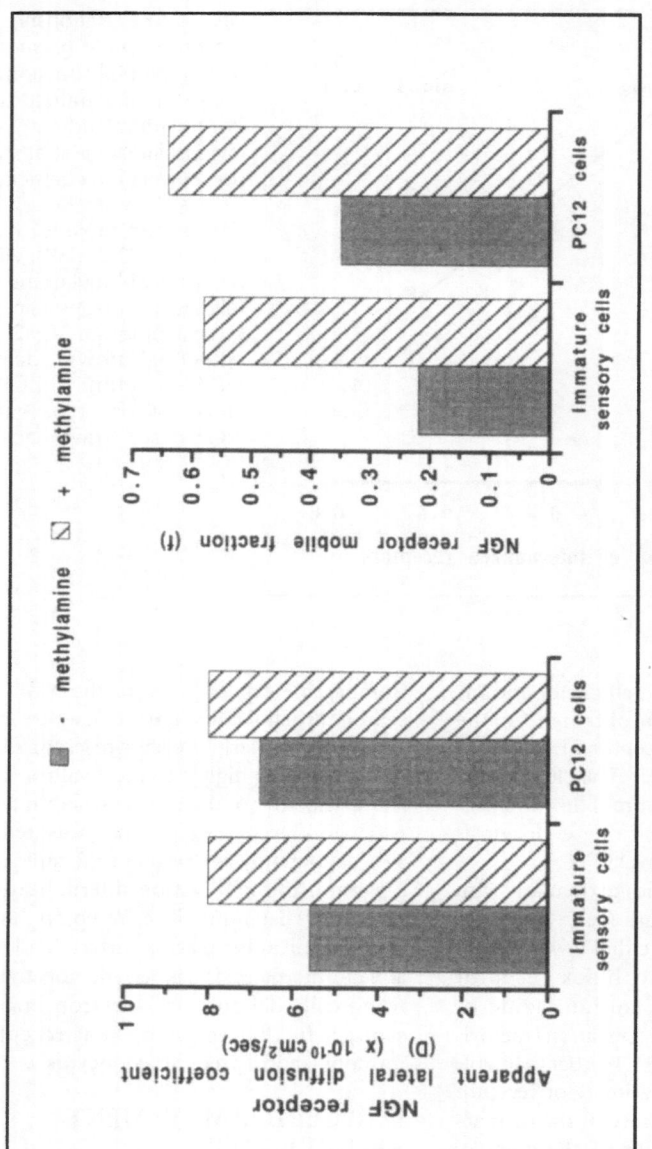

Fig. 6.5. Nerve growth factor (NGF) receptor lateral mobility in the absence and presence of methylamine. Measurements were made of NGF receptor lateral mobility in chicken embryonal immature sensory cells and rat PC12 pheochromocytoma cells; incubation with ligand in the absence or presence of methylamine was at 37°C, while photobleaching recovery measurements were carried out at room temperature.[41] Methylamine inhibits receptor clustering[27,41,56] and concomitantly reduces receptor immobilization prior to endocytosis.

contact with the pit structure;[33,34,37,40,58,59] once contact is made, receptors become essentially immobile prior to endocytosis, the rate of immobilization being important in determining the kinetics of internalization.[36] This has been shown directly for the transferrin receptor, where the receptor mobile fraction is reduced rapidly upon transferrin addition.[33] Interestingly, short term (1h) pretreatment with the microtubule depolymerizing agent colchicine (which also inhibits tubulin polymerization) decreases both receptor immobilization (mobile fraction of 0.63 at 30 min after transferrin addition at 21°C, compared to < 0.2 in the absence of colchicine pretreatment)[33] and internalization (see ref. 33). Longer (3h) pretreatment with either colchicine or vinblastin, also a microtubule perturbing agent, in fact, reduces the mobile fraction to 0.17, whereby the microtubule stabilizer taxol (or paclitaxel, a natural product from *Taxus brevifolia* tree bark), which promotes microtubule assembly and blocks tubulin disassembly, is able to reverse the effect in terms of both receptor immobilization (mobile fraction of 0.5) and internalization.[40] The clear implication is that receptor immobilization upon ligand addition, as well as ensuing endocytosis, is dependent on the integrity of the cellular microtubular assembly.

In the case of the gonadotropin releasing hormone (GNRH) receptor of cultured pituitary cells, cold pretreatment was found to decrease membrane fluidity as determined by fluorescence polarization measurements and concomitantly reduce hormone induced-removal of receptors from the membrane through endocytosis.[57] The temperature-induced effect on desensitization was reversed when the membrane fluidity was increased by incubating cells with the membrane mobilizing agent 2-(2-methoxy-ethoxy)-ethyl-8-(cis-2-n-octyl-cyclopropyl)-octanoate,[57] clearly implying that movement within the plane of the membrane is necessary to bring about receptor internalization.

Increasing the apparent lateral diffusion coefficient of the insulin receptor about 3-fold (see Table 3.2) by "optimizing" the transmembrane α-helical structure using site-directed mutagenesis increases the rate constant of internalization (K_e) by almost 2-fold, as well as leading to an almost 100% increase in receptor degradation subsequent to endocytosis.[58,59] The fact that kinase negative insulin[59,60] and Neu[36] and EGF[61] receptor mutants appear to diffuse normally but are internalized at much slower rates,[36,58,59,60] implies that receptor activation may be necessary for endocytosis to be elicited (see also section G below). Consistent with this, increased receptor immobilization of constitutively active and other Neu receptor mutant derivatives correlates with accelerated antibody-induced internalization,[36] although it should be remembered that the natural hormone ligand for the Neu receptor is not known and that antibody-induced receptor immobilization/internalization may not approximate the physiological situation very well. Receptor immobilization appears to occur more rapidly through facilitated association with coated pits and particular immobile components thereof, such as the AP-2 adaptor protein complex and adaptin subunits (see ref. 36). As discussed in chapter 3, stable association with immobile structures results in a reduction of the mobile fraction whereas transient interaction results in a decrease in the apparent lateral diffusion coefficient (see ref. 35). It thus seems reasonable to assert that the former is the mechanism of immobilization at coated pits of the Neu receptor, as well as of a variety of other receptors,[13,16,36-38,50] as a prelude to internalization.

The results of measurements of the lateral mobility of the GTP-binding protein-activating LH receptor occupied by either LH or chorionic gonadotropin (cGN) support the idea that receptor lateral movement is a critical parameter in the desensitization of response subsequent to hormonal addition and particularly in receptor endocytosis. LH and cGN have distinct steroidogenic effects on luteal and Leydig tumor cells, particularly with respect to the duration of the response elicited.[62,63] While the production of progesterone induced by LH appears to be largely dependent on the latter's continued presence, short incubations with cGN stimulate long-term responses (over six hours) of much greater magnitude.[62,63] This increased potency of cGN compared to LH appears to be directly attributable to an up to 50-fold difference in the kinetics of internalization;[64,65] whereas the half-life of LH-occupied receptor at the membrane of luteal cells is about 40 minutes, that of cGN-occupied receptor is about 17 hours.[64,65] Hormone and receptor degradation is also much slower in the case of cGN.[65] The mechanistic basis of the differences between the two ligands in terms of signaling capabilities mediated by the same receptor appears to be directly related to the LH receptor's lateral mobility properties.[66] Whereas the LH occupied receptor is largely mobile even at physiological temperature, the cGN-occupied receptor is essentially immobile (see Table 4.1).[66-69] The clear implication is that receptor lateral movement is required for receptor internalization; because the cGN-occupied receptor does not move rapidly to coated pits to become internalized, the signal it elicits at the membrane is of much longer duration.

The basis of immobility of the cGN-occupied LH receptor, relative to LH occupied receptor, appears to be attachment to cytoskeletal elements, since treatment with cytochalasins B and D (actin filament perturbing agents) and the microtubule perturber colchicine increase the lateral mobility of human cGN-occupied LH receptors on luteal cells (see also chapter 8 section C; Table 8.1). The carbohydrate moiety of the hormone also appears to be important, since deglycosylation of cGN leads to an 11-fold higher apparent lateral diffusion coefficient compared to that of the LH receptor occupied by nondeglycosylated hormone.[66] To what extent the nature of the glycosylation of the hormone ligand may affect association with the cytoskeleton is unclear.

E. KINETIC CONSIDERATIONS WITH RESPECT TO LATERAL MOBILITY MEASUREMENTS

It should be mentioned in passing that since receptor internalization relates strongly to receptor immobilization, the kinetics of internalization are a critical parameter in determining the extent of the mobile fraction of receptors at physiological temperature. The differences observed at 37°C between tyrosine kinase and GTP-binding protein-activating receptor classes discussed at length in chapter 4 (see Fig. 4.4), must be viewed critically in this context. The EGF receptor is internalized much more rapidly ($t_{1/2}$ = 6 min at 37°C on mouse 3T3 fibroblasts),[37,42,43,55,65] for example, than the vasopressin V_2-receptor ($t_{1/2}$ = 14 min)[45] or the LH-occupied LH receptor, as mentioned above ($t_{1/2}$ = 42 min).[64,65] It accordingly seems reasonable to assert that the relatively low mobile fraction at 37°C measured for tyrosine kinase receptors may in part result from rapid receptor internalization kinetics. Rapid receptor internalization, of course, is not unique to tyrosine kinase receptors;[31,47] both the vasopressin V_1- of rat aortic smooth muscle cells ($t_{1/2}$ = 2 min)[44] and CCK_A-receptors exhibit very rapid internalization at 37°C. It thus seems clear that while rapid receptor endocytosis is not the sole mechanistic basis of low tyrosine kinase receptor mobility at physiological temperature (see chapter 4, section D), it is a significant factor modulating receptor lateral mobility.

F. RECEPTOR PHOSPHORYLATION

As mentioned both above and in chapter 5 (section D), a key event in desensitization subsequent to hormonal stimulation and GTP-binding protein activation is receptor phosphorylation, either by specific receptor[18-22,70-85] or other kinases,[20,23-26] occurring prior to endocytosis. This parallels the kinetics of receptor immobilization and thus conceivably embodies the receptor immobilizing step itself, through effecting association with cytoskeletal or membrane skeletal components (see refs. 86-88; see also chapter 3, section C). That agonist-induced receptor phosphorylation by specific receptor kinases is integral in the process of receptor endocytosis and desensitization is implicated by the study of a number of diverse GTP-binding protein activating receptors, including the parathyroid hormone[26] and the N-formyl peptide,[70] delta opioid,[71] adrenergic,[20,22,23,72,76,78,79,83-85] glucagon,[73] muscarinic,[18,21,74] secretin,[75] CCK$_A$[77,80,81] and vasopressin V$_2$-[82] receptors, as well as rhodopsin.[78,83]

Receptor phosphorylation at specific sites in the third intracellular loop,[23,25,72,74,77,83] and especially at the C-terminus (refer to schematic representation of Fig. 4.10),[23,25,70,74,75,77,83] effects desensitization of response through reducing direct interaction with GTP-binding proteins (see ref. 83). This appears to occur in part through induction of the binding of specific proteins, called arrestins, which "arrest" or prevent GTP-binding protein activation by hormone-bound receptors.[76,78,83,85] In the case of rhodopsin, light-stimulated phosphorylation induces binding of the 48 kDa protein arrestin (also known as S-antigen)[83,89,90] or arrestin 3,[91] whereas activated, phosphorylated β-adrenergic and other receptors appear to bind analogous β-arrestin 1 or 2 molecules.[76,78,83,85,92,93] The molecular basis of arrestin-mediated prevention of receptor interaction with GTP-binding proteins is not clear (see ref. 91), but one can speculate that arrestin binding to phosphorylated receptor may effect receptor immobilization, thereby preventing further activation of GTP-binding proteins. As mentioned in chapter 5 (section E), the GTP-binding protein Gβγ subunits appear to be able to activate receptor kinases specifically, inducing their translocation and binding to the membrane,[18,19,21,22,24,79,94] thus bringing the kinases into the close vicinity of the membrane integral receptor molecules to facilitate their phosphorylation. The receptor phosphorylation state has also been shown to be modulated by dephosphorylation both by protein phosphatase type

2A (PP2A)[95] and the cytosolic protein phosducin, which appears to compete with receptor kinases for the Gβγ subunits,[94] thus preventing receptor kinase association with Gβγ.

In most systems, at least 50% of receptor phosphorylation is due to kinases other than the specific receptor kinases mentioned above (e.g., refs. 23,83,91). PK-A[18,23,26,83,91] and the Ca^{2+},phospholipid-dependent protein kinase (PK-C),[18,21,24-26,83] activated upon adenylate cyclase and phospholipase C activation, respectively, appear to be the major kinases involved. Receptor phosphorylation by these kinases correlates closely with receptor sequestration and internalization (see ref. 82),[20,96-98] and even degradation, as well as with the abrogation of GTP-binding protein activation.[83,91]

Again, precise mechanisms are unknown, and it can be postulated that receptor phosphorylation by other kinases may also play a direct role in receptor immobilization prior to endocytosis. Receptor immobilization mediated by PK-A or PK-C phosphorylation may be the basis of the process of heterologous desensitization.[20] Homologous desensitization is where the downregulation process only affects the receptor directly activated by hormone binding,[20,94] whereas heterologous desensitization is where activation of one receptor can lead to desensitization or refractability to stimulation of a distinct receptor (see ref. 20). In the latter, one can speculate that PK-A or PK-C triggered phosphorylation in response to a particular agonist could effect immobilization of heterologous receptors prior to internalization so that their response to their own specific ligands would be severely reduced. Receptor internalization triggered by nonreceptor-mediated stimulation of PK-A has indeed been shown to lead to internalization of nonligand-occupied vasopressin V$_2$-receptor,[82] although prior receptor immobilization has not been assessed.

How phosphorylation by receptor specific or other kinases may mediate receptor immobilization is not clear, but there are indications that the cytoskeleton may be involved. An example is that of the CCK_A-receptor of acinar pancreatic cells which has been demonstrated using colloidal gold conjugates to undergo rapid immobilization on the cell surface at 37°C prior to endocytosis, concomitant with an inability of the bound ligand to be removed from the membrane by low pH treatment.[31,47] This process of "insulation" of the receptor-ligand complex on the membrane surface is unique to intact cells not occurring in isolated membranes.[47] It is postulated to occur through association of the receptor with membrane skeletal or cytoskeletal components and/or associated protein(s);[47] candidate molecules which may be involved include the monomeric microtubule-associated GTPases, dynamin-2 (which plays a central role in endocytotic coated vesicle formation; see ref. 99) and Rab family members (which are involved in intracellular vesicle transport).[100] CCK_A-receptor phosphorylation may play a role in regulating the cytoskeletal association.[77,80,81] The N-formyl-peptide (chemotactic factor) receptor of neutrophils appears to display quite similar properties in terms of insulation of receptor on the membrane surface and association with cytoskeletal elements and guanine nucleotide regulatory proteins, with quite good evidence for receptor immobilization through the actin cytoskeleton.[101-104] Cytochalasin B treatment slows N-formyl-peptide receptor immobilization induced by ligand occupation of the receptor.[104] Despite the somewhat limited array of receptors for which measurements of this nature have been made, the results imply that association with specific cytoskeletal elements, possibly mediated through phosphorylation, may be a general mechanism of receptor immobilization prior to endocytosis. This will discussed again in a little more detail in chapter 8 (section C).

It should be mentioned in this context that there is evidence for the CCK_A-receptor that immobilization/internalization pathways may be cell-type specific;[47] in contrast to pancreatic acinar cells which possess endogenous receptors, Chinese Hamster Ovary (CHO) cells transfected with the CCK_A-receptor complementary DNA do not display receptor insulation. It seems clear that, just as isolated membranes are essentially useless systems to examine such processes as desensitization and receptor immobilization, especially in the absence of cytoskeleton, cytosolic factors, etc., studies using cells into which a receptor, normally absent from a particular cell type (as are the normal downregulatory mechanisms to desensitize response at the level of the particular receptor), is introduced by transfection should be treated with caution.

G. STUDIES WITH RECEPTOR ANTAGONISTS—RECEPTOR IMMOBILIZATION IS AGONIST-DEPENDENT

Fluorescence photobleaching recovery measurements of GTP-binding protein activating receptors occupied by specific receptor antagonists have yielded further insight into the desensitization process and the central role of receptor lateral mobility.[104,105] Almost all of the fluorescence photobleaching recovery measurements for polypeptide hormone receptors (e.g., those in Table 4.1) have used fluorescently-labeled agonists, ligands which bind a specific receptor and activate it, resulting in initiation of signal transduction. Antagonists, in contrast, are hormone analogs that are capable of specifically binding to receptors with affinities comparable to those of agonists but are unable to stimulate signal transduction. Antagonists have a wide range of pharmacological applications, since they can be used to occupy receptors and thus block responses such as the growth of particular tumors, e.g., prolactinomas can be inhibited by prolactin antagonists, certain breast tumors by estrogen antagonists, etc. Structures of fluorescently-labeled vasopressin agonist and antagonist analogs for the vasopressin V_2-receptor are shown in Figure 6.6.[45,105]

Significantly, antagonist-occupied vasopressin V_2-[105] and N-formyl peptide[104] receptors appear to have lateral mobility properties very similar to those of agonist-occupied receptors (see Table 6.2). The V_2-receptor shows essentially identical temperature dependence of the apparent lateral diffusion coefficient whether occupied by agonist or antagonist.[105] Clearly, antagonist-occupied receptors are not immobile, and accordingly, the basis of antagonism is clearly not receptor immobilization. Rather, lateral diffusion of the receptor in the cell membrane appears to be a constant process in the presence or absence of ligand, with the possibility of a collision with the membrane-associated GTP-binding protein

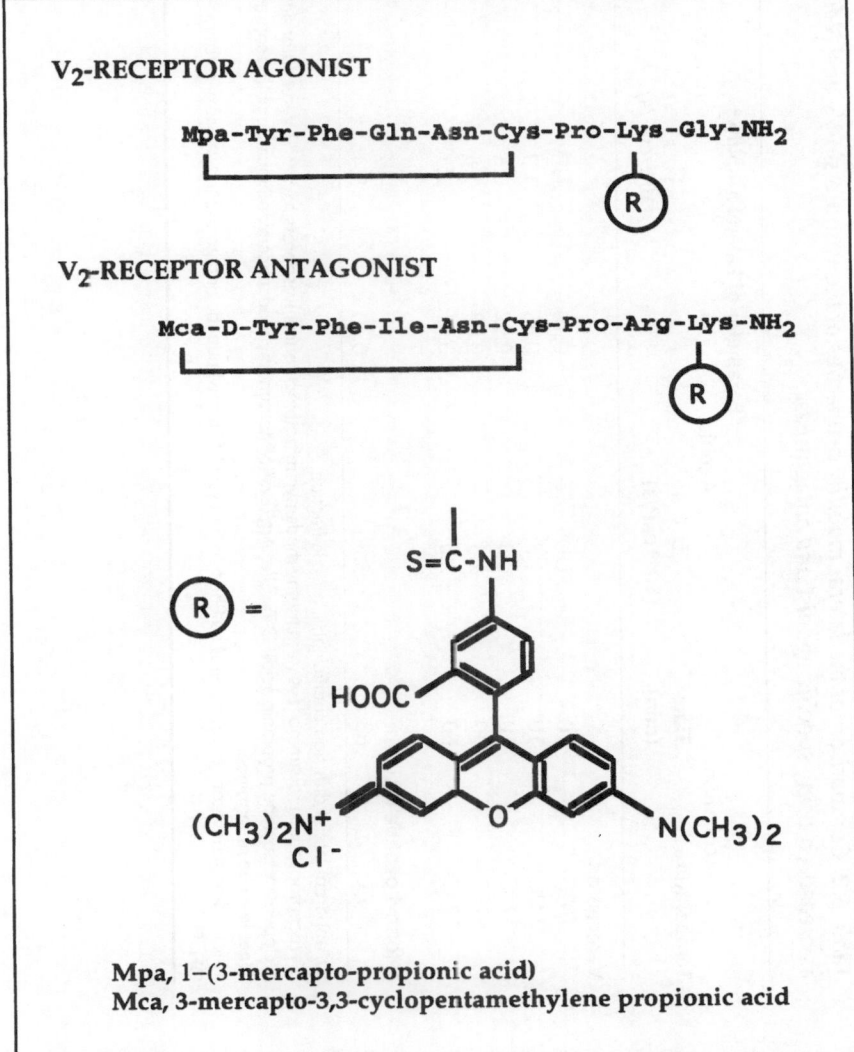

Fig. 6.6. Structures of tetramethyl-rhodamine-labeled vasopressin agonist and antagonist for the vasopressin V_2-receptor.[45,105] The agonist also binds with high-affinity to the V_1-receptor.[44]

Table 6.2. *Comparison of the lateral mobility properties of the vasopressin V_2 and N-formyl-peptide receptors as measured using specific agonist and antagonists.*

| Incubation | | Parameter of Lateral Mobility# | | | | Ref. |
| Temperature | Time (min) | Agonist | | Antagonist | | |
		D (10^{-10} cm²/s)	f	D (10^{-10} cm²/s)	f	
V_2-receptor+						
10°C	10	UDˣ	0.10	UDˣ	0.10	105
23°C	10	1.5	0.65	1.9	0.62	
37°C	10	2.5	0.92	2.6	0.77	
37°C	30	2.8	0.68	2.8	0.76	
37°C	60	2.0	0.43	2.3	0.67	
N-formyl peptide receptor˙						
14°C	60	5.5	0.40	5.4	0.63	104

#D, apparent lateral diffusion coefficient; f, mobile fraction.

+V_2-receptor agonist: deamino-[Lys⁸(tetramethylrhodamylaminothiocarbonyl)]vasopressin;[3,39,45,105] V_2-receptor antagonist: [(β-mercapto-β,β-cyclopentamethylene propionic acid)¹,D-Tyr²,Ile⁴,Arg⁸,Lys⁹(Nᵋ-tetramethylrhodamylaminothiocarbonyl)] vasopressin.[105]

ˣUnable to be determined.

˙N-formyl-peptide agonist: formyl-Nle-Leu-Phe-Nle-Tyr-Lys;[104] N-formyl-peptide antagonist: tert-butyloxy-carbonyl-Phe(D)-Leu-Phe(D)-Leu-Phe-OH.[104]

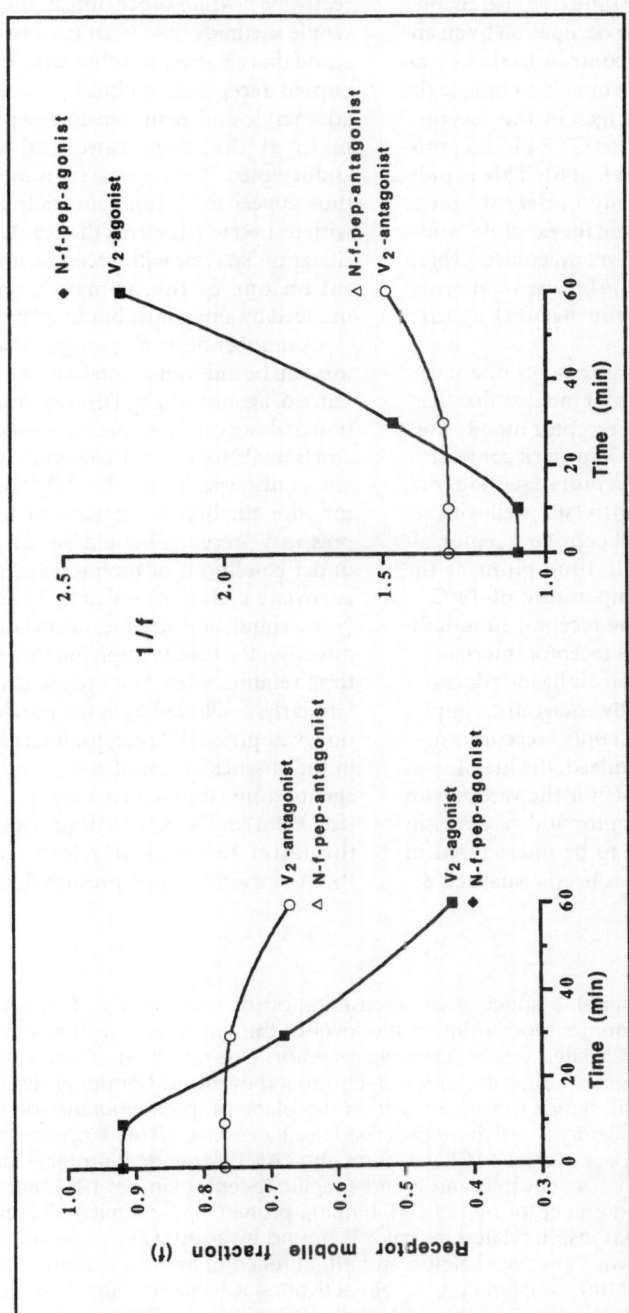

Fig. 6.7. Comparison of vasopressin V₂- and N-formyl-peptide (N-f-pep) receptor agonists and antagonists with respect to time-dependent receptor immobilization. Results for the mobile fraction (see Table 6.2) are plotted against time at 37°C for the V₂-receptor[39,105] and at 14°C for the N-formyl-peptide receptor (left panel).[104] The right panel shows the data plotted reciprocally with respect to the receptor mobile fraction to highlight the differences between agonist-and antagonist-occupied receptors.

complex given all the time (see also chapter 5, section B). Receptor occupation by an antagonistic ligand, in contrast to that by an agonist, appears to be unable to induce the conformational changes in the receptor which are necessary for GTP-binding protein activation (see ref. 106). This is indicated by findings using nuclear magnetic resonance spectroscopy, for example, where vasopressin V_1-receptor antagonists exhibit much slower conversion between conformational states than the natural agonist [Arg⁸]vasopressin.

Significantly, measurements of antagonist-occupied V_2-receptor indicate that there is no reduction in the receptor mobile fraction with time at 37°C, in stark contrast to agonist occupied receptors (see Fig. 6.7, Table 6.2).[39,105] This correlates well with results for the N-formyl peptide receptor, albeit for only a single time point at the nonphysiological temperature of 14°C.[104] This indicates that the receptor immobilization which precedes receptor internalization in the case of agonistic ligands does not appear to be induced by antagonists, implying that antagonist-occupied receptors may not be internalized. Indeed, the inability of fluorescent antagonists for the vasopressin related vasotocin-receptor and vasopressin V_1- and V_2-receptors to be internalized in several cell systems has been visualized directly,[107,108] while biochemical and microscopic methods have been used to demonstrate that a variety of other antagonist-occupied receptors, including muscarinic, adrenergic and neurotensin receptors, remain at the membrane and are not endocytosed.[109-116] Agonists and antagonists thus appear to be fundamentally different with respect to triggering the cellular desensitization system, with receptor immobilization one of the primary events not induced by antagonist binding.[104,105]

A number of studies suggest that receptors can be induced to internalize in the absence of agonistic ligand through the induction of downstream signaling events. Eggena and Buku,[108] for example, demonstrated that antagonists specific for the vasotocin receptor (the amphibian correlate of the vasopressin V_2-receptor) could be internalized under conditions of receptor-independent adenylate cyclase stimulation by forskolin (which binds and activates adenylate cyclase directly), the results implying that endocytosis requires adenylate cyclase activation. Similarly, as alluded to in the previous section, vasopressin V_2-receptor internalization in the absence of ligand can be induced by agents stimulating either adenylate cyclase (forskolin) or PK-A (cAMP analogs),[82] with the latter result clearly implying that PK-A activation, and presumably PK-A-

Fig. 6.8. (opposite page) A speculative representation of some of the desensitizatory/downregulatory phenomena occurring at the level of the membrane with respect to the immobilization of adenylate cyclase activating receptors (see ref. 2). After stimulation with hormone (H)—see Figure 5.5 for the diffusion-controlled events of hormonal stimulation—continued receptor (R) lateral movement within the plane of the membrane (denoted by D_1) amplifies the stimulatory signal through activating successive GTP-binding protein trimers (abg) (I). Upon adenylate cyclase (AC) activation, the cAMP-dependent protein kinase (PK-A) catalytic subunit (C) is activated and it and specific receptor kinases (RK) (attracted to the membrane and the receptor by the GTP-binding protein Gβγ subunits) phosphorylate the receptor (II). The phosphorylated receptor is bound by arrestin-like molecule(s) which mediate association with the cytoskeleton and effect receptor immobilization to abrogate the stimulatory signal (III). GTP-binding protein activation is halted through GTP hydrolysis by $G_α$ (not shown; see Figure 4.7 for the GTP-binding protein cycle). Receptor internalization into endosomes occurs (IV), leading to hormone dissociation from the receptor and either degradation in lysosomes or recycling to the membrane to bring about a return to the basal unstimulated state. For details of the GTP-binding protein cycle in this context (not shown), see refs. 20, 83 and117-119 (see also Figure 4.7 and chapter 5, section E).

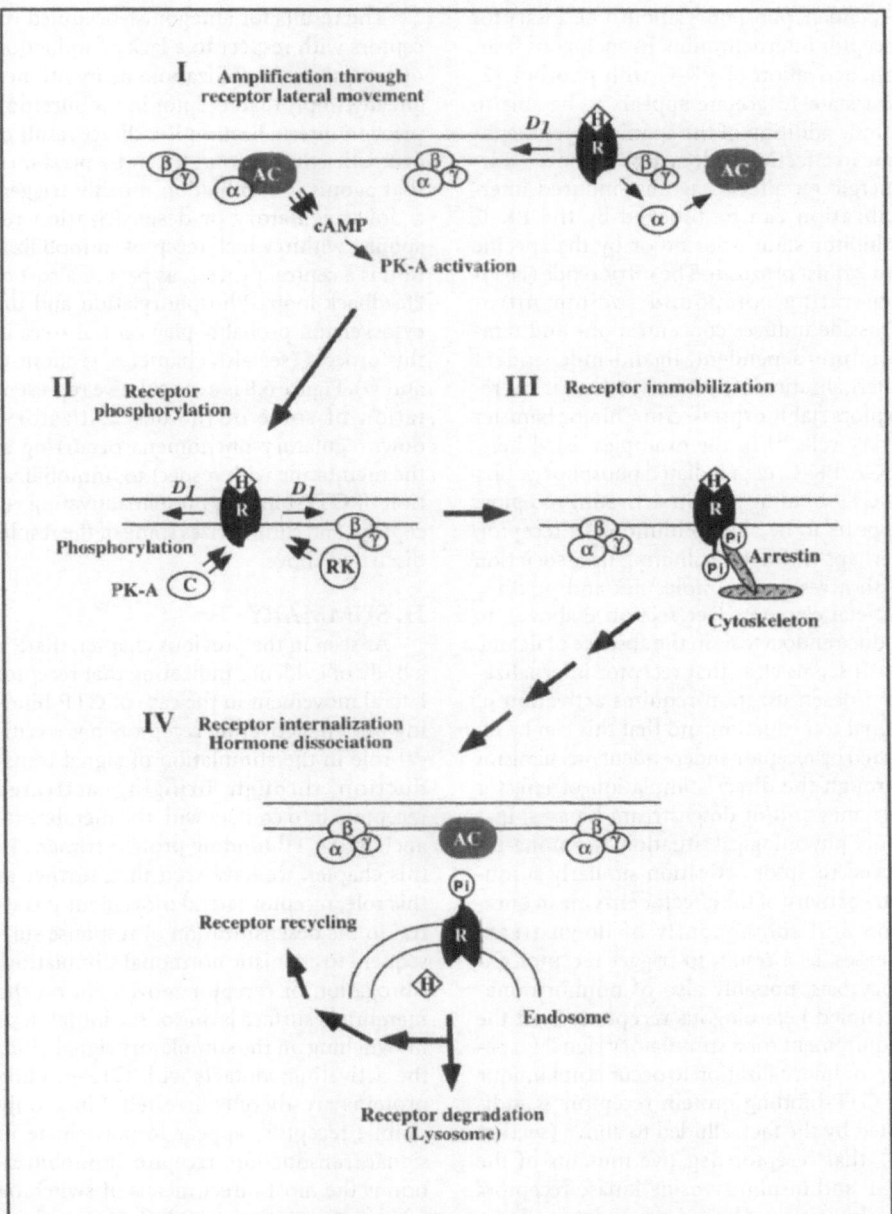

Fig. 6.8. (see caption, opposite page)

dependent phosphorylation, is necessary for receptor internalization. In analogous fashion, activation of PK-C with phorbol 12-myristate 13-acetate appears to be able to mimic addition of the agonist norepinephrine in effecting endocytosis of the α_{1B}-adrenergic receptor;[109] agonist-induced internalization can be blocked by the PK-C inhibitor staurosporine or by the specific antagonist prazosin. The nitric oxide (NO)-generating compound sodium nitroprusside induces concentration- and temperature-dependent, ligand-independent internalization of human M4 muscarinic receptors stably expressed in Chinese hamster ovary cells.[110] In the examples listed here, PK-A/PK-C, *etc.* mediated phosphorylation of GTP-binding protein-activating receptors appears to be able to immobilize receptor, perhaps through facilitating its association with arrestin-like molecules and/or cytoskeletal elements (see section F above), to induce endocytosis in the absence of ligand.

It seems clear that receptor internalization/desensitization requires activation of signal transduction and that this can be effected by receptor-independent mechanisms through the direct stimulation of effector enzymes and/or downstream kinases. In a more physiological situation, hormone-induced receptor activation similarly stimulates activity of the effector enzyme in question and subsequently of downstream kinases, as a result, to trigger receptor endocytosis, possibly also of nonhormone-occupied heterologous receptors. That the requirement for a stimulatory signal for receptor internalization to occur is not unique to GTP-binding protein receptors is indicated by the fact, alluded to above (section D), that receptor negative mutants of the EGF and insulin tyrosine kinase receptors exhibit impaired ligand-dependent endocytosis.[36,58-61] This phenomenon of signal transduction-mediated regulation of downregulation, possibly largely through modulation of receptor immobilization prior to internalization, will be discussed at more length in chapter 8 (section D).

The results for antagonist-occupied receptors with respect to a lack of induction of receptor immobilization, as mentioned already, imply that receptor immobilization prior to internalization is a direct result of agonistic stimulation. It can be presumed that agonistic stimulation directly triggers a downregulatory or desensitization response, within which receptor immobilization is a central process, as part of a sort of "feedback loop". Phosphorylation and the cytoskeleton probably play central roles in this process (see also chapter 8, sections C and D). Figure 6.8 is a speculative representation of some of the desensitizatory/downregulatory phenomena occurring at the membrane with respect to immobilization of GTP-binding protein activating receptors, and summarizes some of the results discussed above.

H. SUMMARY

As seen in the previous chapter, there is a body of evidence indicating that receptor lateral movement in the case of GTP-binding protein activating receptors has a critical role in the stimulation of signal transduction through bringing activated receptors into contact with the membrane-anchored GTP-binding protein trimers. In this chapter, we have seen that, further to this role, receptor lateral movement is central to the desensitization of response subsequent to agonistic hormonal stimulation. Abrogation of receptor movement on the membrane surface is one of the initial steps in switching off the stimulatory signal, since the activating contacts with GTP-binding proteins are thereby arrested. Since only mobile receptors appear to participate in signal transduction, receptor immobilization is the most direct means of switching off the stimulatory signal. Receptor phosphorylation, recognition by arrestin-like proteins, and linkage to the cytoskeleton appear to be integrally involved in this process. Receptor immobilization is a prelude to endocytosis, whereby the stimulatory signal is "permanently" (at least in the short-term in the case of the receptor recycling pathway) switched off, since activated receptor is taken away from the membrane and its complement of GTP-binding proteins. Although GTP-binding proteins and effector enzymes are also downregulated, together with the activation of antagonistic enzymic activities (phosphodiesterase, phosphatases, etc.) as distal desensitizatory events, it is clear that receptor movement and modulation thereof is the key event at the level of the membrane. Chapter 8 shall discuss how response to a single hormone and effector system can affect the responses of a variety of interacting signal transduction pathways in the context of the whole cell.

REFERENCES

1. Jans DA. The mobile receptor hypothesis revisited: a mechanistic role for hormone receptor lateral mobility in signal transduction. Biochim Biophys Acta 1992; 1113:271-276.

2. Jans DA, Pavo I. A mechanistic role for polypeptide hormone receptor lateral mobility in signal transduction. Amino Acids 1995; 9:93-109.

3. Jans DA, Peters R, Jans P et al. Vasopressin V_2-receptor mobile fraction and ligand-dependent adenylate cyclase-activity are directly correlated in LLC-PK$_1$ renal epithelial cells. J Cell Biol 1991; 114(1):53-60.

4. Jans DA, Peters R, Jans P et al. Ammonium chloride affects receptor number and lateral mobility of the vasopressin V_2-type receptor in the plasma membrane of LLC-PK$_1$ renal epithelial cells: role of the cytoskeleton. Exper Cell Res 1990; 191:121-128.

5. Zakharova OM, Rosenkranz AA, Sobolev AS. Modification of fluid lipid and mobile protein fractions of reticulocyte plasma membranes affects agonist-stimulated adenylate cyclase. Application of the percolation theory. Biochim Biophys Acta 1995; 1236:177-184.

6. Jans DA, Hemmings BA. cAMP metabolism in the porcine epithelial cell line LLC-PK$_1$: the central role of the cAMP-dependent protein kinase in cAMP-mediated gene induction. Advances in Second Messenger and Phosphoprotein Res 1988; 21:109-121.

7. Jans DA, Resink TJ, Hemmings BA. A novel LLC-PK$_1$ renal epithelial cell mutant impaired in in vivo down-regulation of cAMP-mediated hormonal response. Arch Biochem Biophys 1991; 285:377-381.

8. Pearson D, Nigg EA, Nagamine Y et al. Mechanisms of cAMP-mediated gene induction; examination of renal epithelial cell mutants affected in the catalytic subunit of the cAMP-dependent protein kinase. Exper Cell Res 1991; 192:315-318.

9. Hemmings BA. cAMP-mediated proteolysis of the catalytic subunit of the cAMP-dependent protein kinase. FEBS Lett 1986; 196(1):126-130.

10. Jans DA, Gajdas EL, Dierks-Ventling C et al. Long term stimulation of cAMP production in LLC-PK$_1$ cells by salmon calcitonin or a photoactivatable analogue of vasopressin. Biochim Biophys Acta 1987; 930:392-400.

11. Luzius H, Jans DA, Fahrenholz F. Use of a UV-activatable analogue of vasopressin to select mutants of LLC-PK$_1$ cells affected in hormonal responsiveness. J Receptor Res 1989; 10:61-80.

12. Lutz W, Salisbury J, Kumar R. Vasopressin receptor-mediated endocytosis: current view. Am J Physiol 1991; 261:F1-F13.

13. Goldstein JL, Brown MS, Anderson RGW et al. Receptor-mediated endocytosis: concepts emerging from the LDL receptor system. Annu Rev Cell Biol 1989; 1:1-139.

14. Carpentier J-L. The cell biology of the insulin receptor. Diabetologia 1989; 32:627-635.

15. Thibonnier M. Signal transduction of V_1-vascular vasopressin receptors. Regulatory Peptides 1992; 38:1-11.

16. De Diego JG, Gorden P, Carpentier J-L. The relationship of ligand receptor mobility to internalization of polypeptide hormones and growth factors. Endocrinol 1991; 128(4):2136-2140.

17. Fan JY, Carpentier J-L, Gorden P et al. Receptor-mediated endocytosis of insulin: role of microvilli, coated pits and coated vesicles. Proc Natl Acad Sci USA 1982; 79:7788-7791.

18. Haga T, Haga K, Kameyama K et al. Phosphorylation of muscarinic receptors: regulation by G proteins. Life Sci 1993; 52(5-6):421-428.

19. Inglese J, Koch WJ, Caron MG. Isoprenylation in regulation of signal transduction by G-protein-coupled receptor kinases. Nature 1992; 359(6391):147-150.

20. Hausdorff WP, Caron MG, Lefkowitz RJ. Turning off the signal: desensitisation of the β-adrenergic receptor function. FASEB J 1990; 4:2881-2889.

21. Hosey MM. Diversity of structure, signaling and regulation within the family of muscarinic cholinergic receptors. FASEB J 1992; 6(3):845-852.

22. Kim CM, Dion SB, Benovic JL. Mechanism of beta-adrenergic receptor kinase activation by G proteins. J Biol Chem 1993; 268(21):15412-15418.

23. Okamoto T, Murayama Y, Hayashi Y et al. Identification of a Gs activator region of the β2-adrenergic receptor that is autoregulated via protein kinase A-dependent phosphorylation. Cell 1991; 67:723-730.

24. Leberer E, Dignard D, Harcus D et al. The protein kinase homologue Ste20p is required to link the yeast pheromone response G-protein βγ subunits to downstream signalling components. EMBO J 1992; 11(13):4815-4824.

25. Haga K, Kameyama K, Haga T et al. Phosphorylation of human m1 muscarinic acetylcholine receptors by G protein-coupled receptor kinase 2 and protein kinase C. J Biol Chem 1996; 271(5):2776-2782.

26. Blind E, Bambino T, Nissenson RA. Agonist-stimulated phosphorylation of the G protein-coupled receptor for parathyroid hormone (PTH) and PTH-related protein. Endocrinology 1995; 136(10):4271-4277.

27. Maxfield FR, Willingham MC, Haigler HT et al. Binding, surface mobility, internalization, and degradation of rhodamine-labeled α_2-macroglobulin. Biochemistry 1981; 20(18):5353-5358.

28. Carpentier J-L, Gorden P, Freychat P et al. The fate of [^{125}I]iodoepidermal growth factor in isolated hepatocytes: a quantitative electron microscopic autoradiographic study. Endocrinol 1981; 109:768-775.

29. Jans DA, Jans P, Luzius H et al. Monensin resistant LLC-PK$_1$ mutants are affected in recycling of the adenylate cyclase-stimulating vasopressin V$_2$-receptor. Mol Cell Endocrinol 1991; 81:165-174.

30. Fishman JB, Dickey BF, Butcher NLR et al. Internalization, recycling, and redistribution of vasopressin receptors in rat hepatocytes. J Biol Chem 1985; 260:12641-12646.

31. Roettger BF, Rentsch RU, Pinon D et al. Dual pathways of internalization of the cholecystokinin receptor. J Cell Biol 1995; 128(6):1029-1041.

32. Schwartz AL, Fridovich SE, Lodish HF. Kinetics of internalization and recycling of the asialoglycoprotein receptor in a hepatoma cell line. J Biol Chem 1982; 257(8):4230-4237.

33. Thatte HS, Bridges KR, Golan DE. Microtubule inhibitors differentially affect translational movement, cell surface expression, and endocytosis of transferrin receptors in K562 cells. J Cell Physiol 1994; 160:345-357.

34. Thatte HS, Bridges KR, Golan DE. ATP depletion causes translational immobilization of cell surface transferrin receptors in K562 cells. J Cell Physiol 1996; 166:446-452.

35. Fire E, Zwart DE, Roth MG et al. Evidence from lateral mobility studies for dynamic interactions of a mutant influenza hemagglutinin with coated pits. J Cell Biol 1991; 115:1585-1594.

36. Gilboa L, Ben-Levy R, Yarden Y et al. Roles for a cytoplasmic tyrosine and tyrosine kinase activity in the interactions of Neu receptors with coated pits. J Biol Chem 1995; 270:7061-7067.

37. Hillman GM, Schlessinger J. Lateral diffusion of epidermal growth factor complexed to its surface receptors does not account for the thermal sensitivity of patch formation and endocytosis. Biochemistry. 1982; 21(7):1667-1672.

38. Ullrich A, Schlessinger J. Signal transduction by receptors with tyrosine kinase activity. Cell 1990; 61:203-212.

39. Jans DA, Peters R, Fahrenholz F. An inverse relationship between receptor internalization and the fraction of laterally

mobile receptors for the vasopressin renal-type V_2-receptor; an active role for receptor immobilization in down-regulation? FEBS Lett 1990; 274:223-226.

40. Paccaud JP, Reith W, Johansson B et al. Role of internalization signals and receptor mobility. J Biol Chem 1993; 268(31):23191-23196.

41. Levi A, Schechter Y, Neufeld EJ et al. Mobility, clustering and transport of nerve growth factor in embryonal sensory cells and in a sympathetic neuronal cell line. Proc Natl Acad Sci USA 1980; 77:3469-3473.

42. Schlessinger J, Schechter Y, Willingham MC et al. Direct visualization of binding, aggregation, and internalization of insulin and epidermal growth factor on living fibroblastic cells. Proc Natl Acad Sci USA 1978; 75:2659-2663.

43. Zidovetzki R, Yarden Y, Schlessinger, J et al. Rotational diffusion of epidermal growth factor complexed to its surface receptor the rapid microaggregation and endocytosis of occupied receptors. Proc Natl Acad Sci USA 1981; 78:6981-6985.

44. Jans DA, Peters R, Fahrenholz F. Lateral mobility of the phospholipase-C-activating vasopressin V_1-type receptor in A7r5 smooth muscle cells: a comparison with the adenylate cyclase-coupled V_2-receptor. EMBO J 1990; 9(9):2693-2699.

45. Jans DA, Peters R, Zsigo J et al. The adenylate cyclase-coupled vasopressin V_2-receptor is highly laterally mobile in membranes of LLC-PK$_1$ renal epithelial cells at physiological temperature. EMBO J 1989; 8(9):2431-2438.

46. Henis YI, Katzir Z, Shia MA et al. Oligomeric structure of the human asialoglycoprotein receptor: nature and stoichiometry of mutual complexes containing H1 and H2 polypeptides assessed by fluorescence photobleaching recovery. J Cell Biol 1990; 111:1409-1418.

47. Roettger BF, Rentsch RU, Hadac EM et al. Insulation of a G protein-coupled receptor on the plasmalemmal surface of the pancreatic acinar cell. J Cell Biol 1995; 130:579-590.

48. Carraway III KL, Koland JG, Cerione RA. Visualization of epidermal growth factor (EGF) receptor aggregation in plasma membranes by fluorescence energy transfer. J Biol Chem 1989; 264:8699-8707.

49. King AC, Cuatrecasas P. Peptide hormone-induced receptor mobility, aggregation and internalization. N Engl J Med 1981; 305:77-88.

50. Maxfield R, Schlessinger J, Shechter Y et al. Collection of insulin, EGF and α_2-macroglobulin in the same patches on the surface of cultured fibroblasts and common internalization. Cell 1978; 14(4):805-810.

51. Luborsky L, Slater W, Behrman H. Luteinizing hormone (LH) receptor aggregation: modification of ferritin-LH binding and aggregation by prostaglandin F2α and ferritin-LH. Endocrinol 1984; 115:2217-2226.

52. Mao SY, Varin-Blank N, Edidin M et al. Immobilization and internalization of mutated IgE receptors in transfected cells. J Immunol 1991; 146(3):958-966.

53. Srinivasan Y, Guzikowski AP, Haugland RP et al. Distribution and lateral mobility of glycine receptors on cultured spinal cord neurons. J Neurosci 1990; 10(3):985-995.

54. Velazquez JL, Thompson CL, Barnes EM Jr et al. Distribution and lateral mobility of GABA/benzodiazepine receptors on nerve cells. J Neurosci 1989; 9(6):2163-2169.

55. Schlessinger J, Schechter Y, Cuatrecasas P et al. Quantitative determination of the lateral diffusion coefficients of the hormone-receptor complexes of insulin and epidermal growth factor on the plasma membrane of cultured fibroblasts. Proc Natl Acad Sci USA 1978; 75:5353-5357.

56. Maxfield FR, Willingham MC, Pastan I et al. Binding and mobility of the cell surface receptors for 3,3',5-triiodo-L-thyronine. Science 1981; 211:63-65.

57. Gorospe WC, Conn PM. Membrane fluidity regulates development of gonadotrope desensitization to GnRH. Mol Cell Endocrinol 1987; 53(1-2):131-140.

58. Yamada K, Carpentier JL, Cheatham B et al. Role of the transmembrane domain and flanking amino acids in internalization and down-regulation of the insulin receptor. J Biol Chem 1995; 270(7):3115-3122.

59. Goncalves E, Yamada K, Thatte HS et al. Optimizing transmembrane domain helicity accelerates insulin receptor internalization and lateral mobility. Proc Natl Acad Sci USA 1993; 90:5762-5766.

60. McClain DA, Maegawa H, Lee J et al. A mutant insulin receptor with defective tyrosine kinase displays no biologic activity and does not undergo endocytosis. J Biol Chem 1987; 262:14663-14671.

61. Livneh E, Benveniste M, Prywes R et al. Large deletions in the cytoplasmic kinase domain of the epidermal growth factor receptor do not affect its lateral mobility. J Cell Biol 1986; 103:327-331.

62. Segaloff DL, Puett D, Ascoli M. The dynamics of the steroidogenic response of perfused Leydig tumor cells to human chorionic gonadotropin, ovine luteinizing hormone, cholera toxin, and adenosine 3',5'-cyclic monophosphate. Endocrinology 1981; 108(2):632-638.

63. Bourdage RJ, Fitz TA, Niswender GD. Differential steroidogenic responses of ovine luteal cells to ovine luteinizing hormone and human chorionic gonadotropin. Proc Soc Exp Biol Med 1984; 175(4):483-486.

64. Mock EJ, Niswender GD. Internalization of ovine luteinizing hormone/human chorionic gonadotropin recombinants: differential effects of the alpha- and beta-subunits. Endocrinology 1983; 113(1):265-269.

65. Mock EJ, Niswender GD. Differences in the rates of internalization of ^{125}I-labeled human chorionic gonadotropin, luteinizing hormone, and epidermal growth factor by ovine luteal cells. Endocrinology 1983; 113(1):259-264.

66. Niswender GD, Roess DA, Sawyer HR et al. Differences in the lateral mobility of receptors for luteinizing hormone (LH) in the luteal plasma membrane when occupied by ovine LH versus human chorionic gonadotropin. Endocrinol 1985; 116:164-169.

67. Roess DA, Niswender GD, Barisas BG. Cytocholasins and colchicine increase the lateral mobility of human chorionic gonadotropin-occupied luteinizing hormone receptors on ovine luteal cells. Endocrinol 1988; 122:261-269.

68. Roess DA, Rahman NA, Kenny N. Molecular dynamics of luteinizing hormone receptors on rat luteal cells. Biochim Biophys Acta 1992; 1137:309-316.

69. Philpott CJ, Rahman NA, Kenny N et al. Rotational dynamics of luteinizing hormone receptors and MHC class I antigens on murine Leydig cells. Biochim Biophys Acta 1995; 1235(1):62-68.

70. Prossnitz ER, Kim CM, Benovic JL et al. Phosphorylation of the N-formyl peptide receptor carboxyl terminus by the G protein-coupled receptor kinase, GRK2. J Biol Chem 1995; 270(3):1130-1137.

71. Pei G, Kieffer BL, Lefkowitz RJ et al. Agonist-dependent phosphorylation of the mouse delta-opioid receptor: involvement of G protein-coupled receptor kinases but not protein kinase C. Mol Pharmacol 1995; 48(2):173-177.

72. Eason MG, Moreira SP, Liggett-SB. Four consecutive serines in the third intracellular loop are the sites for β-adrenergic receptor kinase-mediated phosphorylation and desensitization of the α2A-adrenergic receptor. J Biol Chem 1995; 270(9):4681-4688.

73. Heurich RO, Buggy JJ, Vandenberg MT et al. Glucagon induces a rapid and sustained phosphorylation of the human glucagon receptor in Chinese hamster ovary cells. Biochem Biophys Res Commun 1996; 220(3):905-910.

74. Pals-Rylaarsdam R, Xu Y, Witt-Enderby P et al. Desensitization and internalization of the m2 muscarinic acetylcholine receptor are directed by independent mechanisms. J Biol Chem 1995; 270(48):29004-29011.

75. Ozcelebi F, Holtmann MH, Rentsch RU et al. Agonist-stimulated phosphorylation of the carboxyl-terminal tail of the secretin receptor. Mol Pharmacol 1995; 48(5):818-824.

76. Sohlemann-P, Hekman-M, Puzicha-M et al. Binding of purified recombinant β-arrestin to guanine-nucleotide-binding-protein-coupled receptors. Eur J Biochem 1995; 232(2):464-472.

77. Ozcelebi F, Miller LJ. Phosphopeptide mapping of cholecystokinin receptors on agonist-stimulated native pancreatic acinar cells. J Biol Chem 1995; 270(7):3435-3441.

78. Lohse MJ, Andexinger S, Pitcher J et al. Receptor-specific desensitization with purified proteins. Kinase dependence and receptor specificity of β-arrestin and arrestin in the β2-adrenergic receptor and rhodopsin systems. J Biol Chem 1992; 267(12):8558-8564.

79. Garcia-Higuera I, Mayor F Jr. Rapid agonist-induced β-adrenergic receptor kinase translocation in C6 glioma cells. FEBS Lett 1992; 302(1):61-64.

80. Klueppelberg UG, Gates LK, Gorelick FS et al. Agonist regulated phosphorylation of the pancreatic cholecystokinin receptor. J Biol Chem 1991; 266:2403-2408.

81. Lutz MP, Pinon DI, Gates LK et al. Control of cholecystokinin receptor dephosphorylation in pancreatic acinar cells. J Biol Chem 1993; 268:12136-12142.

82. Jans DA, Hemmings BA. cAMP-dependent protein kinase activation affects vasopressin V₂-receptor number and internalization in LLC-PK₁ renal epithelial cells. FEBS Lett 1991; 281:267-271.

83. Johnson GL, Dhanasekaran N. The G-protein family and their interactions with receptors. Endocrine Reviews 1992; 10(3):317-331.

84. Sibley DR, Strasser RH, Caron MG et al. Regulation of transmembrane signaling by receptor phosphorylation. Cell 1986; 48:913-922.

85. Benovic JC, Kuhn H, Weyand I et al. Functional desensitization of the isolated β-adrenergic receptor by the β-adrenergic receptor kinase: potential role of an analog of the retinal protein arrestin (48-kDa protein). Proc Natl Acad Sci USA 1987; 84:8879-8882.

86. Meier T, Perez GM, Wallace BG. Immobilization of nicotinic acetylcholine receptors on mouse C2 myotubes by agrin-induced protein tyrosine phosphorylation. J Cell Biol 1995; 131(2):441-451.

87. Smith PR, Stoner JC, Viggiano SC et al. Effects of vasopressin and aldosterone on the lateral mobility of epithelial Na⁺ channels in A6 epithelial cells. J Memb Biol 1995; 147(2):195-205.

88. Griffin FM Jr, Mullinax PJ. Effects of differentiation in vivo and of lymphokine treatment in vitro on the mobility of C3 receptors of human and mouse mononuclear phagocytes. J Immunol 1985; 135(5):3394-3397.

89. Wilden U, Hall SW, Kuhn H. Phosphodiesterase activation by photoexcited rhodopsin is quenched when rhodopsin is phosphorylated and binds the intrinsic 48-kDa protein of rod outer segments. Proc Natl Acad Sci USA 1986; 83:1174-1178.

90. Bennett N, Sitaramayya A. Inactivation of photoexcited rhodopsin in retinal rods: the roles of rhodopsin kinase and 48-kDa protein (arrestin). Biochemistry 1988; 27:1710-1715.

91. Sterne-Marr R, Gurevich VV, Goldsmith P et al. Polypeptide variants of β-arrestin and arrestin 3. J Biol Chem 1993; 268(21):15640-15648.

92. Freedman NJ, Liggett SB, Drachman DE et al. Phosphorylation and desensitization of the human β1-adrenergic receptor. Involvement of G protein-coupled receptor kinases and cAMP-dependent protein kinase. J Biol Chem 1995; 270(30):17953-17961.

93. Dawson TM, Arriza JL, Jaworsky DE et al. β-adrenergic receptor kinase-2 and β-arrestin-2 as mediators of odorant-induced desensitization. Science 1993; 259(5096):825-829.

94. Hekman M, Bauer PH, Sohlemann P et al. Phosducin inhibits receptor phosphorylation by the β-adrenergic receptor kinase in a PKA-regulated manner. FEBS Lett 1994; 343(2):120-124.

95. Pitcher JA, Payne ES, Csortos C et al. The G-protein-coupled receptor phosphatase: a protein phosphatase type 2A with a distinct subcellular distribution and substrate specificity. Proc Natl Acad Sci USA 1995; 92(18):8343-8347.

96. Kelleher DJ, Pessin JE, Ruoho AE et al. Phorbol ester induces desensitization of adenylate cyclase and phosphorylation of the β-adrenergic receptor in turkey erythrocytes. Proc Natl Acad Sci USA 1984; 81:4316-4320.

97. Kwatra MM, Hosey MM. Phosphorylation of the cardiac muscarinic receptor in intact chick neart and its regulation by muscarinic agonist. J Biol Chem 1986; 261:12429-12432.

98. Kwatra MM, Leung E, Maan AC et al. Correlation of agonist-induced phosphorylation of chick heart muscarinic receptors with receptor desensitization. J Biol Chem 1987; 262:16314-16321.

99. Robinson MS. The role of clathrin, adaptors and dynamin in endocytosis. Curr Opin Cell Biol 1994: 6(4):538-544.

100. Fischer-von-Mollard G, Stahl B, Li C et al. Rab proteins in regulated exocytosis. Trends Biochem Sci 1994; 19(4):164-168.

101. Jesaitis AJ, Bokoch GM, Tolley JO et al. Lateral segregation of neutrophil chemotactic receptors into actin- and fodrin-rich plasma membrane microdomains depleted in guanyl nucleotide regulatory proteins. J Cell Biol 1988; 107:921-928.

102. Jesaitis AJ, Tolley JO, Bokoch GM et al. Regulation of chemoattractant receptor interaction with transducing proteins by organizational control in the plasma membrane of human neutrophils. J Cell Biol 1989; 109:2783-2790.

103. Jesaitis AJ, Tolley JO, Allen RA. Receptor-cytoskeleton interactions and membrane traffic may regulate chemoattractant-induced superoxide production in human granulocytes. J Biol Chem 1986; 261:13662-13669.

104. Johansson B, Wymann MP, Holmgren-Peterson K et al. N-formyl peptide receptors in human neutrophils display distinct membrane distribution and lateral mobility when labeled with agonist and antagonist. J Cell Biol 1993; 121:1281-1289.

105. Pavo I, Jans DA, Peters R et al. A vasopressin antagonist that binds to the V_2-receptor of LLC-PK$_1$ renal epithelial cells is highly laterally mobile but does not effect ligand-induced receptor immobi-

lization. Biochim Biophys Acta 1994; 1223:240-246.

106. Schmidt JM, Ohlenschlager O, Ruterjans H et al. Conformation of [8-arginine]vasopressin and V1 antagonists in dimethyl sulfoxide solution derived from two-dimensional NMR spectroscopy and molecular dynamics simulation. Eur J Biochem 1991; 201:355-371.

107. Lutz W, Londowski JM, Sanders M et al. A vasopresin analog that binds but does not activate V1 or V2 vasopressin receptors is not internalized into cells that express V1 or V2 receptors. J Biol Chem 1992; 267:1109-1115.

108. Eggena P, Lu M, Buku A. Internalization of fluorescent vasotocin-receptor agonist and antagonist in the toad bladder. Am J Physiol 1990; 259:C462-C470.

109. Fonseca MI, Button DC, Brown RD. Agonist regulation of α1B-adrenergic receptor subcellular distribution and function. J Biol Chem 1995; 270(15):8902-8909.

110. Maggio R, Barbier P, Toso A et al. Sodium nitroprusside induces internalization of muscarinic receptors stably expressed in Chinese hamster ovary cell lines. J Neurochem 1995; 65(2):943-946.

111. Hermans E, Octave JN, Maloteaux JM. Receptor mediated internalization of neurotensin in transfected Chinese hamster ovary cells. Biochem Pharmacol 1994; 47(1):89-91.

112. Gerard NP, Gerard C. Receptor-dependent internalization of platelet-activating factor. J Immunol 1994; 152(2):793-800.

113. Hunyady L, Merelli F, Baukal AJ et al. Agonist-induced endocytosis and signal generation in adrenal glomerulosa cells. A potential mechanism for receptor-operated calcium entry. J Biol Chem 1991; 266(5):2783-2788.

114. Mantey S, Frucht H, Coy DH et al. Characterization of bombesin receptors using a novel, potent, radiolabeled antagonist that distinguishes bombesin receptor subtypes. Mol Pharmacol 1993; 43(5):762-774.

115. Mantyh PW, Allen CJ, Ghilardi JR et al. Rapid endocytosis of a G protein-coupled receptor: substance P evoked in-

ternalization of its receptor in the rat striatum in vivo. Proc Natl Acad Sci USA 1995; 92(7):2622-2626.

116. Widmann C, Dolci W, Thorens B et al. Agonist-induced internalization and recycling of the glucagon-like peptide-1 receptor in transfected fibroblasts and in insulinomas. Biochem J 1995; 310(1):203-214.

117. Conklin BR, Bourne HR. Structural elements of Gα subunits that interact with Gβγ, receptors, and effectors. Cell 1993; 73:631-641.

118. Birnbaumer L. G proteins in signal transduction. Annu Rev Pharacol Toxicol 1990; 30:675-705.

119. Clapham DE, Neer EJ. New roles for G-protein βγ-dimers in transmembrane signalling. Nature 1993; 365:403-406.

EVIDENCE FOR THE ROLE OF IMMOBILIZATION OF LIGAND-OCCUPIED MEMBRANE RECEPTORS IN SIGNAL TRANSDUCTION

A. INTRODUCTION

As we saw in chapter 6, immobilization of hormone-occupied receptors plays an important role in downregulating response subsequent to hormonal stimulation in GTP-binding protein activating systems.[1-3] However, receptor immobilization also appears to play an important role in eliciting the stimulatory signal in other receptor systems. As will be discussed below, these include tyrosine kinase receptors, as already hinted at in chapter 4, as well as receptors mediating binding to the cell substratum or cell-cell interaction. In these cases, immobilization of receptors through dimerization/aggregation and/or complexation with cytosolic, cytoskeletal or extracellular components appears to represent the primary stimulus triggering signal transduction. Receptor movement within the plane of the membrane before ligand engagement, however, is required to effect receptor dimerization/aggregation and/or bring receptors to the sites of interaction with extracellular components or components on the surfaces of other cells and hence is also essential to signal transduction in adhesion/cell-cell recognition responses.

B. RECEPTOR IMMOBILIZATION IN TYROSINE KINASE RECEPTOR SIGNALING

Based on direct measurements, tyrosine kinase receptors appear to be largely immobile at physiological temperature (see chapter 4; Fig. 4.4; Table 4.1).[3-8] While the kinetics of receptor internalization probably play a role (see chapter 6, section E),[2,3,5,7,9-11] this appears to relate largely to the tyrosine kinase receptor signal transduction mechanism, which requires both receptor dimerization occurring within the plane of the membrane (see Fig. 4.5) and the association of a variety of cytosolic signaling components, such as phospholipase Cγ, phosphatidylinositol 3-kinase, etc. (see Figs. 4.6 and 4.11).[12-14] Dimerization is required in order to enable intermolecular receptor phosphorylation of the associated monomers to

The Mobile Receptor Hypothesis: The Role of Membrane Receptor Lateral Movement in Signal Transduction, by David A. Jans. © 1997 R.G. Landes Company.

occur.[15-17] Ligand binding by the epidermal growth factor (EGF) receptor leads to rapid temperature-dependent receptor microclustering on the cell surface, as shown using donor photobleaching fluorescence resonance energy transfer and imaging techniques[18] and, consistent with this, constitutively active mutant forms of the EGF and EGF-related p185[neu] (Neu, gene product of the HER2 or c-ErbB-2 protooncogene) receptors are predimerized.[19-21] Analogously, clustering of insulin receptors occurs upon insulin addition, and antibodies that induce receptor aggregation concomitantly activate receptor kinase activity.[22,23]

That tyrosine kinase receptor dimerization/microaggregation effects receptor immobilization (see also chapter 4, section D) has been shown by direct measurements for:

i.   the Neu receptor, where activation induced either by dimerizing mutations or by agonistically functioning antibodies results in a marked reduction in the receptor mobile fraction,[20] whereby association with cytoskeletally-linked structures appears to be involved[20] (see also ref. 24).

ii. the EGF receptor, where hormone-occupied receptor has been shown to diffuse at a rate almost four times slower than the unoccupied receptor,[25] and the high-affinity receptor, responsible for full biological response on A431 cells, appears to be immobile (see Table 4.1)[8] due to the fact that it is in a predimerized or oligomerized state.[18]

iii. the nerve growth factor (NGF) receptor, which is preclustered and immobile on responsive cells, in contrast to those on nonresponsive cells, implying that immobilization of NGF receptors prior to ligand binding may be essential to signal transduction.[26]

iv. the platelet-derived growth factor (PDGF) receptor, where the addition of PDGF to Fab-antibody-labeled receptor reduces the receptor mobile fraction (see Table 4.1 and legend), indicating that receptor immobilization occurs subsequent to, and as a result of, ligand-dependent receptor dimerization.[27] High PDGF concentrations have also been shown to result in a reduced mobile fraction, presumably resulting from increased receptor aggregation relative to that at lower concentrations.[28]

It seems clear that receptor immobilization is central to tyrosine kinase receptor-mediated signal transduction. The kinetics of signal transduction in tyrosine kinase receptor cells systems are slower at lower temperatures (see refs. 9 and 29-34) despite the fact that the receptor mobile fractions are very high (see Fig. 4.4; Table 4.1), which presumably reflects the fact that receptor mobility in this specific context is not essential to signal transduction activity. This is in contrast to the situation with GTP-binding protein coupling receptors, where only mobile receptors appear to participate in signaling (see chapter 5; Figs. 5.1, 5.2 and 5.4).[2,3,35-38] It should be stressed that immobile receptors as such, of course, do not represent the stimulatory signal in the case of tyrosine kinase receptors, but receptor immobilization appears to be a direct consequence of ligand-induced dimerization and is probably also essential for the subsequent association of signaling molecules from the cytosolic phase, which is necessary for signal transduction.

It is clear that lateral movement of hormone-occupied receptor is of course required to bring receptors into contact with one another as a prelude to the immobilizing signal transduction step of dimerization (see, however, refs. 8,26).[13,39,40] In vitro experiments, for example, show conclusively that immobilization of the EGF receptor prior to hormone addition prevents both dimerization and receptor kinase activation upon ligand addition.[41] Thus, movement of the ligand-occupied tyrosine kinase receptor is necessary to achieve receptor dimerization initially, but subsequent receptor immobilization appears to be mechanistically integral to communication of the stimulatory signal represented by hormone binding.

C. RECEPTOR IMMOBILIZATION IN SIGNALING BY Fc RECEPTORS

Fc receptors—receptors which specifically bind the Fc portion of antibodies—play a crucial role in the activation of a variety of cells of the immune system—being integrally involved in functions such as antigen presentation and phagocytosis by macrophages—mast cell degranulation and histamine release, and induction of cytokine production by natural killer cells. In their signaling role in mast cell/ basophil degranulation, Fc receptors appear to resemble tyrosine kinase receptors quite closely in terms of their requirement for receptor aggregation in order for the stimulatory signal represented by ligand binding to be transduced at the level of the membrane.[42-49] In the case of the receptors for immunoglobulin (Ig) G (Fcγ R) and IgE (Fcε R), receptor dimerization/oligomerization triggers a variety of signaling events (see ref. 50),[42,44-46,48-59] including the association of soluble tyrosine kinases such as p53/56[lyn], p59[fyn], and p72[syk], the initiation of tyrosine phosphorylation kinase cascades, an increase in intracellular Ca^{2+}, stimulation of inositol 1,4,5-trisphosphate production through phosphorylation/activation of phospholipases Cγ and A_2, and activation of forms of the Ca^{2+}, phospholipid-dependent protein kinase (PK-C), including the δ isoform. Tyrosine and serine/threonine phosphorylation of receptor subunits effected by soluble tyrosine kinases (above)[48,49,55,57,59] and PK-C and other serine/threonine kinases[48,50,52,55] is an early response to and direct consequence of receptor aggregation (see also ref. 60). Signaling events subsequent to receptor dimerization thus appear to occur exclusively through the participation of cytosolic signaling components, GTP-binding protein activation not appearing to be involved.[44,52]

That receptor dimerization/aggregation is a key step in inducing the stimulatory signal has been shown for the low-affinity IgG (Fcγ RII/RIII) and high-affinity IgE (Fcε RI) receptors in experiments in which monomeric ligand (IgG or IgE, respectively) effects essentially negligible activation, but induction of receptor aggregation with multivalent anti-receptor antibody in the absence of ligand can stimulate cells (see ref. 50; see Table 7.1; Figs. 7.1 and 7.2).[42,47,49,61-63] Tyrosine phosphorylation of receptor-associated proteins, including phospholipase Cγ and Shc, initiates the subsequent cascade of biochemical changes.[49,50] Clearly, a number of the signaling events and molecules involved are identical to those for tyrosine kinase receptors (see chapter 4; Figs. 4.6 and particularly 4.11); the most obvious difference is that the Fc receptors do not possess intrinsic tyrosine kinase activity, relying instead on the association of cytosolic tyrosine kinases to communicate the signal generated by the initial dimerization step and hence in this regard more closely resemble cytokine receptors. In structural terms, Fc receptors comprise 2-4 subunits, of which the α-subunit contains a single transmembrane domain and large extracellular domain with 2-3 Ig-like domains involved in ligand binding. The human Fcε RI, for example (see also Table 3.2), comprises one α-, one β- and two γ-subunits, of which the α-subunit contains a 20 amino acid cytoplasmic domain, the β-subunit possesses four transmembrane domains and two large intracellular loops of 59 and 43 amino acids, respectively, and the γ-subunit has a single transmembrane domain with a 36 amino acid cytoplasmic domain.[44,45]

Receptor immobilization upon dimerization and activation has been demonstrated directly for the Fcε RI receptor using the fluorescence photobleaching recovery technique (see Table 7.1; Figs. 7.1 and 7.2 below).[63-66] Schlessinger et al[66] were the first to show that the aggregated Fcε RI receptor was essentially immobile. This has since been confirmed by others (e.g., refs. 63-67), while a number of studies have shown that receptors occupied by monomeric (nonactivating, nondimerizing) IgE molecules are highly mobile, at least at room temperature (mobile fraction 0.72-0.85; see Table 2.3).[63-65,67,68] Receptor immobilization upon cell activation has been estimated to

Table 7.1. Lateral mobility of high-affinity IgE (Fcε RI) receptors as influenced by ligand-occupation and effect on degranulation activity and Ca^{2+} mobilization in rat basophilic leukemia cells[63]

Ligand	Fcε RI Mobile Fraction[x] (f) (25°C)[#]	[Ca^{2+}] response Activity (% max)	Degranulation [^{3}H]5HT release (% max)
none	0.85	5	< 10
100 nM anti-IgE mAB (B1E3)	0.57	6	7
20 nM (DCT)$_2$-cys (bivalent ligand)	0.58	8	
100 nM B1E3 + 20 nM (DCT)$_2$-cys	0.13	94	88
DNP$_{27}$-BSA (multivalent ligand)	0.20	100	100

[^{3}H]5HT, [5-1,2-^{3}H(N)]hydroxytryptamine binoxalate; DCT, [(2,4-DNP)amino]caproyl]-L-tyrosyl; (DCT)$_2$-cys, N,N'-bis[DCT]-L-cystine; DNP$_{27}$-BSA, DNP conjugated to bovine serum albumen (molar ratio of 1 mol BSA to 27 mol DNP).
[x]The Fcε RI receptor was labeled with a fluorescein-labeled monoclonal anti-DNP (dinitrophenol) IgE.
[#]The apparent lateral diffusion coefficient in all cases was about 2×10^{-10} cm^2/sec.[63]

Fig. 7.1. Relationship between degranulation activity ([5-1,2-^{3}H(N)]hydroxytryptamine binoxalate - 5HT - release) and the mobile fraction of the high-affinity IgE receptor Fcε RI in rat basophilic leukemia cells (l). Results for the activating antibody (A2) correspond to concentrations of 300, 60, 12 and 2.4 nM antibody, respectively, for increasing mobile fraction and that for the non-activating antibody (B5), 300 nM (see ref. 65). Lateral mobility measurements were performed using rhodamine-labeled monomeric IgE and the fluorescence photobleaching recovery technique.[65]

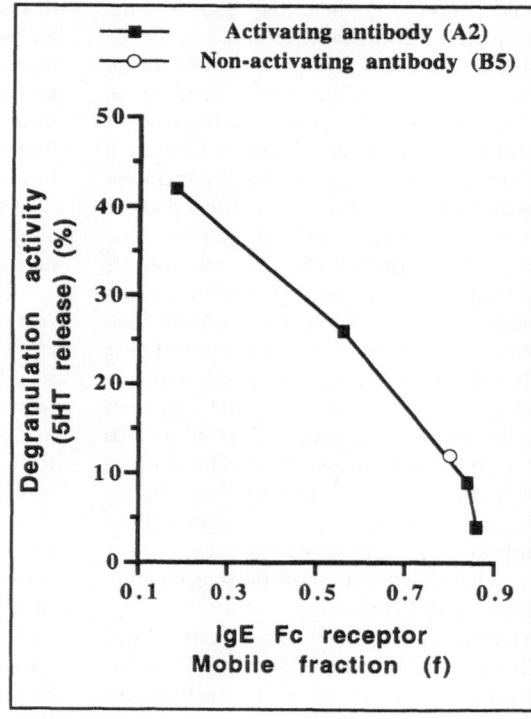

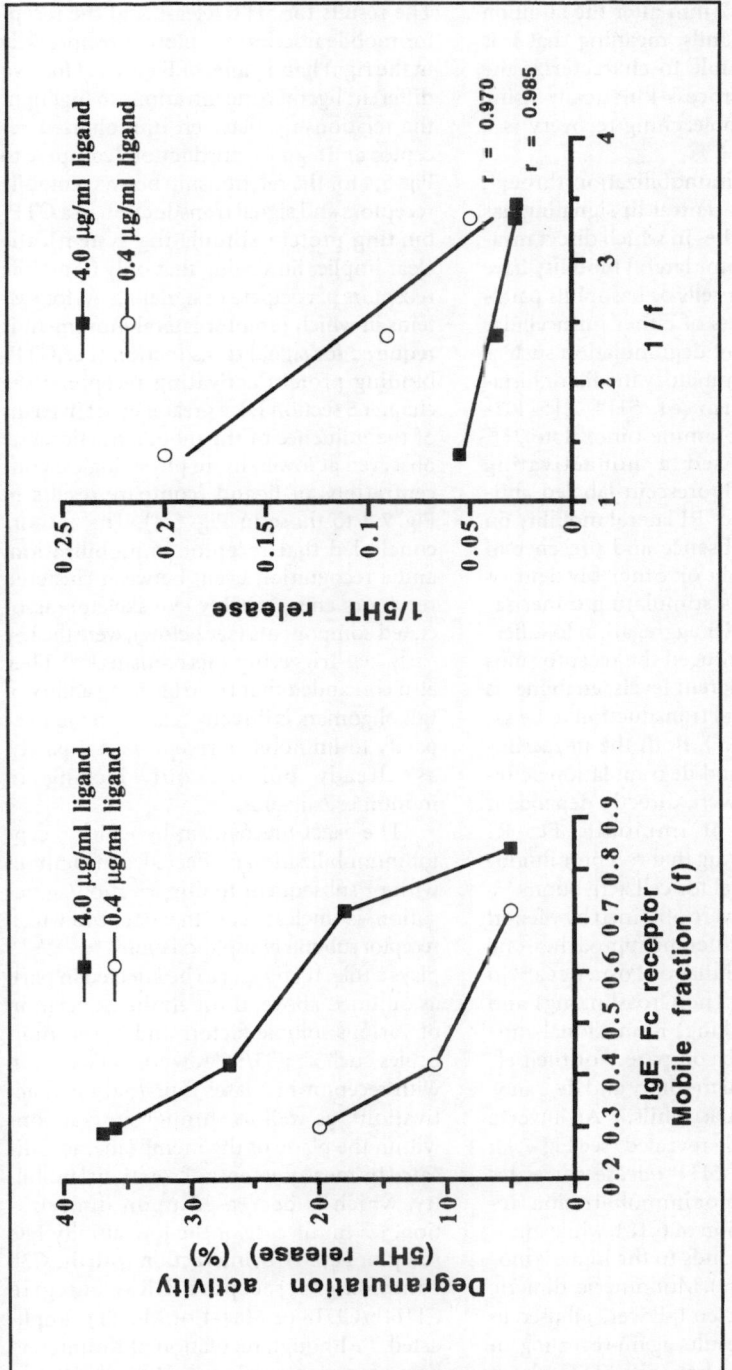

Fig. 7.2. Relationship between degranulation activity ([[5-1,2-^{3}H(N)]]hydroxytryptamine binoxalate - 5HT - release) and the mobile fraction of the high-affinity IgE receptor Fcϵ RI in rat basophilic leukemia cells (II). Rhodamine-labeled monomeric (two different preparations), dimeric and oligomeric (two different preparations) IgE ligands were used at two different concentrations as indicated.[64] Results in the right panel are plotted reciprocally (compare to Fig. 5.4), with the regression coefficients indicated. It should be noted that the highest mobile fraction (0.84, corresponding to 5% 5HT release at 4.0 µg/ml ligand) is not included in the double reciprocal plot; since there appears to be a threshold with respect to the number of immobilized receptors only above which signal transduction can occur.

occur in less than 2 min after the addition of crosslinking ligands, meaning that it is essentially impossible to characterize the immobilization process kinetically using fluorescence photobleaching recovery (see ref. 65).

That receptor immobilization through oligomerization is central in signaling has been shown in studies in which direct measurements of receptor lateral mobility have been made in mast cells or basophils parallel to measurements of either intracellular Ca^{2+} or indicators of degranulation such as release of the inflammatory mediator histamine in the form of 5HT ([5-1,2-H(N)]hydroxytryptamine binoxalate).[63-65] Posner et al[63] used a nonactivating (nondimerizing) fluorescein-labeled antibody to measure Fcε RI lateral mobility on basophils in the absence and presence of additional antibody or other bivalent or multivalent ligands stimulating dimerization and higher order aggregation to differing extents. This reduced the receptor mobile fraction to different levels, enabling its importance in signal transduction to be assessed (see Table 7.1). Both the intracellular Ca^{2+} response and degranulation activity (5HT release) were directly dependent on the fraction of immobile Fcε RI (Table 7.1),[63] implying that receptor immobilization is essential for cell activation.

Similar results were obtained by Menon et al[64,65] using two different approaches, one of which resembled that of Posner et al[63] in that nonactivating (noncrosslinking) and activating (crosslinking) monoclonal anti-IgE antibodies were compared for their effect on Fcε RI lateral mobility and degranulation activity in basophils.[65] An inverse relationship is clearly revealed (see Fig. 7.1); the most extensive 5HT release correlates with maximal receptor immobilization (receptor mobile fraction of 0.18), while minimal release corresponds to the highest mobile fraction (0.8-0.9). Monomeric, dimeric and oligomeric labeled IgE were all used in a parallel study,[64] results again revealing an inverse relationship between 5HT release and the receptor mobile fraction (Fig. 7.2).

The results for 5HT release and the receptor mobile fraction are plotted reciprocally in the right hand panel of Figure 7.2 for two different ligand concentrations to highlight the relationship between immobilized receptor and signal transduction (compare to Fig. 5.4 for the relationship between mobile receptors and signal transduction in a GTP-binding protein stimulating system), the clear implication being that only immobile receptors participate in signaling. As for systems in which receptor lateral movement is required for signal transduction (i.e. ,GTP-binding protein activating receptors; see chapter 5 section D), a greater effect in terms of the influence of the mobile fraction was observed at lower, more physiological concentrations of ligand (compare results in Fig. 7.2 to those in Fig. 5.4). The authors concluded that receptor immobilization, and a recognition event between clustered receptors and probably cytoskeleton-associated components (see below), were the key early cell-triggering mechanisms.[64,65] They also concluded that the triggering ability of IgE oligomers is directly related to their capacity to immobilize receptors, a capacity, as already pointed out, lacking in monomeric ligands.

The exact mechanism by which receptor immobilization is effected, concomitant with or subsequent to dimerization/aggregation, is unclear, as is the extent to which receptor subunit phosphorylation[48-50,52,55,57,59,60] plays a role. It appears to be effected in part, as outlined above, through the association of various soluble factors and kinase molecules such as p53/56[lyn], whose association with receptor increases four-fold upon activation[49] as well as through interactions within the plane of the membrane, as indicated by measurements of rotational mobility, which is decreased upon dimerization.[69,70] In the case of the low-affinity IgG receptor FcγRIIIB, interaction with the C3b complement receptor CR3 (integrin CD11b/CD18 or Mac-1 or Mo-1) is implicated.[71] Although regulation of the interaction by phosphorylation is implicated in reducing Fcγ RIIIB lateral mobility (see chapter 3 section A), this has not been demonstrated under conditions of receptor activation (see also chapter 3, section G).[71] A role for linkage to the cytoskeleton in the receptor immobilization process is also implicated by several observations, although inhibitors of cytoskeletal component biosynthesis such as cytochalasin D (actin filament perturbing agent) do not appear to significantly affect immobilization[64,65] except in the case of the study of Schlessinger et al[66] where treatment with cytochalasin B appeared to reduce mobility. Ligand induced aggregation of IgE receptors does not appear to be dependent on temperature or metabolic energy being able to occur at 4°C in the presence of sodium azide or deoxyglucose.[64]

Biochemical approaches have demonstrated that actin binding proteins (ABPs) directly link the high-affinity IgG receptor Fcγ RI, as well as other cell surface glycoproteins, to cytoskeletal actin microfilaments.[72] There is also clear evidence for association of Fcε RI with the membrane skeleton,[54,73,74] whereby receptor activation by multivalent ligands induces a significant increase in association with the detergent insoluble membrane skeleton in rat basophilic leukemia (RBL) cells[54,73,74] or in COS or P815 cells transfected with Fcε RI receptor subunits.[73] The critical event in initiating this interaction between the receptor and the membrane skeleton is the extent of aggregation of the receptor on the membrane surface.[54,73,74] The fact that this interaction appears to be retained within purified plasma membrane preparations indicates that all of the components necessary for linkage to the membrane skeleton are membrane-associated and that intracellular components are not required.[54] A model has been proposed in which the crosslinking of membrane proteins leads to their becoming nonspecifically enmeshed in a network of membrane skeletal proteins on either the outside and/or the inside of the membrane. It seems feasible to conclude that receptor dimerization/aggregation effects rapid receptor immobilization[65] largely through initiating tight linkage to the membrane skeleton, both of which are integral to signal transduction by Fc receptors (see above; Figs. 7.1 and 7.2; Table 7.1).[63-65]

Although signaling by Fc receptors is integrally dependent on receptor immobilization as we have just seen, the latter also seems to correspond well with receptor internalization as discussed at length in chapter 6 (see Fig. 6.4 in particular) with respect to GTP-binding protein activating, tyrosine kinase and other receptors.[2-12,75-78] Receptor internalization is known to be enhanced by crosslinking (immobilizing) antibodies (see ref. 67). Poo et al[68] used mouse and rat IgE ligands and the technique of electromigration to estimate receptor lateral movement and show that the mouse ligand induced both higher immobilization (mobile fraction of 0.56) compared to rat IgE (mobile fraction of 0.92) and significantly higher receptor internalization. Mao et al[67] used a variety of mutant derivatives, comprising mutations in each of the total of five extensive cytoplasmic domains of the four subunit Fcε RI. Although the mutations generally did not markedly affect lateral mobility of the monomeric or oligomeric-ligand occupied receptor (see chapter 3, section E, Table 3.2), the mutations which affected the extent of receptor internalization (see Fig. 7.3)[67] concomitantly increased the receptor mobile fraction (that is, reduced receptor immobilization) in the case of oligomer-occupied receptor. The selected results shown in Figure 7.3 indicate an inverse relationship between the fraction of mobile receptors and the extent of receptor internalization, concurring with the findings presented in chapter 6 (Fig. 6.4). As for the receptor classes discussed in chapter 6, Fc receptor immobilization is implicated to be a prerequisite for internalization; however, these results with respect to the Fcε RI system should not be interpreted purely in the context of desensitization subsequent to hormonal stimulation, since there may also be a specific intracellular signaling role of the internalized ligand-receptor complex (see ref. 67).[43,44,79-81] This aspect will not be

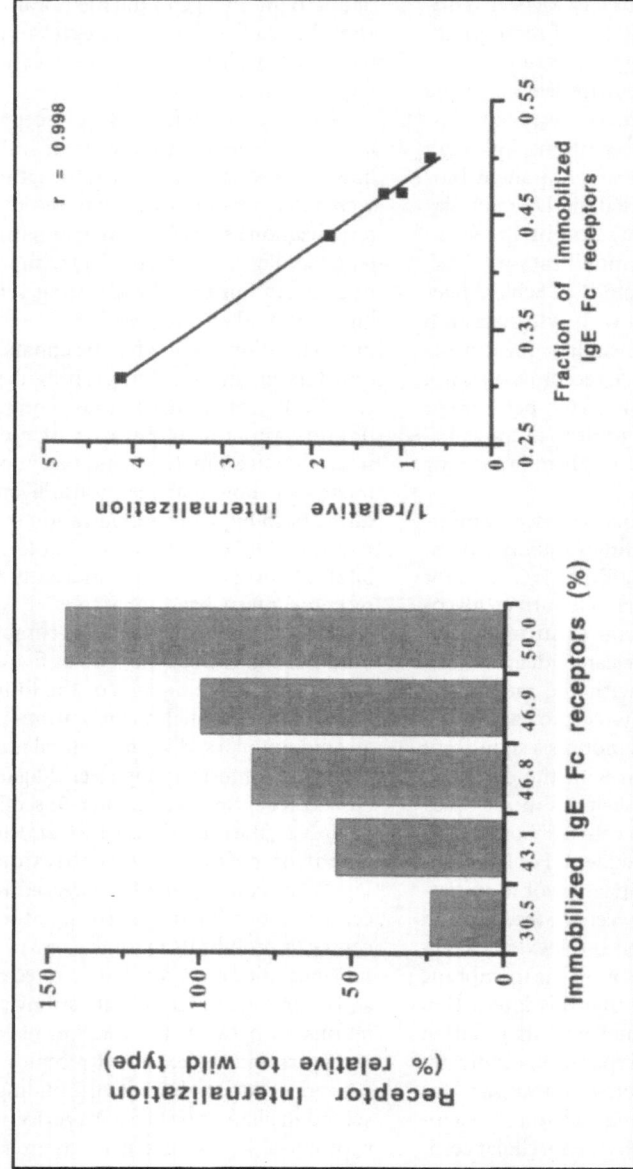

Fig. 7.3. Relationship between receptor immobilization and internalization of wild type and mutant forms of the high-affinity IgE receptor Fcε RI. Data plotted is for the following combinations of receptor structure (comprising α-, β- and two γ-subunits, with the wild type structure $\alpha_{20}\beta_{59,43}\gamma_{36}$, where the numbers represent the length in amino acids of the cytoplasmic domains: the β–subunit has two extensive amino-terminal and carboxy-terminal cytoplasmic domains of 59 and 43 amino acids, respectively): $\alpha_4\beta_{59,43}\gamma_{36}$; $\alpha_4\beta_{59,5}\gamma_{36}$; $\alpha_{20}\beta_{59}\gamma_6$; $\alpha_{20}\gamma_6$ (the carboxy terminal cytoplasmic domain of the β-subunit is replaced by the α-subunit cytoplasmic domain); and $\alpha_{20}\beta_{5,43}\gamma_{36}$ (see also Table 3.2).[67] The fraction of immobilized receptors is the percentage of mobile receptors (measured using monomeric ligand) becoming immobilized upon the addition of oligomeric ligand; for relative internalization, see ref. 67. The panel on the right shows the linear relationship when results are plotted reciprocally with respect to the relative internalization, with the regression coefficient (r) indicated. Compare the data in the left panel with that in Figure 6.4 for GTP-binding protein activating and tyrosine kinase receptors.

addressed here since these signaling events do not impinge on those at the level of the plasma membrane, but may be quite relevant in Fc-receptor-mediated phagocytosis by macrophages and neutrophils. Fcε RI internalization is cytochalasin sensitive,[67] again implying an active role of the cytoskeleton in mediating receptor immobilization as a prelude to endocytosis, while phosphorylation is also implicated as having a role by the fact that phorbol ester treatment appears to be able to trigger Fcε RI internalization in the absence of receptor aggregation.[80]

The fact that lateral movement of unoccupied Fc receptors is required to bring receptors into contact with one another as a prelude to the immobilizing signal transduction step of dimerization upon ligand binding is implied by several studies.[66,82-84] Using direct measurements, Schlessinger et al[66] showed that excessive Fcε RI receptor aggregation and concomitant immobility[66] inhibits the degranulation response of peritoneal mast cells.[82-84] This clearly implies that complete receptor immobility blocks signal transduction. The sequence of events in signal transduction mediated by Fc receptors, where receptor movement in the absence of ligand precedes receptor immobility in its presence, is summarized in equations [1.10]-[1.12] in chapter 1 (section C). As we have seen, [1.10] is strictly effected through diffusion within the plane of the membrane; hence the requirement for lateral movement of receptor in the unbound state. Upon ligand binding/dimerization ([1.11]), the receptor dimer-ligand complex becomes immobilized to allow subsequent association of soluble signaling components, which occurs in the aqueous phase. While tyrosine kinase receptor dimerization and immobilization upon ligand binding is analogous (as is the subsequent association of signaling components from the cytosol), the fundamental difference appears to be that movement of the receptor in its hormone-bound rather than nonligand form is required (see equations [1.1], [1.3] and [1.4] in chapter 1, section C). As alluded to above,

the role of Fc receptors in cell-cell recognition (antigen presentation, etc.) is analogous to that of the CD2/LFA-3 receptor/coreceptor pair, dealt with in the next section (see below). The role of Fc receptors mediating adhesion reactions as in phagocytosis mediated by macrophages/neutrophils is analogous to interactions discussed in section E below with particular reference to the fibronectin receptor.

D. RECEPTOR IMMOBILIZATION IN CELL-CELL INTERACTION

Receptor immobilization appears to be mechanistically important in signal transduction by receptor/cell surface molecules which mediate cell-cell interaction. The adhesive interactions of cells with other cells and with the extracellular matrix are crucial to developmental processes and have a central role in the immune system.[85] Several families of cell-surface molecules are involved in these interactions, including receptors of the Ig superfamily such as CD2, LFA-3, CD4, CD8, the T-cell receptor (CD3) major histocompatibility complex (MHC) class I and II molecules, integrins—central to the dynamic regulation of adhesion and migration (see section E, below)—and the selectins which are prominent in lymphocyte and neutrophil interaction with vascular endothelium.[85]

While experimental results with respect to the lateral movement of these types of molecules are largely restricted to members of the Ig class of molecules (see Table 2.3), it seems reasonable to postulate that all of these types of molecules are similar with respect to the dynamics of their lateral movement in the plane of the membrane and its importance in cell-cell recognition. As already indicated, members of the first class of molecules belong to the Ig receptor superfamily, each comprising one or more molecules containing one to six Ig-like domains and a single transmembrane domain (or a glycosylphosphoinositol - GPI - anchor, in the case of certain LFA-3 isoforms).[85] Of the molecules discussed below, CD2, LFA-3, ICAM-1 and CD4 are all single subunit receptors; CD8 comprises two identical subunits and CD3 consists eight subunits of which at least six represent distinct forms (see ref. 85).

The best characterized proteins with respect to lateral mobility properties are the T-cell adhesion glycoprotein CD2 (or lymphocyte function-associated antigen-2 - LFA-2), involved in T cell adhesion and activation, and its specifically recognized glycoprotein ligand/coreceptor LFA-3 (lymphocyte function-associated antigen-3 or CD58),[86-90] which is present on the surface of interacting cells such as target cells (e.g., erythrocytes in rosette assays) or antigen-presenting cells such as macrophages.[85] Similar receptor-coreceptor pairs involved in cytotoxic T-cell-target cell or helper T cell-antigen presenting cell interaction include the T cell receptor and CD4 or CD8, which interact in concert with MHC class I or II, respectively, and LFA-1 which recognizes ICAM-1 or -2 (see refs. 88,91).[85] Many of these receptor/coreceptor pairs are present on the same cells and function cooperatively; e.g., CD2 signal transduction appears to require CD3 in mature T lymphocytes, although it appears to be able to signal independently of the T cell receptor in natural killer cells and thymocytes. All of these receptors are required to move laterally to the cellular site of cell-cell contact, prior to rapid immobilization in an anchoring role (see chapter 1, Fig. 1.4). Hence in this case, mobility of unliganded receptor is required both prior to response and in order to accentuate it, but receptor immobilization is essential to elicit the response itself. That receptor crosslinking—often in complexes with heterologous cell surface molecules—leads to immobilization and/or stimulation of signal transduction, in similar fashion to the Fc receptors discussed in the previous section, has been demonstrated for CD2,[86,89,92] the T cell receptor,[91] CD4[93,94] and LFA-3.[95]

That CD2 crosslinking leads to a marked reduction in the receptor mobile fraction has been shown in Jurkat cells, whereby the mobile fraction is less than 0.1, in contrast to a mobile fraction of 0.7 of Fab fragment-labeled receptor.[86] Consistent with this, bivalent ligand occupied receptor exhibits reduced mobility (apparent lateral diffusion coefficient of 1.8×10^{-10} cm^2/sec)[96] compared to Fab fragment-labeled CD2 (apparent lateral diffusion coefficient of 7.2×10^{-10} cm^2/sec),[86] although it should be pointed out that the former study[96] was carried out on transfected Chinese hamster ovary cells and hence not directly comparable to experiments on Jurkat cells. Several studies have demonstrated the direct role of mobilization of intracellular free Ca^{2+} upon T cell activation in this process (see also chapter 3, section G; Table 3.3).[86,89] The increase of intracellular Ca^{2+} appears to effect CD2 immobilization directly, and the kinetics of this subsequent to activation by paired antibodies are shown in Figure 7.4. As intracellular Ca^{2+} rises (Fig. 7.4, left panel), the receptor mobile fraction drops to levels of near total immobility (Fig. 7.4, right panel). Crosslinking and stimulation of the T cell receptor (CD3) of Jurkat cells has also been shown both to elevate intracellular Ca^{2+} and to immobilize CD2 (the mobile fraction is reduced to 0.54).[89]

Treatment with various agents such as the metal ion chelator, EGTA (ethylene glycol-bis[β-amino ethyl ether] N,N,N',N'-tetraacetic acid), and the Ca^{2+} ionophore, ionomycin, partially reverses the increase in intracellular Ca^{2+} and concomitantly reduces CD2 immobilization (see Fig. 7.4) resulting from antibody stimulation (see also Table 3.3).[86] Figure 8.4 (see chapter 8) shows the relationship between CD2 and intracellular Ca^{2+} with respect to various treatments perturbing the intracellular Ca^{2+} level. Trifluoperazine, a specific inhibitor of the Ca^{2+} binding protein, calmodulin, also reduces activation-induced CD2 immobilization effected by increased intracellular Ca^{2+}, implying that calmodulin activation is required.[90] CD2 immobilization is reversed about two hours after activation, but this can be slowed (by about 2-fold) by the

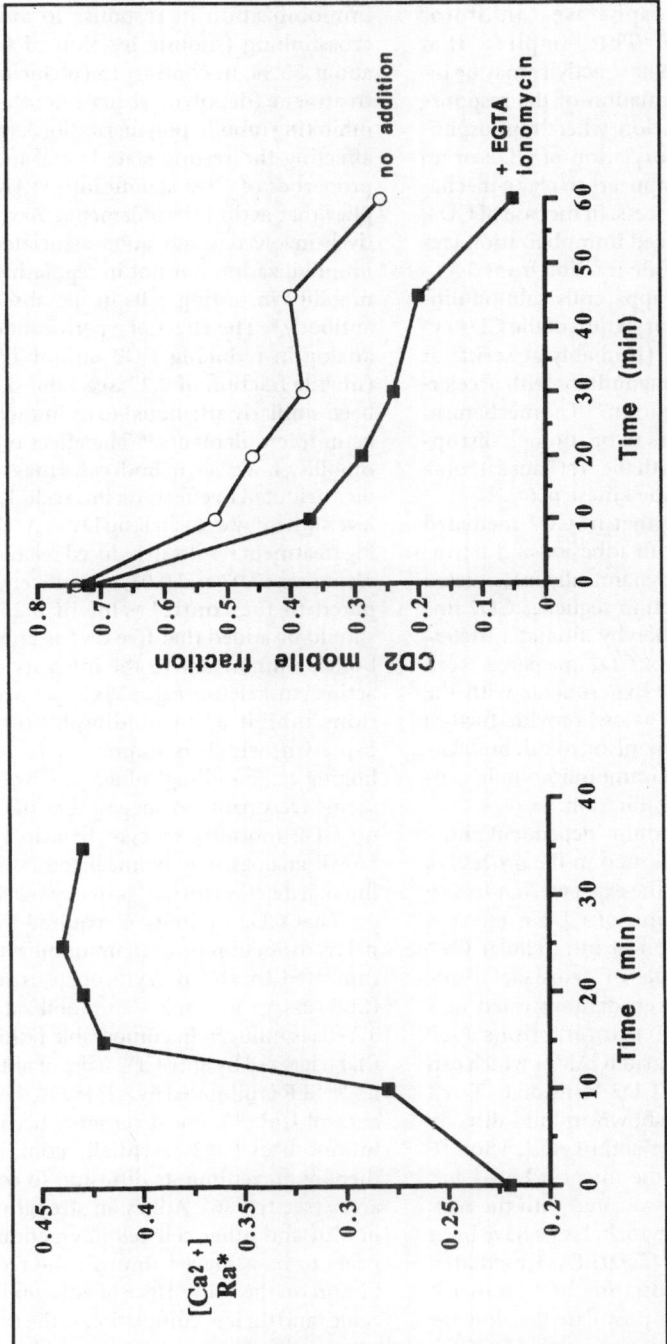

Fig. 7.4. Time course of CD2 immobilization (right panel) and Ca²⁺ mobilization (left panel) upon activation of Jurkat T cells using paired antibodies.[86] *Ra is the fraction of cells with [Ca²⁺] significantly greater than control resting cells (see ref. 86). The right panel also depicts accelerated CD2 immobilization in the presence of the Ca²⁺ chelator, EGTA (ethylene glycol-bis[β-amino ethyl ether] N,N,N',N'-tetraacetic acid), and the Ca²⁺ ionophore, ionomycin.

calcineurin phosphatase inhibitor cyclosporin A.[90] This implies that calcineurin phosphatase activity may be involved in downregulation of the response of CD2 immobilization, whereby phosphorylation/dephosphorylation of CD2 or an associated protein appears to play a mechanistic role in the process. In the case of CD4, phorbol ester induced immobilization (reduction of the mobile fraction from 0.8 to 0.5),[94] through an apparently calmodulin-dependent phosphorylation of the CD4 cytoplasmic domain (probably at serine at position 408), is concomitant with accelerated CD4 internalization.[97] The mechanism of this effect appears to be through disruption of a complex with the *src* (Raus sarcoma virus) family tyrosine kinase p56[lck].[97]

It thus appears that in CD2-mediated effects on T cells, both adhesion and activation appear to be dynamically interrelated in that T cell activation regulates CD2 immobilization and thereby ultimately determines the strength of CD2-mediated T cell adhesion to LFA-3. Experiments with the latter,[87,88] discussed below, provide further insight into the mechanistic role of both lateral movement and immobilization in cell-cell interaction and adhesion.

Although calmodulin-dependent phosphorylation is implicated in the process as mentioned above,[90] the exact mechanism by which immobilization of CD2 is effected through mobilization of intracellular Ca^{2+} is not clear. The role of cytoskeletal microfilaments has been demonstrated in a number of adhesive interactions (see ref. 86).[98,99] The integrin ICAM-1 which can act in concert with CD2 to mediate T-cell adhesion has been shown to bind directly to α-actinin and the actin cytoskeleton,[100] for example, while the integrin LFA-1 has been shown to be associated with the ABP talin.[101,102] Further, cytochalasins have been shown to decrease CD2/LFA-3-mediated sheep erythrocyte rosetting,[103,104] so that it seems reasonable to postulate that linkage to the cytoskeleton may be involved in CD2 immobilization. Consistent with this, cytochalasin D pretreatment reduces CD2

immobilization in response to antibody crosslinking (mobile fraction of 0.3) by about 50 %, in contrast to colchicine pretreatment (depolymerizing microtubules/inhibiting tubulin polymerization), without affecting the resting state lateral mobility properties of CD2 significantly.[86] This implies that actin microfilaments may be actively involved in activation-associated CD2 immobilization but not in regulating CD2 mobility in resting cells in the absence of antibody.[86] The effect of a permeant cAMP analog in reducing CD2 immobilization (mobile fraction of 0.29; see Table 3.3) has been similarly attributed to its influence on actin microfilaments.[86] The effect on CD2 mobility, however, in both cases may be better attributed to effects on intracellular Ca^{2+} levels whereby cytochalasin D or cAMP analog treatment results in reduced relative Ca^{2+} responses (0.07 and 0.03, respectively, compared to the control value of 0.25).[86] It should be added that free Ca^{2+} is known to both be important to the integrity of the actin cytoskeleton; e.g., µM Ca^{2+} concentrations inhibit actin binding by the ABP L-plastin, which is important for crosslinking or "bundling" of actin.[105,106] It thus seems reasonable to suggest that the effect on CD2 mobility of cytochalasin D and cAMP analogs may be mediated indirectly through the effects of Ca^{2+} on the cytoskeleton.

That CD2 mobility is required for signal transduction prior to immobilization is indicated by sheep erythrocyte rosetting (SER) assays, in which Ca^{2+} mobilization in T-cells is induced in comparable fashion to that triggered by anti-CD2 pairs of antibodies.[107] SER is inhibited by A23187 in the presence of Ca^{2+},[104] almost certainly because it immobilizes CD2 essentially completely, thereby preventing its diffusion to contact areas (see ref. 86). Adhesion strengthening in SER and other cell-cell interactions appears to be achieved through the redistribution on the cell surface of adhesion molecules and their accumulation at the cell-cell contact site. The kinetic results of Liu et al[86] (see Fig. 7.4) suggest that there is an interval between maximal Ca^{2+} response (at about 15 minutes) and maximal CD2 immobilization (between 45 and 60 minutes). This interval allows anisotropic redistribution of CD2 molecules, enabling them to migrate to the contact area and thus increase the numbers of CD2 molecules at the site, prior to their immobilization there. Subsequent immobilization of CD2 serves to maintain it at high density at the contact site, thereby increasing cellular adhesion strength.[86] The timing of CD2 immobilization in relationship to T cell activation modulated through cAMP-mediated or other signals may determine whether adhesion is enhanced or weakened under a specific set of conditions.

Experiments using an adhesion assay system to test the binding of Jurkat cells to egg phosphatidylcholine (PC) bilayers reconstituted with the two distinct anchorage isoforms of the CD2 ligand/coreceptor LFA-3 (see ref. 88)[87,108] support the idea that both receptor lateral mobility and receptor immobilization are essential for adhesion. The distinct transmembrane domain (TM)- and GPI-anchor-containing LFA-3 form, are essentially immobile (mobile fraction less than 0.1) and highly mobile (mobile fraction of 0.73), respectively, as shown by fluorescence photobleaching recovery measurements (see Table 7.2).[87] Jurkat cell adhesion to LFA-3-containing PC membranes was tested using static and laminar flow assays and found to be dependent on LFA-3 density and contact time. It was much more efficient on PC membranes containing the GPI-LFA-3 than those containing the TM-LFA-3 isoform (Table 7.2), with differences between the two less evident at high LFA-3 concentrations and longer contact times. The extent of maximal binding in static assays was about 10 times lower, and the LFA-3 density yielding half-maximal binding 15 times higher for the TM-isoform, relative to the GPI-isoform (Table 7.2). Similarly, the shearing force necessary to remove 50% of cells in laminar flow assays was about 10-fold higher for the GPI form, relative to the TM-derivative at LFA-3 densities up to 250 mol/μm^2

(Table 7.2).[87] At a higher density of 1500 mol/μm^2, the difference between the two isoforms was about 4-fold, with a greater than 40-fold difference if the adhesion assay was reduced to 5°C (see Table 7.2).[87] Both LFA-3 forms were identical in terms of cell binding (both low in terms of strength) when immobilized on plastic,[87] clearly implying that the differences in adhesion strengthening capabilities are directly attributable to the differences in lateral mobility properties. The results imply that the ability of a membrane receptor and its membrane-bound coreceptor to diffuse laterally enhances cell adhesion both by allowing accumulation of ligands in the cell contact area and by increasing the rate of receptor-ligand/coreceptor bond formation. As for the Fc receptor experiments described below (Fig. 7.2), and results for the vasopressin V$_2$-receptor (Fig. 5.4), lateral mobility appears to be a more critical factor in terms of adhesion strengthening under the more physiological conditions of lower coreceptor concentrations and weaker interactions, e.g., compare the differences in adhesion strength between the mobile and immobile LFA-3 isoforms at 50 and 250 mol/μm^2 (Table 7.2); see ref. 87.

A comparable type of study[88] using several adhesion molecules, including both CD2/LFA-3 and LFA-1/ICAM-1 combinations, led to similar conclusions; since a number of adhesive bonds between cells are necessary for stable interaction, movement of unliganded receptor is required to enable accumulation at the site of cell contact, while receptor immobilization at the site upon ligand binding ensures a stable bond. As a model for contact-induced receptor redistribution upon cell adhesion, McCloskey et al[109] employed an artificial system in which RBL cells, carrying fluorescent anti-DNP IgE bound to their surface Fcε RI receptors, were allowed to adhere to phospholipid vesicles containing a dinitrophenyl (DNP)-lipid, hapten. Time-dependent increase in the strength of adhesion paralleled the kinetics of accumulation of the adhesion-mediating antibody molecules at the zone of

Table 7.2. Lateral mobility of the adhesion ligand/coreceptor LFA-3 and its influence on adhesion by CD2-expressing Jurkat T-cells.[86]

| Protein | Treatment | Lateral mobility* | | Jurkat cell Adhesion° | | Shearing force removing 50% of cells (dynes/cm²) LFA-3 density (mol/mm²) | | | |
		D (x 10^{-10} cm²/s)	f	Maximum binding (%)	Density for 50% max binding (mol/mm²)	50^x	250^x	1500#	1500@
GPI LFA-3+	None	23	0.73	67	12	5	7.8	11.6	8
	crosslinking antibody	0	0	<0.10					
TM LFA-3+	None	0	0	7	180	<0.5	0.8	2.9	<0.2

*Values are for the apparent lateral diffusion coefficient (D) and mobile fraction (f) measured at 22°C at an LFA-3 density of 50-250 molecules/mm².
°Measurements at an LFA-3 density of 80 molecules/mm².
^{x}Force exerted after 20 minutes binding at 22°C.
#Force exerted after 3 minutes binding at 22°C.
@Force exerted after 3 minutes binding at 5°C.
+The two LFA-3 forms have either a glycosyl phosphoinositol (GPI) or transmembrane (TM) domain membrane anchor.

membrane-membrane contact as determined by various microscopic techniques,[109] results thus correlating well with those above.

As early as 1978, Bell[110,111] defined equations to describe the various receptor-mediated interactions in cell adhesion (see ref. 112) based on the observation that membrane-bound antibodies diffuse much more slowly than those in solution (by two orders of magnitude). These equations have been used to describe the series of events in cell-cell interaction/adhesion as two consecutive and reversible steps (see also equations [1.13]-[1.15] in section C of chapter 1) and have been modified slightly here:

$$R + CR \underset{d^-}{\overset{d^+}{\Longleftrightarrow}} R.CR'' \underset{i^-}{\overset{i^+}{\Longleftrightarrow}} R.CR * \quad [7.1],$$

where R, receptor, and CR, coreceptor are on different cells. $R.CR$ is the transient (denoted by '') receptor-coreceptor encounter complex and $R.CR^*$ is a stabilized (immobilized) complex; d^+ and d^- are the rates of formation and breakdown respectively of the encounter complex, essentially determined directly by the rates of diffusion of R and CR, respectively, while i^+ and i^- are the forward and reverse rate constants for the intrinsic reaction of formation of $R.CR^*$ from $R.CR''$, which is essentially the immobilization step. Since the concentration of the transient $R.CR''$ complex is small compared to R, CR and $R.CR^*$, the overall reaction of $R + CR \rightarrow R.CR^*$ can be expressed in terms of the rate constants:

$$k^+ \Longleftrightarrow \frac{d^+ i^+}{d^- + i^+} \quad [7.2]$$

and for the reverse reaction:

$$k^- \Longleftrightarrow \frac{d^- i^-}{d^- + i^+} \quad [7.3]$$

It is evident that if $i^+ >> d^-$, the transient encounter complex R.CR'' is more likely to proceed to $R.CR^*$ than to dissociate, k^+ tends to d^+, and the forward reaction becomes diffusion-limited.[87,110-112] It follows that under these circumstances the critical parameter for complex formation, and hence cell-adhesion, is lateral diffusion of R and CR to bring them into contact with one another

to allow initial interaction (i.e., formation of encounter complex $R.CR''$). Subsequent stabilization (formation of the immobile complex $R.CR^*$) is not directly influenced by the rates of diffusion of R and CR but is dependent on i^+, which is determined by the affinity of the receptor for its ligand/coreceptor and the cytoskeleton or other components in the cytosolic aqueous phase. Adhesion strengthening would appear to be directly proportional to the number of immobilized receptor-coreceptor complexes at the interaction site (see ref. 87 and Table 7.2), occurring through repetitions of the sequence of events in [7.1] (or [1.13]-[1.15]).

The conclusion from the kinetic analysis above is that, through its reduced mobility, the TM-LFA-3 isoform is about two to three orders of magnitude slower than the GPI-form in terms of the initial rate of encounter complex formation. The analysis also predicts a greater temperature dependence of adhesion mediated by the former, completely consistent with the experimental results (Table 7.2 and see ref. 87). It should not be forgotten that the receptor and coreceptor lateral movement required for interaction is of unliganded receptors; upon ligation the receptor-coreceptor complex becomes rapidly immobile which is essential for the signal transduction to proceed. Clearly, since lateral movement of the receptor-coreceptor complex would not assist in stabilizing the cell-cell interaction, receptor-coreceptor immobilization is a prerequisite for adhesion. Thus, although details differ, the requirement of both lateral mobility before and immobilization after ligation for signaling by cell-cell interaction molecules, such as CD2 and LFA-3, functionally resembles that for signal transduction by Fc receptors (section C above) and receptors mediating cell adhesion to the extracellular matrix (next section).

It seems reasonable to conclude that the lateral mobility properties of CD2 and LFA-3 on their respective cell types, as well as of the other similar molecules mentioned above, are instrumental in determining the rate of cell adhesion and adhesion strengthening. The two forms of LFA-3 with radically different mobility properties presumably play differential roles in cell-cell adhesion; one can speculate that the GPI-form is critical in mediating adhesion where long-range receptor movement is required, while the TM-form may be primarily important in situations where high receptor densities mean that an enhanced rate of bond formation would dominate and only short-range diffusion would be required for collision with the partner receptor (see ref. 87).

E. RECEPTOR IMMOBILIZATION IN CELL-ADHESION TO AN EXTRACELLULAR SUBSTRATUM

Cell adhesion to an immobile ligand/substrate is a biological process comparable to cell-cell interaction/recognition, the basic difference being that in the case of the latter, as we have just seen, the mobilities of the molecules mediating the interaction on each interacting cell need to be taken into account, whereas in the latter case only lateral mobility of the receptor molecule in the adhering cell needs to be considered. Cell adhesion can be seen as the reverse of cell migration, which is important in a variety of processes such as embryonic development (see refs. 113 and 114), as well as tumor cell metastasis.[115-117] Extracellular matrix components involved in cell adhesion include laminin, fibronectin (also known as LETS—large external transformation-sensitive - protein, or CSP—cell surface protein) and collagen.[85,101,115-118] Receptors such as integrins (mentioned in the previous section) mediating adhesion to the extracellular matrix possess multiple binding sites for one or more of these components.[85,119,120] Integrins comprise an α/β-subunit structure, whereby the α-subunit retains a single transmembrane domain and extensive extracellular domain, while the β-subunit is completely extracellular and noncovalently associated with the α-subunit.[85]

Cell anchorage to the extracellular matrix occurs at restricted sites of the membrane called focal or close contacts, where very little of the ventral membrane surface is actually in contact with the substratum. In contrast, motile cells interact uniformly with the substrate, with broad close contacts and very few focal contact sites. Adhesion stimulates a variety of changes in cell structure and shape, largely through the formation of contractile stress fibers including actin and other components.[121]

Duband et al[113] compared the lateral mobility of the fibronectin receptor (probably the $\alpha_4\beta_1$ integrin VLA-4 or LPAM-1 or CD49d/CD29 - or αVβ_1 - CD51/CD29; the "fibronectin receptor" $\alpha_5\beta_1$ - VLA-5 or ECMRV1 or CD-/CD29 - is primarily found on fibroblasts, platelets, and epithelial and endothelial cells)[85,122,123] using a labeled antibody probe to the β_1-receptor subunit in avian embryonic neural crest, somitic, dermal and heart cells at various locomotory stages, including motile and stationary cells cultured on fibronectin substrates. A clear finding was that the receptor mobile fraction was greatly reduced in focal contacts (mobile fraction of 0.13-0.18) in all types of cells, compared to in other areas of the cell membrane (mobile fraction of 0.57-0.84).[113] The apparent lateral diffusion coefficient did not vary significantly (values of 2.0-3.4 x 10^{-10} cm^2/sec),[113] while whether cells were motile or stationary did not appear to affect the results. It could be concluded that the receptor was immobilized at the points of contact with the substrate but mobile in other parts of the membrane. Schlessinger et al[124] produced concurring results using labeled fibronectin (CSP) to show essential immobility (mobile fraction of about 0.1 and apparent lateral diffusion coefficient of less than 5 x 10^{-12} cm^2/sec) of the fibronectin receptor in chicken embryo fibroblasts.

Duband et al[113] found that the fibronectin receptor mobile fraction varied according to temperature in neural crest cells and somitic fibroblasts, being highest (mobile fraction of 0.84 in the case of neural crest cells) at 37°C and lowest (0.54) at 15°C, correlating with increased motility at higher temperatures (see Table 2.3).[113] When stronger binding to the substratum and concomitant reduced motility was effected by culturing cells on polyclonal antibodies to the fibronectin receptor, strong receptor immobilization (51-63% of mobile receptors) was observed at all temperatures tested (see Fig. 7.5 for results for 37°C), the clear implication being that receptor immobilization was concomitant with cell adhesion (or alternatively, receptor mobility was required for motility).[113] As above, the apparent lateral diffusion coefficient was unchanged. A monoclonal antibody competing with fibronectin binding by the receptor was found to reduce fibronectin receptor immobilization (mobile fraction of 0.58 compared to the control of 0.13 in the absence of the inhibiting antibody, while not affecting the apparent lateral diffusion coefficient) and concomitantly inhibited cell adhesion.[113]

The results imply that most fibronectin receptors readily diffuse in the plane of the membrane in locomoting embryonic cells, this degree of lateral mobility correlating with the labile nature of the adhesions to the substratum required for high motility. In contrast, receptors in stationary cells are immobilized in focal contacts and fibrillar streaks which are in close, stable association with both extracellular and cytoskeletal structures (e.g., linkage to talin).[101,102,125,126] This provides firm anchorage to the substratum. That cytoskeletal components are immobilized at focal contacts has been shown directly for actin[127] and α-actinin,[128] and it seems reasonable to postulate that fibronectin receptor association with the actin cytoskeleton is the main mechanism of its immobilization.

Thus, in similar fashion to the CD2 and LFA-3 receptors described above, and adhesion receptors in general, unliganded fibronectin receptors are required to move laterally to the site of the membrane in contact with the substratum, prior to ligand engagement and rapid immobilization in their anchoring role. Receptor mobility is thus required both prior to interaction with the immobile ligand and in order to accentuate it (i.e., strengthen binding), but receptor immobilization is essential to elicit the response itself. This is reflected in the equations [1.16] - [1.18] listed in chapter 1 (section D) and obviously comparable to signal transduction mediated by the receptors discussed in the previous two sections of this chapter.

Similar results using indirect methods have been reported for the protease precursor pro-urokinase and its GPI-anchored receptor (Mo3) in human fibroblasts and rhabdomyosarcoma cells.[129] Unoccupied urokinase receptors are diffusely distributed and relatively mobile, while ligand occupied receptors are localized largely at focal contacts and apparently immobile. In the presence of ligand, receptors redistribute to adhesion sites where they immobilize, apparently through linkage to the

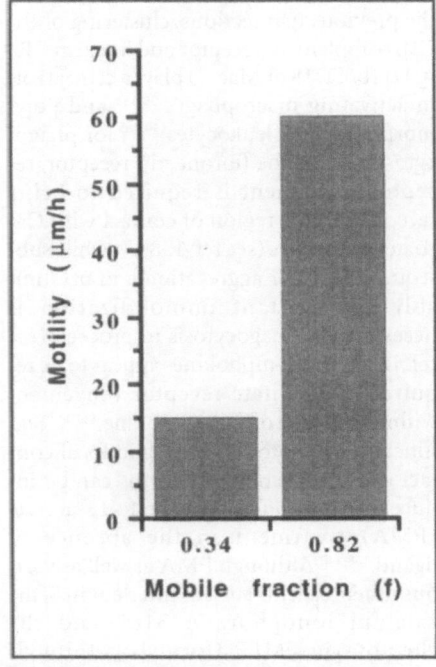

Fig. 7.5. Relationship of fibronectin receptor mobility to cell motility.[113] Results are shown for avian neural crest cells grown on either fibronectin or an anti-fibronectin receptor (fibronectin receptor immobilizing) antibody as substrates.

cytoskeletal component vinculin at focal contacts.[129] Urokinase and its receptor thus appear to be able to function in analogous fashion to fibronectin and the fibronectin receptor(s) in terms of mediating cell adhesion/motility distinct from urokinase's proteolytic activity;[129-132] significantly, both urokinase and its receptor are strongly implicated in cell motility in metastasis,[133-135] where it is quite likely that tumor cell motility and adhesion both play an active role.

Receptor lateral movement plays a role in other adhesive interactions such as phagocytosis. As already alluded to above (section C), Fc-receptor-mediated binding of antibody-coated bacteria, etc. by macrophages and neutrophils prior to phagocytosis is functionally equivalent to the process of adhering to an immobile substrate (see also ref. 109). In similar fashion to the adhesive molecules discussed above and in the previous two sections, clustering of the C3b complement receptor and integrin CR3 (CD11b/CD18 or Mac-1) plays a central role in activating macrophages[136-138] and polymorphonuclear leukocytes[139,140] for phagocytosis. As for the fibronectin receptor, receptor movement is required to bring receptors to the region of contact with C3-coated substrates (see ref. 138),[136] while subsequent receptor aggregation and presumably concomitant immobilization is necessary for phagocytosis to proceed (see ref. 138).[137,139] Lymphokines appear to be required to facilitate receptor movement within the plane of the membrane.[141,142] Significantly, receptor clustering in focal contact sites in several cell systems can be induced by phorbol-12-myristate-13-acetate (PMA) treatment in the absence of ligand.[138-140] Although PMA as well as various other reagents such as interleukin 8, the calcium ionophore A23187 and the chemotaxin FMLP (formyl-methionyl-leucyl-phenylalanine) rapidly increase surface CR3 through inducing fusion of intracellular vesicles containing CR3 with the plasma membrane in several systems,[139,140,142-145] this does not appear to be

the basis of the adhesion response. Rather, it appears to be the receptors already at the membrane which mediate the adhesive interaction prior to phagocytosis, as demonstrated by kinetic studies,[144,145] and the exploitation of systems in which receptor expression remains constant.[139,140] A role of the actin cytoskeleton (see ref. 139) in mediating CR3 clustering (immobilization) is implicated by studies using cytochalasin D.[138,139]

F. SUMMARY

We have seen in previous chapters that lateral movement of ligand-occupied receptor within the plane of the membrane is both instrumental in evoking the stimulatory signal in GTP-binding protein activating receptor and other systems, and that receptor immobilization is central to the process of abrogating the stimulatory signal in downregulation subsequent to hormonal stimulation. In this chapter we have seen that receptor immobilization plays a central role in evoking stimulatory signals in cell adhesion and other signal transduction phenomena. This should not be seen as evidence against the Mobile Receptor Hypothesis. Firstly, it is clear that in all of the receptor systems discussed where receptor immobilization actively participates in signal transduction, receptor movement in the absence of ligand is required to enable receptor aggregation to occur (in analogous fashion to the role of receptor dimerization in tyrosine kinase receptor activation) and clearly is essential for phenomena such as adhesion strengthening (signal amplification) to occur; lack of receptor movement prior to binding to ligand, coreceptor or immobile substrate completely blocks response (see refs. 13, 39-41, 66 and 82-84). Secondly, it underlines the central role of receptor lateral movement in signal transduction of all types, whether it be receptor mobility or immobilization that is specifically required in a particular system for signaling/signal amplification. The dynamics of receptor lateral movement are clearly key elements in signaling at the level of the membrane. The final chapter will put this in a wider context, concentrating on the central roles of the cytoskeleton, the regulation of signal transduction through receptor phosphorylation and interaction between the two, leading to the postulate that regulation of receptor lateral mobility may be the single most important factor determining primary cellular response to hormonal stimulation.

REFERENCES

1. Fahrenholz F, Jans DA, Peters R. Lateral mobility of the V_1- and V_2-receptors in plasma membranes: a role in signal transduction and receptor down-regulation. Colloques INSERM: Vasopressin 1991; 208:49-56.

2. Jans DA. The mobile receptor hypothesis revisited: a mechanistic role for hormone receptor lateral mobility in signal transduction. Biochim Biophys Acta 1992; 1113:271-276.

3. Jans DA, Pavo I. A mechanistic role for polypeptide hormone receptor lateral mobility in signal transduction. Amino Acids 1995; 9:93-109.

4. Levi A, Schechter Y, Neufeld EJ et al. Mobility, clustering and transport of nerve growth factor in embryonal sensory cells and in a sympathetic neuronal cell line. Proc Natl Acad Sci USA 1980; 77:3469-3473.

5. Schlessinger J, Schechter Y, Cuatrecasas P et al. Quantitative determination of the lateral diffusion coefficients of the hormone-receptor complexes of insulin and epidermal growth factor on the plasma membrane of cultured fibroblasts. Proc Natl Acad Sci USA 1978; 75:5353-5357.

6. Zidovetzki R, Yarden Y, Schlessinger J et al. Rotational diffusion of epidermal growth factor complexed to its surface receptor the rapid microaggregation and endocytosis of occupied receptors. Proc Natl Acad Sci USA 1981; 78:6981-6985.

7. Hillman GM, Schlessinger J. The lateral diffusion of epidermal growth factor complexed to its surface receptors does not account for the thermal sensitivity of patch formation and endocytosis. Biochemistry 1982; 21:1667-1672.

8. Rees AR, Gregoriou M, Johnson P et al. High-affinity epidermal growth factor receptors on the surface of A-431 cells have restricted lateral diffusion. EMBO J 1984; 3:1843-1847.

9. Schlessinger J, Schechter Y, Willingham MC et al. Direct visualization of binding, aggregation, and internalization of insulin and epidermal growth factor on living fibroblastic cells. Proc Natl Acad Sci USA 1978; 75:2659-2663.

10. Zidovetzki R, Yarden Y, Schlessinger, J et al. Rotational diffusion of epidermal growth factor complexed to its surface receptor the rapid microaggregation and endocytosis of occupied receptors. Proc Natl Acad Sci USA 1981; 78:6981-6985.

11. Mock EJ, Niswender GD. Differences in the rates of internalization of ^{125}I-labeled human chorionic gonadotropin, luteinizing hormone, and epidermal growth factor by ovine luteal cells. Endocrinology 1983; 113(1):259-264.

12. Panaotou G, Waterfield MD. The assembly of signalling complexes by receptor tyrosine kinases. Bioessays 1993; 15(3):171-177.

13. Ullrich A, Schlessinger J. Signal transduction by receptors with tyrosine kinase activity. Cell 1990; 61:203-212.

14. Noh DY, Shin SH, Rhee SG. Phosphoinositide-specific phospholipase C and mitogenic signaling. Biochim Biophys Acta 1995; 1242(2):99-113.

15. Honegger AM, Kris RM, Ullrich A et al. Evidence that autophosphorylation of solubilized EGF-receptors is mediated by intermolecular cross phosphorylation. Proc Natl Acad Sci USA 1989; 86:925-929.

16. Honegger AM, Schmidt A, Ullrich A et al. Evidence for EGF-induced autophosphorylation of the EGF-receptor in living cells. Mol Cell Biol 1990; 10(8):4035-4044.

17. Ballotti R, Lammers R, Scimeca I-C et al. Intermolecular transphosphorylation between insulin receptors and EGF-insulin receptor chimerae. EMBO J 1989; 8:3303-3309.

18. Gadella TW Jr, Jovin TM. Oligomerization of epidermal growth factor receptors on A431 cells studied by time-resolved fluorescence imaging microscopy. A stereochemical model for tyrosine kinase receptor activation. J Cell Biol 1995; 129(6):1543-1558.

19. Sorokin A. Activation of the EGF receptor by insertional mutations in its juxtamembrane regions. Oncogene 1995; 11(8):1531-1540.

20. Gilboa L, Ben Levy R, Yarden Y et al. Roles for a cytoplasmic tyrosine and tyrosine kinase activity in the interactions of Neu receptors with coated pits. J Biol Chem 1995; 270(13):7061-7067.

21. Weiner DB, Lui J, Cohen JA et al. A point mutation in the neu oncogene mimics ligand induction of receptor aggregation. Nature 1989; 339:230-231.

22. Heffetz D, Zick Y. Receptor aggregation is necessary for activation of the soluble insulin receptor kinase. J Biol Chem 1986; 261:889-894.

23. Kahn CR, Baird KL, Jarrett DB et al. Direct demonstration that receptor crosslinking or aggregation is important in insulin action. Proc Natl Acad Sci USA 1978; 75(9):4209-4213.

24. Fire E, Zwart DE, Roth MG et al. Evidence from lateral mobility studies for dynamic interactions of a mutant influenza hemagglutinin with coated pits. J Cell Biol 1991; 115:1585-1594.

25. Giugni TD, Braslau DL, Haigler HT. Electric field-induced redistribution and postfield relaxation of epidermal growth factor receptors on A431 cells. J Cell Biol 1987; 104(5):1291-1297.

26. Venkatakrishnan G, McKinnon CA, Pilapil CG et al. Nerve growth factor receptors are preaggregated and immobile on responsive cells. Biochemistry 1991; 30(11):2748-2753.

27. Ljungquist-Hoeddelius P, Lirvall M, Wasteson A et al. Lateral diffusion of PDGF-β receptor in human fibroblasts. Bioscience Reports 1991; 11(1):43-52.

28. Ljungquist P, Wasteson A, Magnusson K-E. Lateral diffusion of plasma membrane receptors labelled with either platelet-derived growth factor (PDGF) or wheat germ agglutinin (WGA) in human leukocytes and fibroblasts. Bioscience Reports 1989; 9:63-73.

29. Yarden Y, Schlessinger J. Epidermal growth factor induces rapid, reversible aggregation of the purified epidermal growth factor receptor. Biochem 1987; 26:1443-1451.

30. Kahn CR. Membrane receptors for hormones and neurotransmitters. J Cell Biol 1976; 70:261-286.

31. Kahn CR. The molecular mechanism of insulin action. Annu Rev Med 1985; 36:429-451.

32. Cuatrecasas P. Membrane receptors. Annu Rev Biochem 1974; 43:169-214.

33. Fuchs R, Male P, Mellman I. Acidification and ion permeabilities of highly purified rat liver endosomes. J Biol Chem 1989; 264(4):2212-2220.

34. Contreras I, Caro J. The collisional hypothesis of insulin action, receptor internalization and diabetes. Trends in Biochem Sci 1989; 14:399.

35. Atlas D, Volsky DJ, Levitzki A. Lateral mobility of beta-receptors involved in adenylate cyclase activation. Biochim Biophys Acta 1980; 597(1):64-69.

36. Jans DA, Peters R, Jans P et al. Vasopressin V₂-receptor mobile fraction and ligand-dependent adenylate cyclase-activity are directly correlated in LLC-PK₁ renal epithelial cells. J Cell Biol 1991; 114(1):53-60.

37. Jans DA, Peters R, Jans P et al. Ammonium chloride affects receptor number and lateral mobility of the vasopressin V₂-type receptor in the plasma membrane of LLC-PK₁ renal epithelial cells: role of the cytoskeleton. Exper Cell Res 1990; 191:121-128.

38. Zakharova OM, Rosenkranz AA, Sobolev AS. Modification of fluid lipid and mobile protein fractions of reticulocyte plasma membranes affects agonist-stimulated adenylate cyclase. Application of the percolation theory. Biochim Biophys Acta 1995; 1236:177-184.

39. Schlessinger J. Signal transduction by allosteric receptor oligomerization. Trends Biochem Sci 1988; 13:443-447.

40. Schlessinger J. The epidermal growth factor receptor as a multifunctional allosteric protein. Biochemistry 1989; 27:3119-3123.

41. Yarden Y, Schlessinger J. Self-phosphorylation of epidermal growth factor: evidence for a model of intermolecular allosteric activation. Biochem 1987; 26:1434-1442.

42. Metzger H, Kinet JP. How antibodies work: focus on Fc receptors. FASEB J 1988; 2(1):3-11.

43. Blank U, Ra C, Miller L et al. Complete structure and expression in transfected cells of high affinity IgE receptor. Nature 1989; 337(6203):187-189.

44. Metzger H. The receptor with high affinity for IgE. Immunol Rev 1992; 125:37-48.

45. Fridman WH, Bonnerot C, Daeron M et al. Structural bases of Fcγ receptor functions. Immunol Rev 1992; 125:49-76.

46. Hogarth PM, Hulett MD, Ierino FL et al. Identification of the innumoglobulin binding regions (IBR) of FcγRII and FcεRI. Immunol Rev 1992; 125:21-35.

47. Posner RG, Lee B, Conrad DH et al. Aggregation of IgE-receptor complexes on rat basophilic leukemia cells does not change the intrinsic affinity but can alter the kinetics of the ligand-IgE interaction. Biochemistry 1992; 31(23):5350-5356.

48. Mao SY, Yamashita T, Metzger H. Chemical cross-linking of IgE-receptor complexes in RBL-2H3 cells. Biochemistry 1995; 34(6):1968-1977.

49. Yamashita T, Mao SY, Metzger H. Aggregation of the high-affinity IgE receptor and enhanced activity of p53/56[lyn] protein-tyrosine kinase. Proc Natl Acad Sci USA 1994; 91(23):11251-11255.

50. Zhang F, Yang B, Odin JA et al. Lateral mobility of Fcγ RIIa is reduced by protein kinase C activation. FEBS Lett 1995; 376(1-2):77-80.

51. Trinchieri G, Valiante N. Receptors for the Fc fragment of IgG on natural killer cells. Nat Immun 1993; 12(4-5):218-234.

52. Rosales C, Brown EJ. Signal transduction by neutrophil immunoglobulin G Fc receptors. Dissociation of intracytoplasmic calcium concentration rise from inositol 1,4,5-trisphosphate. J Biol Chem 1992; 267(8):5265-5271.

53. Sarmay G, Rozsnyay Z, Koncz G et al. Interaction of signaling molecules with human Fcγ RIIb1 and the role of various Fcγ RIIb isoforms in B-cell regulation. Immunol Lett 1995; 44(2-3):125-131.

54. Apgar JR. Association of the crosslinked IgE receptor with the membrane skeleton is independent of the known signaling mechanisms in rat basophilic leukemia cells. Cell Regul 1991; 2(3):181-191.

55. Paolini R, Numerof R, Kinet JP. Phosphorylation/dephosphorylation of high-affinity IgE receptors: a mechanism for coupling/uncoupling a large signaling complex. Proc Natl Acad Sci USA 1992; 89(22):10733-10737.

56. Kent UM, Mao SY, Wofsy C et al. Dynamics of signal transduction after aggregation of cell-surface receptors: studies on the type I receptor for IgE. Proc Natl Acad Sci USA 1994; 91(8):3087-3091.

57. Pribluda VS, Metzger H. Transmembrane signaling by the high-affinity IgE receptor on membrane preparations. Proc Natl Acad Sci USA 1992; 89(23):11446-11450.

58. Wofsy C, Kent UM, Mao SY et al. Kinetics of tyrosine phosphorylation when IgE dimers bind to Fcε receptors on rat basophilic leukemia cells. J Biol Chem 1995; 270(35):20264-20272.

59. Li W, Deanin GG, Margolis B et al. Fcε R1-mediated tyrosine phosphorylation of multiple proteins, including phospholipase Cγ 1 and the receptor βγ2 complex, in RBL-2H3 rat basophilic leukemia cells. Mol Cell Biol 1992; 12(7):3176-3182.

60. Gergely J, Sarmay G. B-cell activation-induced phosphorylation of FcγRII: a possible prerequisite of proteolytic receptor release. Immunol Rev 1992; 125:1-19.

61. Ishizaka T, Ishizaka K. Triggering of histamine release from rat mast cells by divalent antibodies against IgE-receptors. J Immunol 1978; 120:800-805.

62. Isersky C, Taurog JD, Poy G et al. Triggering of cultured mastocytoma cells by antibodies to the receptor for IgE. J Immunol 1978; 121:549-558.

63. Posner RG, Subramanian K, Goldstein B et al. Simultaneous cross-linking by two nontriggering bivalent ligands causes synergistic signaling of IgE Fc epsilon RI complexes. J Immunol 1995; 155(7):3601-3609.

64. Menon AK, Holowka D, Webb WW et al. Clustering, mobility, and triggering activity of small oligomers of immunoglobulin E on rat basophilic leukemia cells. J Cell Biol 1986; 102:534-540.

65. Menon AK, Holowka D, Webb WW et al. Cross-linking of receptor-bound IgE

to aggregates larger than dimers leads to rapid immobilization. J Cell Biol 1986; 102:541-550.

66. Schlessinger J, Webb WW, Elson EL. Lateral motion and valence of Fc receptors on rat peritoneal mast cells. Nature 1976; 264:550-552.

67. Mao SY, Varin-Blank N, Edidin M et al. Immobilization and internalization of mutated IgE receptors in transfected cells. J Immunol 1991; 146(3):958-966.

68. McCloskey MA, Liu ZY, Poo MM. Lateral electromigration and diffusion of Fcε receptors on rat basophilic leukemia cells: effects of IgE binding. J Cell Biol 1984; 99(3):778-787.

69. Chang EY, Mao SY, Metzger H et al. Effects of subunit mutation on the rotational dynamics of Fc epsilon RI, the high affinity receptor for IgE, in transfected cells. Biochemistry 1995; 34(18):6093-6099.

70. Myers JN, Holowka D, Baird B. Rotational motion of monomeric and dimeric immunoglobulin E-receptor complexes. Biochemistry 1992; 31(2):567-575.

71. Poo H, Krauss JC, Mayo-Bond L et al. Interaction of Fcγ receptor type IIIB with complement receptor type 3 in fibroblast transfectants: evidence from lateral diffusion and resonance energy transfer studies. J Mol Biol 1995; 247(4):597-603.

72. Leedman PJ, Faulkner-Jones B, Cram DS et al. Cloning from the thyroid of a protein related to actin binding protein that is recognized by Graves disease immunoglobulins. Proc Natl Acad Sci USA 1993; 90(13):5994-5998.

73. Mao SY, Alber G, Rivera J et al. Interaction of aggregated native and mutant IgE receptors with the cellular skeleton. Proc Natl Acad Sci USA 1992; 89(1):222-226.

74. Robertson D, Holowka D, Baird B. Cross-linking of immunoglobulin E-receptor complexes induces their interaction with the cytoskeleton of rat basophilic leukemia cells. J Immunol 1986; 136:4565-4572.

75. Jans DA, Peters R, Fahrenholz F. An inverse relationship between receptor internalization and the fraction of laterally mobile receptors for the vasopressin renal-type V₂-receptor; an active role for receptor immobilization in down-regulation? FEBS Lett 1990; 274:223-226.

76. Maxfield FR, Willingham MC, Haigler HT et al. Binding, surface mobility, internalization, and degradation of rhodamine-labeled α₂-macroglobulin. Biochemistry 1981; 20(18):5353-5358.

77. Roettger BF, Rentsch RU, Hadac EM et al. Insulation of a G protein-coupled receptor on the plasmalemmal surface of the pancreatic acinar cell. J Cell Biol 1995; 130:579-590.

78. Carraway III KL, Koland JG, Cerione RA. Visualization of epidermal growth factor (EGF) receptor aggregation in plasma membranes by fluorescence energy transfer. J Biol Chem 1989; 264:8699-8707.

79. Furuichi K, Rivera J, Isersky C. The fate of IgE bound to rat basophilic leukemia cells. III. Relationship between antigen-induced endocytosis and serotonin release. J Immunol 1984; 133:1513-1520.

80. Ra C, Furuichi K, Mullins JM et al. Internalization of IgE receptors on rat basophilic leukemia cells by phorbol ester: comparison with endocytosis induced by receptor aggregation. Eur J Immunol 1989; 19:1771-1777.

81. Furuichi K, Rivera J, Triche T et al. The fate of IgE bound to rat basophilic leukemia cells. IV. Functional association between the receptors for IgE. J Immunol 1985; 134:1766-1773.

82. Becker KE, Ishizaka T, Metzger H et al. Surface IgE on human basophils during histamine release. J Exp Med 1973; 138:394-409.

83. Lawson D, Fewtrell C, Gomperts B et al. Anti-immunoglobulin-induced histamine secretion by rat peritoneal mast cells studied by immunoferritin electron microscopy. J Exp Med 1975; 142(2):391-402.

84. Magro AM, Alexander A. Histamine-release - in vitro studies of inhibitory region of dose-response curve. J Immunol 1974; 112:1762-1765.

85. Springer TA. Adhesion receptors of the immune system. Nature 1990; 346:425-434.

86. Liu SJ, Hahn WC, Bierer BE et al. Intracellular mediators regulate CD2 lateral diffusion and cytoplasmic Ca²⁺ mobilization upon CD2-mediated T cell activation. Biophys J 1995; 68(2):459-470.

87. Chan PY, Lawrence MB, Dustin ML et al. Influence of receptor lateral mobility on adhesion strengthening between membranes containing LFA-3 and CD2. J Cell Biol 1991; 115(1):245-255.

88. Carpen O, Dustin ML, Springer TA et al. Motility and ultrastructure of large granular lymphocytes on lipid bilayers reconstituted with adhesion receptors LFA-1, ICAM-1, and two isoforms of LFA-3. J Cell Biol 1991; 115(3):861-871.

89. Liu SJ, Golan DE. Mobilization of intracellular free Ca^{2+} upon T cell activation via TCR/CD3 is required to induce a decrease in CD2 lateral mobility in Jurkat T cell membranes. Biophys J 1994; 66:149a.

90. Liu SJ, Golan DE. Regulation of CD2 lateral mobility by calmodulin and calcineurin in Jurkat T cell membranes. Biophys J 1994; 66:149a.

91. Blotta MH, Marshall JD, DeKruyff RH et al. Cross-linking of the CD40 ligand on human CD4+ T lymphocytes generates a costimulatory signal that up-regulates IL-4 synthesis. J Immunol 1996; 156(9):3133-3140.

92. Arulanandam AR, Koyasu S, Reinherz EL. T cell receptor-independent CD2 signal transduction in FcR+ cells. J Exp Med 1991; 173(4):859-868.

93. Pal R, Nair BC, Hoke GM et al. Lateral diffusion of CD4 on the surface of a human neoplastic T-cell line probed with a fluorescent derivative of the envelope glycoprotein (gp120) of human immunodeficiency virus type 1 (HIV-1). J Cell Physiol 1991; 147(2):326-332.

94. Grebenkamper K, Tosi PF, Lazarte JE et al. Modulation of CD4 lateral mobility in intact cells by an intracellularly applied antibody. Biochem J 1995; 312(1):251-259.

95. Bierer BE, Golan DE, Brown CS et al. A monoclonal antibody to LFA-3, the CD2 ligand, specifically immobilizes major histocompatibility complex proteins. Eur J Immunol 1989; 19(4):661-665.

96. Corcao G, Sutcliffe RG, Kusel JR et al. Lateral diffusion of human CD2 wild type and mutants with large deletions in the transmembrane domain. Biochem Biophys Res Commun 1995; 208:1131-1136.

97. Sleckman BP, Shin J, Igras VE et al. Disruption of the CD4-p56^{lck} complex is required for rapid internalization of CD4. Proc Natl Acad Sci USA 1992; 89:7566-7570.

98. Chopra H, Hatfield JS, Chang YS et al. Role of tumor cell cytoskeleton and membrane glycoprotein IRgpIIb/IIIa in platelet adhesion to tumor cell membrane and tumor cell-induced platelet aggregation. Cancer Res 1988; 48:3787-3800.

99. Jaffe SH, Friedlander F, Matsuzaki F et al. Differential effects of the cytoplasmic domains of cell adhesion molecules on cell aggregation and sorting-out. Proc Natl Acad Sci USA 1990; 87:3589-3593.

100. Carpen O, Pallai P, Staunton DE et al. Association of intercellular adhesion molecule-1 (ICAM-1) with actin-containing cytoskeleton and alpha-actinin. J Cell Biol 1992; 118:1223-1234.

101. Burridge K. Substrate adhesions in normal and transformed fibroblasts: organization and regulation of cytoskeletal, membrane, and extracellular matrix components at focal contacts. Cancer Rev 1987; 4:18-78.

102. Kupfer A, Singer SJ. The specific interaction of helper T-cells and antigen-presenting B cells. IV. Membrane and cytoskeletal reorganizations in the bound T cell as a function of antigen dose. J Exp Med 1989; 170:1697-1713.

103. Freed BM, Lempert N, Lawrence DA. The inhibitory effects of N-ethylmaleimide, colchicine and cytochalasins on human T-cell functions. Int J Immunopharmacol 1989; 11:459-465.

104. Ishijima SA, Asakura H, Suzuta T. Participation of cytoplasmic organelles in E-rosette formation. Immunol Cell Biol 1991; 69:403-409.

105. Namba Y, Ito M, Zu Y et al. Human T-cell L-plastin bundles actin filaments in a calcium-dependent manner. J Biochem (Tokyo) 1992; 112:503-507.

106. Pacaud M, Derancourt I. Purification and further characterization of macrophage 70-kDa protein, a calcium-regulated, actin-binding protein identical to L-plastin. Biochemistry 1993; 32:3448-3455.

107. Ledbetter JA, Rabinovitch PS, Hellstrom I et al. Role of CD2 cross-linking in cytoplasmic calcium responses and T cell activation. Eur J Immunol 1988; 18:1601-1608.

108. Ferguson LM, Dustin ML, Chan P-Y et al. Redistribution of GPI-linked LFA-3 to regions of contact between LFA-3 reconstituted planar bilayer membranes and CD2+ T-lymphoblasts. J Cell Biol 1991; 115:71a.

109. McCloskey MA, Poo MM. Contact-induced restribution of specific membrane components: local accumulation and development of adhesion. J Cell Biol 1986; 102:2185-2196.

110. Bell GI. Models for the specific adhesion of cells to cells: a theoretical framework for adhesion mediated by reversible bonds between cell surface molecules. Science 1978; 200:618-627.

111. Bell GI. Theoretical models for cells-cell interactions in immune responses. Dev Cell Biol 1979; 4:371-392.

112. Helmreich EJM, Elson EL. Protein and lipid mobility. Adv in Cyclic Nucleotide and Prot Phosphor Res 1984; 18:1-62.

113. Duband J-L, Nuckolls GH, Ishihara A et al. Fibronectin receptor exhibits high lateral mobility in embryonic locomoting cells but is immobile in focal contacts and fibrillar streaks in stationary cells. J Cell Biol 1988; 107:1385-1396.

114. Duband JL. Extracellular matrix and embryonal morphogenesis: role of fibronectin in cell migration. Reprod Nutr Dev 1990; 30(3):379-395.

115. Thiery JP, Duband J-L, Tucker GC. Cell migration in the vertebrate embryo. Annu Rev Cell Biol 1985; 1:91-113.

116. Zetter BR. Adhesion molecules in tumor metastasis. Semin Cancer Biol 1993; 4(4):219-229.

117. McCarthy JB, Basara ML, Palm SL et al. The role of cell adhesion proteins - laminin and fibronectin - in the movement of malignant and metastatic cells. Cancer Metastasis Rev 1985; 4(2):125-152.

118. Mecham RP. Receptors for laminin on mammalian cells. FASEB J 1991; 5(11):2538-2546.

119. Albelda SM, Buck CA. Integrins and other cell adhesion molecules. FASEB J 1990; 4(11):2868-2880.

120. Sonnenberg A. Integrins and their ligands. Curr Top Microbiol Immunol 1993; 184:7-35.

121. Tucker RP, Edwards BF, Erickson CA. Tension in the culture dish: microfilament organization and migratory behaviour of quail neural crest cells. Cell Motil 1985; 5:225-237.

122. Shimizu Y, van-Seventer GA, Horgan KJ et al. Roles of adhesion molecules in T-cell recognition: fundamental similarities between four integrins on resting human T cells (LFA-1, VLA-4, VLA-5, VLA-6) in expression, binding, and costimulation. Immunol Rev 1990; 114:109-143.

123. Buck CA, Horwitz AF. Integrin, a transmembrane glycoprotein complex mediating cell-substratum adhesion. J Cell Sci Suppl 1987; 8:237-250.

124. Schlessinger J, Barak LS, Hammes GG et al. Mobility and distribution of a cell surface glycoprotein and its interaction with other membrane components. Proc Natl Acad Sci USA 1977; 74(7):2909-2913.

125. Horwitz A, Duggan K, Buck C et al. Interaction of plasma membrane fibronectin receptor with talin. A transmembrane linkage. Nature 1986; 320:531-532.

126. Burridge K, Molony L, Kelly T. Adhesion plaques: sites of transmembrane interaction between the extracellular matrix and the actin cytoskeleton. J Cell Sci Suppl 1987; 8:211-229.

127. Kreis TE, Geiger B, Schlessinger J. Mobility of microinjected rhodamine actin within living chicken gizzard cells determined by fluorescence photobleaching recovery. Cell 1982; 29:835-845.

128. Stickel SK, Wang Y-L. Alpha-actinin-containing aggregates in transformed cells are highly dynamic structures. J Cell Biol 1987; 101:1521-1526.

129. Myohanen HT, Stephens RW, Hedman K et al. Distribution and lateral mobility of the urokinase-receptor complex at the cell surface. J Histochem Cytochem 1993; 41(9):1291-1301.

130. Gudewicz PW, Gilboa N. Human urokinase-type plasminogen activator stimulates chemotaxis of human neutrophils. Biochem Biophys Res Commun 1987; 147:1176-1181.

131. Fibbi G, Ziche M, Morbidelli M et al. Interaction of urokinase with specific re-

ceptors stimulates mobilization of bovine adrenal capillary endothelial cells. Exp Cell Res 1988; 179:385-395.

132. Del Rosso M, Fibbi G, Dini G et al. Role of specific membrane receptors in urokinase-dependent migration of human keratinocytes. J Invest Dermatol 1990; 94:310-316.

133. Ossowski L, Clunie G, Masucci MT et al. In vivo interaction between urokinase and its receptor: effect on tumor cell invasion. J Cell Biol 1991; 115:1107-1112.

134. Bruckner A, Filderman AE, Kirchheimer JC et al. Endogenous receptor-bound urokinase mediates tissue invasion of the human lung carcinoma cell lines A549 and Calu-1. Cancer Res 1992; 52:3043-3047.

135. Kobayashi H, Ohi H, Sugimura M et al. Inhibition of in vitro ovarian cancer cell invasion of modulation of urokinase type plasminogen activator and cathepsin B. Cancer Res 1992; 52:3610-3614.

136. Griffin Jnr FM, Mullinax PJ. Augmentation of macrophage complement receptor function in vitro. III. C3b receptors that promote phagocytosis migrate within the plane of macrophage plasma membrane. J Med Exp 1981; 154:291-305.

137. Hermanowski-Vosatka A, Detmers PA, Gotze O et al. Clustering of ligand on the surface of a particle enhances adhesion to receptor-bearing cells. J Biol Chem 1988; 263(33):17822-17827.

138. Ross GD, Reed W, Dalzell JG et al. Macrophage cytoskeletal association with CR3 and CR4 regulates receptor mobility and phagocytosis of iC3b-opsonized erythrocytes. J Leukoc Biol 1992; 51(2):109-117.

139. Detmers PA, Wright SD, Olsen E et al. Aggregation of complement receptors on human neutrophils in the absence of ligand. J Cell Biol 1987; 105:1137-1145.

140. Pryzwansky KB, Wyatt T, Reed W et al. Phorbol ester induces transient focal concentrations of functional, newly expressed CR3 in neutrophils at sites of specific granule exocytosis. Eur J Cell Biol 1991; 54(1):61-75.

141. Griffin FM Jr, Mullinax PJ. Effects of differentiation in vivo and of lymphokine treatment in vitro on the mobility of C3 receptors of human and mouse mononuclear phagocytes. J Immunol 1985; 135(5):3394-3397.

142. Griffin FM Jr, Mullinax PJ. High concentrations of bacterial lipopolysaccharide, but not microbial infection-induced inflammation, activate macrophage C3 receptors for phagocytosis. J Immunol 1990; 145(2):697-701.

143. Detmers PA, Powell DE, Walz A et al. Differential effects of neutrophil-activating peptide 1/IL-8 and its homologues on leukocyte adhesion and phagacytosis. J Immunol 1991; 147(12):4211-4217.

144. Vedder NB, Harlan JM. Increased surface expression of CD11b/CD18 (Mac-1) is not required for stimulated neutrophil adherence to cultured endothelium. J Clin Invest 1988; 81:676-682.

145. Siu K, Detmers PA, Levin SM et al. Transient adhesion of neutrophils to endothelium. J Exp Med 1989; 169:1779-1793.

CHAPTER 8

THE MOBILE RECEPTOR HYPOTHESIS: A GLOBAL VIEW

A. INTRODUCTION

This chapter attempts to put the findings of the previous seven chapters into a wider context and to expand on the implications of the Mobile Receptor Hypothesis. As we have seen in the last three chapters, receptor movement, as well as the "reverse side of the coin," receptor immobilization, plays a central role in signaling at the level of the membrane in a variety of signal transduction and receptor systems[1,2] and thereby in the regulation of cellular processes such as immune responses, differentiation/transformation, cell proliferation/death, growth, metabolism and motility/metastasis. From a variety of observations in the previous chapters, and in chapters 6 and 7 in particular, it should be clear that there is a body of evidence for the role of the cytoskeleton in modulating membrane protein lateral mobility as well as for the ability of signaling pathways to interact with and regulate other signal transduction cascades through modulation of the lateral mobilities of heterologous membrane components. Both of these aspects will be discussed in some detail below, with the conclusion that signal transduction pathways are able to modulate the lateral movement of heterologous membrane proteins and receptors largely through regulating their respective membrane skeletal/cytoskeletal linkages. Possible pharmacological applications of knowledge of the factors regulating lateral mobility will be discussed in section E.

B. RECEPTOR LATERAL MOVEMENT IN SIGNAL TRANSDUCTION

We have seen that receptor lateral mobility plays a central role in signal transduction, its role being to bring molecules into contact with one another within the plane of the membrane. Since membrane protein lateral movement is much slower than diffusion in the aqueous phase of the cytosol (see chapter 2), events within the plane of the membrane are rate-limiting in terms of signaling in a variety of receptor systems (see chapters 4-7). Receptor lateral mobility is central to signal transduction in a number of ways:

1. Lateral movement of hormone-occupied receptor is essential to signal transduction in GTP-binding protein activating receptor systems where only mobile receptors are active in signaling.[1-19] The basis of this appears to be the essential immobility of GTP-binding proteins (see chapter 4, section G and chapter 5, section F),[20,21] which means

that receptor lateral movement is required to bring receptors into contact with and activate GTP-binding proteins within the plane of the membrane.

2. Lateral movement of hormone-occupied receptor is necessary for signal amplification in GTP-binding protein activating receptor systems (see chapters 4, section E and chapter 5, section G).[1,2] Since GTP-binding protein activation by receptors is through a transient nonimmobilizing contact,[22] receptor movement within the plane of the membrane results in successive collisions with and activation of multiple GTP-binding proteins, thus enabling the signal at the level of the membrane to be amplified.[18,23-26]

3. Immobilization of hormone-occupied receptor is central to the process of abrogating the stimulatory signal in downregulation subsequent to hormonal stimulation, as a prelude to receptor internalization, removal from the membrane and elimination of the stimulatory signal represented by hormone-occupied receptor (see chapter 6).[1,2,6] This is of particular importance in GTP-binding protein activating receptor systems, since continued receptor movement effects signal amplification as outlined above.

4. Lateral movement of hormone-occupied receptor is required to bring tyrosine kinase receptors into the dimeric conformation and enable the intermolecular phosphorylation event necessary for signal transduction.[27-37] Analogously, lateral movement of ligand-occupied receptor is needed to bring the respective ligand-binding and nonligand-binding signal transducing receptor subunit(s) together into an active signaling complex in the case of cytokine receptors.

5. Receptor immobilization is central in evoking stimulatory signals in cell adhesion and other signal transduction phenomena (see chapter 7). Receptor

dimerization/aggregation upon ligand binding constitutes the initial stimulus activating signal transduction in many receptor systems, including those of Fc receptors,[38-43] immunoglobulin (Ig) family receptors such as CD2, LFA-3, etc.[44-48] involved in mediating cell-cell adhesion, and integrins and other molecules involved in adhesion to the extracellular matrix.[49,50] Only immobile receptors appear to be active in signal transduction. Tyrosine kinase receptors can also included in this category with respect to the role of the immobile signaling complex. The association of a variety of cytosolic signaling components (see Figs. 4.6 and 4.11),[51-53] including soluble tyrosine kinases such as p59^{lyn} in the case of Fc receptors, etc.[54-59] contributes to receptor immobilization in all cases. Short-term rapid receptor movement of nonligand occupied receptor within the plane of the membrane is required to bring receptors into contact with one another/ligand-coreceptors/immobile extracellular ligand, whereby receptors that are immobile in the absence of ligand are inactive in signaling.[44,52,60-65]

6. Receptor mobility is essential to adhesion strengthening in many adhesion interactions, whereby lateral movement of unliganded receptor is required to bring receptors to the membrane site of cell-cell interaction,[43,45,46] contact with the substratum[49,50] or ligand-coated phagocytic substrate[66-71] (see chapter 7), where they subsequently aggregate and immobilize to effect stronger adhesion.

7. Movement of ligand-occupied receptor is central to the function of receptors, such as the low density lipoprotein (LDL)- and transferrin-receptors whose function is to transport their respective ligands LDL and transferrin from outside to inside the cell through cycles of ligand binding, en-

docytosis, ligand dissociation and recycling back to the membrane. The significance of lateral movement of these receptors (or "ligand transporters") for their function has not really been touched upon in this book, but rapid receptor movement within the plane of the membrane appears to be required to bring the receptors to the coated pit membrane structures, which are the site of endocytosis (see chapter 6, section B).[72,73] The transferrin[74,75] and LDL[76] receptors exhibit apparent lateral diffusion coefficients (10 and 20 x 10^{-10} cm^2/sec at 37°C, respectively) (see, however, Table 2.3, where lower values for the LDL receptor appear to result from the rapid internalization kinetics) significantly higher than those of polypeptide hormone receptors (see Table 4.1), in keeping with this transport role. They also exhibit very rapid immobilization kinetics,[74] paralleling very rapid endocytosis kinetics. Receptor lateral movement as well as its subsequent abrogation appears to be central to the transport role of these receptors (see also below).[72,73]

A key factor in all of the above is the rate of lateral diffusion, since this clearly determines the rate of reaction/interaction with other membrane components and hence the rate of signal transduction (or transport in the case of 7. above).[1,2] Even more important would appear to be the receptor mobile fraction; receptor immobilization (where the apparent lateral diffusion coefficient is less than 10^{-12} cm^2/sec) in most of the biological systems mentioned above appears to takes place exclusively through a reduction in the mobile fraction,[5-7,9,12,29,36,38-40,42,44,47-49,55,77-79] via association with immobile structures (see ref. 80). The mobile fraction regulates the magnitude of the biological response, by determining the number of receptors that are activated; reducing the mobile fraction removes the signal amplification capability of hormone-occupied GTP-binding protein activating

receptors[5,7,9,12] and effects and maintains dimerized/aggregated receptor signaling complexes in the case of tyrosine kinase receptors,[27,29,33,35] Fc receptors[38-40] and adhesion-mediating receptors/integrins,[44,49,79] etc. Most of the evidence (see section C below) suggests that the major mechanism of receptor immobilization is through links to the cytoskeleton,[2,5,7,39,40,42,55,57,66,69,70, 74,78,81-96] with regulation thereof effected by hormonally induced or other signals[35,36,44,47,48,55,66,68,70,71,77-79,97,98] (see section D below). Receptor lateral mobility (see Figs. 5.4 and 7.2; Table 7.2) appears to be a more critical parameter in determining biological response at low ligand concentrations, indicating the importance and relevance of membrane protein lateral mobility to the physiological situation.

C. THE CENTRAL ROLE OF THE CYTOSKELETON

Actin filaments and microtubules are closely associated with membranes through various linkages, including actin/membrane binding proteins such as talin, vinculin and hisactophilin (see refs. 99-102). Since the membrane skeleton components spectrin (arranged in strands of α- and β-subunits, of which there are multiple forms) and ankyrin are known to be able to be linked to membrane proteins, their direct links in turn to the cytoskeleton mean that both cytoskeletal and membrane skeletal components may participate directly or indirectly in diffusion-controlled events within the plane of the membrane. Specific cytoskeletal and other components are particular to regions of cell-cell contact (e.g., cadherins, catenins) and focal contact sites (e.g., integrins).

As alluded to above (see also chapter 7), facilitated association of membrane proteins with the membrane skeleton and cytoskeleton appears to be the means by which receptor immobility/immobilization is effected, with phosphorylation as one important mechanism by which this interaction can be modulated. A simplistic idea is that tight linkage to cytoskeletal or membrane skeletal elements, including structures, such as coated pits,[72,73,75,95,103] immobilizes membrane proteins (see ref. 80). Such interactions are indeed implicated in immobilizing Fc receptors upon their activation[87-89,96] and in the immobilization of other receptors involved in cell-cell recognition/adhesion interactions, such as CD2—where the actin cytoskeleton and α-actinin are involved,[44,91,94,104] the fibronectin receptor, where talin and actin filaments play a role,[92,105,106] and the C3b complement receptor CR3 (integrin CD11b/CD18 or Mac-1 or Mo-1)—where the actin cytoskeleton is implicated.[69,70] General membrane protein mobility is increased by treatments perturbing membrane skeletal/cytoskeletal integrity.[107,108] Examples of specific proteins whose lateral mobility is restricted by cytoskeletal elements are shown in Table 8.1A. Where linkage to the cytoskeleton is the mechanism of immobilization, inhibitors which perturb the cytoskeletal element in question increase the lateral mobility of the particular protein or reduce receptor immobilization in response to stimuli which decrease receptor mobility. This includes the GTP-binding protein activating receptors for luteinizing hormone,[82] cholecystokinin[83] and N-formyl peptide.[118] Specific examples where receptor phosphorylation is implicated are indicated in Table 8.1A. The role of phosphorylation of GTP-binding protein receptors in triggering binding to arrestins[130-133] and purported subsequent immobilization has been discussed in some detail in chapter 6 (section F) and will not be touched upon here.

Although it is largely accepted that linkage to the cytoskeleton generally inhibits membrane protein lateral movement, it is also clear that cytoskeletal integrity is essential to movement within the plane of the membrane of a number of membrane proteins. Antibody induced redistribution and capping of surface antigens, for example, is known to require microtubule integrity as well as be inhibitable by cytochalasin D and colchicine (e.g., refs. 69 and 134-136). In contrast to the examples in Table 8.1A, the

lateral mobility of membrane integral proteins which rely on cytoskeletal elements for their mobility is inhibited by perturbers of cytoskeletal structure (see Table 8.1B). In the case of the transferrin receptor, while short term treatment with colchicine slows ligand-induced receptor immobilization and increases the mobile fraction (Table 8.1A), longer term colchicine treatment reduces the mobile fraction directly (Table 8.1B).[74] This implies that microtubule structural integrity is necessary for receptor movement within the plane of the membrane as well as for receptor immobilization prior to internalization.[74] Cytochalasin pretreatment has no effect on transferrin receptor lateral movement, implying that microtubules rather than actin filaments modulate transferrin receptor mobility.[74] Examples where membrane protein phosphorylation are involved in regulating cytoskeletal association are indicated in Table 8.1B. Aldosterone treatment increases association of the amiloride sensitive epithelial Na^+ channel with spectrin/actin, concomitant with an increase in its mobile fraction, whereby methylation is implicated as being involved in regulating the interaction (see ref. 77).

In similar fashion to the above examples, lateral mobility of the vasopressin V_2-receptor appears to be controlled either directly or indirectly by the cytoskeleton.[5,7] Treatments reducing the mobile fraction of the V_2-receptor and concomitantly ligand-dependent signal transduction (see chapter 5, section D) significantly affect the actin cytoskeleton as illustrated in Figure 8.1. While pretreatment at 4°C (see also ref. 12) reversibly disrupts actin filament structure (Fig. 8.1, Panel F) probably through inhibiting actin polymerization (see ref. 137), extended treatment with the weak base ammonium chloride intensifies actin filament structure, increasing its rigidity (Fig. 8.1, Panel J). Both treatments affecting the actin cytoskeleton reduce the V_2-receptor mobile fraction, while not affecting the apparent lateral diffusion coefficient significantly (see Figs. 5.3 and 8.2).[5,7] Cytochalasin B pretreatment (two hours) also reduces

Table 8.1. Examples of cytoskeletal modulation of receptor mobility in biological membranes

Protein	Membrane Skeletal/Cytoskeletal Element Implicated	Effect on Lateral Mobility
A. Cytoskeletal Association Inhibiting Lateral Movement		
IgE receptor (Fcε RI) (high-affinity)	actin (?)[57,87,88]	Cytochalasin D inhibits receptor immobilization and internalization[42,58,109+]
CD2	ABP/L-plastin[110] α-actinin/actin[91]	Cytochalasin D reduces receptor immobilization (i.e., increases the mobile fraction) in response to activation (cytoskeletal link is positively regulated by intracellular Ca^{2+})[44,110+]
Fibronectin receptor	Talin/α-actinin/actin[93,94,102]	Link to cytoskeleton immobilizes the receptor in focal contacts
GTP-binding proteins	Tubulin[21,111x]	Association with or compartmentalization through the cytoskeleton impedes lateral movement[20]
LH receptor	Actin microfilaments (?)	Cytochalasins B or D or colchicine increase the mobile fraction of chorionic gonadotropin occupied receptor (from 0.1 to c. 0.3) and the apparent lateral diffusion coefficient (over six-fold)[82]
CCK$_A$ receptor	Dynamin-2/Rab family[112,113]	Surface immobilization[114,115+] through cytoskeletal linkage prior to internalization[83]
N-formyl peptide receptor	Actin[116,117]	Cytochalasin B treatment slows ligand-induced receptor immobilization and internalization[118]
acetylcholine receptor (nicotinic)	β-spectrin/dystrophin[67,119-121]	Agrin-induced tyrosine phosphorylation increases association with the cytoskeleton and a reduction in the mobile fraction[67,122]
transferrin receptor	Microtubules	Microtubular attachment immobilizes receptor prior to internalization;[74] short term (one hour) colchicine pretreatment slows immobilization
Na$^+$,K$^+$-ATPase	Ankyrin/Actin[123,124]	Cytochalasin D treatment increases the apparent lateral diffusion coefficient by seven-fold as well as increasing the mobile fraction from 0.47 to 0.63)[125#]

Table continues on next paage

Table 8.1. (continued)

red blood cell band 3	Spectrin	Association with spectrin decreases mobility[126-128]

B. Cytoskeletal Association Enhancing Lateral Movement

Vasopressin V_2-receptor (renal)	Actin microfilaments	Treatments affecting actin filament integrity (e.g., cold pretreatment or treatment with NH_4Cl) reduce the receptor mobile fraction[5,7] Cytochalasin B reduces the mobile fraction from 0.9 to 0.76 and the apparent lateral diffusion coefficient from 3.2 to 2.4 x 10^{-10} cm^2/sec[5,129+]
	Microtubules (?)	Vinblastin reduces the mobile fraction from 0.9 to 0.76 and the apparent lateral diffusion coefficient from 3.2 to 2.3 x 10^{-10} cm^2/sec[5]
transferrin receptor	Microtubules	Microtubular integrity is necessary for receptor movement;[74] long term treatment (3h) with colchicine or vinblastin results in receptor immobilization (mobile fraction of 0.2 compared to 0.58 in the absence of treatment). The microtubule stabilizer taxol reverses this effect (mobile fraction of 0.5).*
epithelial Na^+ channel (amiloride-sensitive)	Spectrin	Association with the cytoskeleton is regulated by aldosterone, resulting in a two-fold increase in the mobile fraction; treatment with the methylation blocker 3-deazaadenosine reverses this effect[77]
	Actin microfilaments (?)	Cytochalasin D pretreatment reduces both the apparent lateral diffusion coefficient (0.99 to 0.82 x 10^{-10} cm^2/sec) and mobile fraction (0.12 to 0.07)[77]
red blood cell band 3	Ankyrin	Association with ankyrin increases mobility[126-128]

LH, luteinizing hormone; ABP, actin binding protein; CCK$_A$, cholecystokinin type-A. *Receptor phosphorylation implicated in regulating the linkage to the cytoskeleton (see Table 8.2). +See also chapter 5 (section F). #Ca^{2+} depletion also enhances mobility;[125] presumably through affecting the cytoskeleton;[110] see also ref. 44. ^ATP-depletion through various treatments also decreases the mobile fraction, presumably through a direct or indirect effect on the cytoskeleton (see ref. 75).

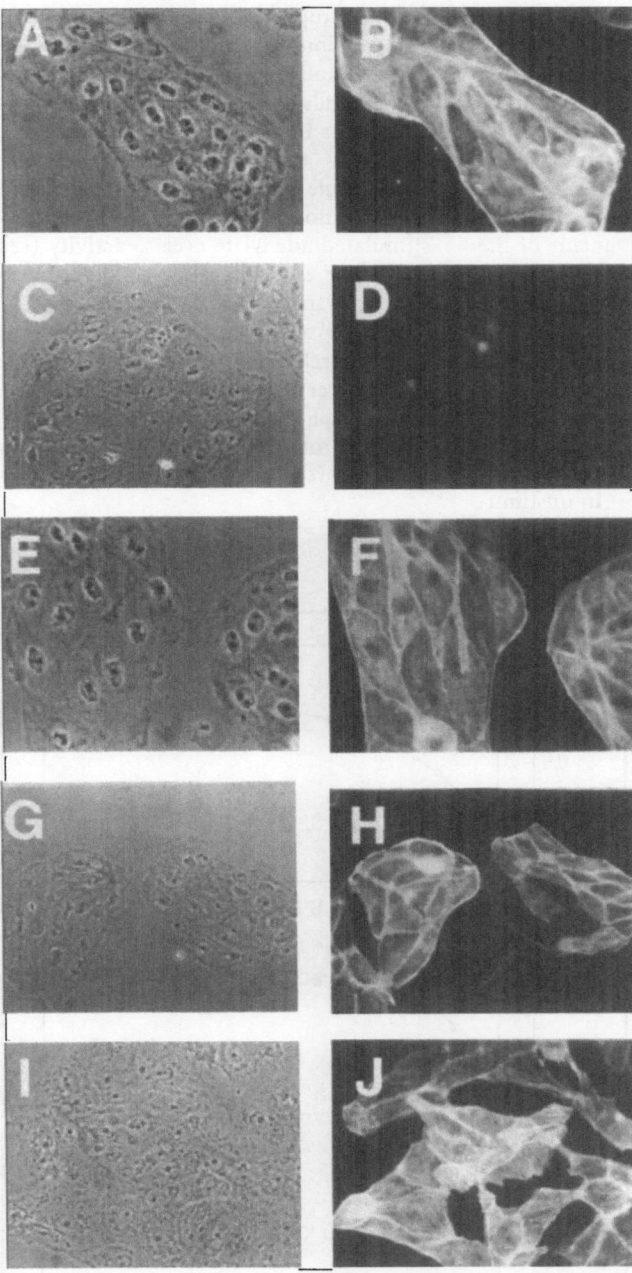

Fig. 8.1. Treatments affecting the lateral mobility of the vasopressin V_2-receptor affect the actin cytoskeleton. Cells of the LLC-PK$_1$ renal epithelial cell line were pretreated for 1h at 37°C (Panels A-D), 4°C (Panels E and F), 4°C followed by 1h at 37°C (Panels G and H) or 2d with 10 mM NH$_4$Cl (at 37°C) (Panels I and J), prior to fixation and staining with a rhodamine-labeled derivative of the actin-filament binding alkaloid phalloidin.[5,7] Phase contrast images are shown in the left panels and fluorescent images of the same field of cells on the right. Staining in the presence of a 100-fold excess of unlabeled phalloidin is shown in panel D to demonstrate specificity of the stain for actin.

the receptor mobile fraction but to a lesser extent (Fig. 8.2), while vinblastin treatment (16 hours) has a similar effect, implying that microtubule structure may also be important to ensure the integrity of the actin cytoskeleton and thereby V_2-receptor lateral movement.[5]

The implication from these and other studies is that distinct components of the cytoskeleton play roles in both restricting and facilitating lateral movement of plasma membrane proteins. As already seen in chapter 4 (section G), trimeric GTP-binding proteins are essentially immobile within the plane of the membrane.[20] The mechanistic basis of immobility appears to be through linkage to the cytoskeleton (see also chapter 5, section E).[21,22,138,139] In unstimu-

lated neutrophils, for example, the $Gs\alpha$ and $G_{i2}\alpha$ subunits appear to be linked to the cytoskeleton but dissociate upon activation,[138] while direct Gs association with microtubules in S49 lymphoma cells is implied by the observation that the microtubule inhibitors colchicine/vinblastin increase nonhydrolyzable GTP analog-stimulated adenylate cyclase activity (see ref. 21).[139] Significantly, the membrane-linked cytoskeletal proteins spectrin/fodrin[140,141] and dynamin,[142] among others, have a "pleckstrin homology domain" (see also chapter 5, section F), which has been clearly implicated in interacting with GTP-binding proteins and the $G\beta\gamma$ subunits, in particular (see refs. 21, 140 and 141).

Fig. 8.2. Treatments affecting the cytoskeleton affect the lateral mobility of the vasopressin V_2-receptor.[5,7] Cells of the LLC-PK$_1$ renal epithelial cell line were pretreated with cytochalasin B (2h) or 10 mM NH$_4$Cl (2d) as indicated, prior to determination of V_2-receptor lateral mobility parameters using the fluorescence photobleaching recovery technique essentially as described in the legend to Figure 4.2. The apparent lateral diffusion coefficient (D) and mobile fraction (f) are indicated.

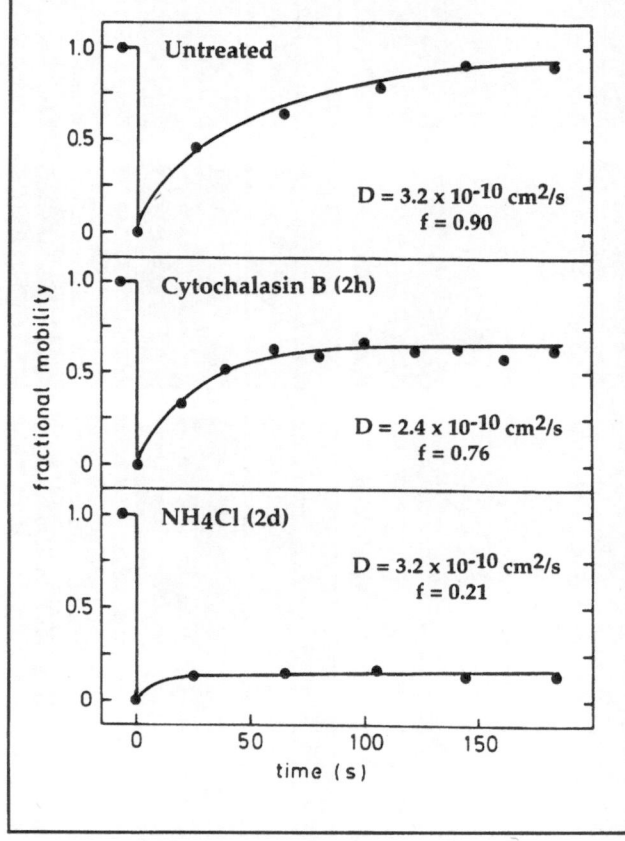

The basis for phosphorylation-mediated regulation of cytoskeletal association in GTP-binding protein activating receptor systems may relate directly to the observation that the region of the β-adrenergic receptor kinase necessary for Gβγ binding, absent from the rhodopsin kinase, also contains a pleckstrin homology domain.[21,141,143 144] One can speculate that the membrane-bound Gβγ subunits are the sites of association of receptor kinases as well as representing a direct link to the membrane skeleton; association of the kinase with the receptor during phosphorylation brings the receptor into this cytoskeletally linked complex. Phosphorylation itself may be secondary to complexation in terms of importance to initial receptor association with the cytoskeleton and concomitant immobilization. Arrestin-like molecules may be critical in stabilizing cytoskeletal association rather than in bringing about the initial complexation itself. Figure 8.3 is a speculative scheme indicating the role of cytoskeletal linkages in mediating GTP-binding protein and GTP-binding protein activating receptor lateral movement and therefore signal transduction. Cytoskeletal linkages are integral to both facilitating movement (e.g., the actin cytoskeleton in the case of the V_2-receptor) and arresting it during the abrogation of the stimulatory response; in the case of the GTP-binding protein trimer, spectrin/microtubule-linkages inhibit movement, meaning that receptor movement is required for signal transduction to occur.

The potential role of membrane skeletal/cytoskeletal-associated proteins such as arrestins actin binding proteins such as hisactophilin/talin/vinculin, proteins—such as agrin which appears to regulate aggregation and association of the nicotinic acetylcholine receptor with the cytoskeleton.[119-122,145,146] and specific interactive components, such as actin/tubulin-binding domains and pleckstrin homology domains in the overall membrane signaling network, should not be underestimated. Receptor kinases and GTP-binding protein subunits are able to be specifically localized to membrane sites in direct contact with the membrane skeleton and cytoskeleton, while the movement of plasma membrane signaling components such as receptors, GTP-binding proteins, etc. appears to be able to be modulated by cytoskeletal components in a precisely controlled fashion. The modulation in turn of these cytoskeletal components through Ca^{2+} (e.g., actin/L-plastin/profilin; see chapter 7 section D),[44,147,148] methylation[77] and phosphorylation (see above and below) means that the cytoskeletal network is a dynamic structure, able to alter rapidly in response to a variety of stimuli. Signals feeding directly or indirectly into the membrane skeleton/cytoskeleton can modulate plasma membrane protein mobilities and hence their activity either in stimulatory or desensitizing responses. Since incoming extracellular signals are initially transduced at the level of the membrane, and protein lateral mobility is integral in this, modulation of plasma membrane protein lateral movement through the membrane skeleton/cytoskeleton may be a central mechanism of signal integration.

D. SIGNAL TRANSDUCTION: RECEPTOR LATERAL MOBILITY MODULATION BY HETEROLOGOUS SIGNALING

It should be clear from the previous chapters, and particularly chapter 3 (see section G) that an important aspect of signaling at the level of the membrane is the fact that the lateral mobility of a number of proteins, and significantly integral membrane receptors, can be modulated by signal transduction through heterologous receptors. Since phosphorylation plays a major role in regulating receptor mobility, and treatments such as those modifying cellular Ca^{2+} affect the cytoskeleton which in turn influences receptor lateral movement, it is clear that stimulation of one receptor can result in the modulation of the lateral mobility of other receptors/membrane proteins. This constitutes part of the basis of long-term response and phenotypic change such as differentiation/transformation, etc. where multiple receptor activities are involved. It may also be the basis of phenomena such as heterologous desensitization, where, in contrast to homologous desensitization where downregulation only affects the receptor directly activated by hormone binding,[149] stimulation of one GTP-binding protein activating receptor can lead to desensitization or refractability to stimulation of other receptors (see ref. 149). In simple terms, cAMP-dependent protein kinase (PK-A) or PK-C (Ca^{2+}, phospholipid-dependent kinase) triggered phosphorylation activity in response to a particular agonist appears to be capable of effecting immobilization of heterologous receptors (see below) so that the response to their own specific ligands is severely reduced.

Table 8.2 lists examples of proteins whose lateral mobility is modulated by signal transduction (see also Tables 3.3 and 3.4). These include those whose mobility can be decreased (Table 8.2A) or increased (Table 8.2B) in response to extracellular signals, respectively. The lateral mobility parameters of a number of receptors, for example, including the integrin/C3b complement receptor CR3, the low-affinity IgG receptor FcγRIIa (CD32) and the nicotinic acetylcholine receptor appear to be able to be regulated by phosphorylation by PK-C in the absence of ligand (see Table 8.2). Lateral mobility of CD32 and the acetylcholine receptor in fact appears to be modulated by synergistic phosphorylation events mediated by several kinases. As mentioned in section E of chapter 3, mutation of tyrosine at position 252 within the CD32 cytoplasmic domain increases the CD32 mobile fraction by about 30%, as well reducing the apparent lateral diffusion coefficient by 50% (see Table 3.3), implying that tyrosine phosphorylation is also important in regulating CD32 lateral mobility.[55] Consistent with this idea, truncation mutants of the cytoplasmic domain which lack tyrosine 252 as well as a second putative tyrosine phosphorylation site are not affected by PMA in terms of lateral mobility properties.[55] As alluded to above (see Tables 8.1 and 8.2), nicotinic acetylcholine receptor immoblization induced by agrin is also strongly regulated by tyrosine phosphorylation. Clearly, heterologous external signals inducing PK-C activation or tyrosine phosphorylation activity are able to modulate the lateral mobility of both the acetylcholine receptor or CD32.

The lateral mobility properties of receptors such as CD2 and the acetylcholine receptor are responsive to intracellular Ca^{2+}, and possibly calmodulin-dependent phosphorylation (see also ref. 79). In the case of CD2 and as shown in Figure 8.4, there is a direct inverse relationship between the receptor mobile fraction and the cellular Ca^{2+} response (see also chapter 7, section D; Fig. 7.4). Intracellular Ca^{2+} is elevated upon T cell activation, and incubation with agents such as the Ca^{2+} ionophore, ionomycin, or Ca^{2+} chelator, EGTA (ethyleneglycol-bis[β-aminoethyl]-N,N,N',N'-tetraacetic acid) which modulate intracellular Ca^{2+} levels, affects CD2 lateral mobility and in particular the mobile fraction (Table 3.3; Fig. 8.4). The effects of cAMP and cytochalasin D on lateral mobility properties (see Tables 8.1 and 8.2) presumably result from indirect modulation of intracellular Ca^{2+} (see section D of chapter 7), which in turn probably stems from changes in cytoskeletal components such as actin, L-plastin or profilin whose structure is known to be influenced by Ca^{2+}.[44,147,148] By directly (or indirectly) controlling CD2 lateral diffusion, and thereby its cell surface distribution, cytoplasmic Ca^{2+} mobilization exerts an integral regulatory role in CD2-mediated cell adhesion phenomena. In analogous fashion to the receptors mentioned above, it is clear that heterologous agents causing changes in intracellular Ca^{2+} (and probably calmodulin-dependent phosphorylation) can affect CD2 lateral mobility.

Other examples of heterologous signals controlling receptor lateral movement include the increase of the apparent lateral diffusion coefficient and mobile fraction of the platelet-derived growth factor (PDGF)

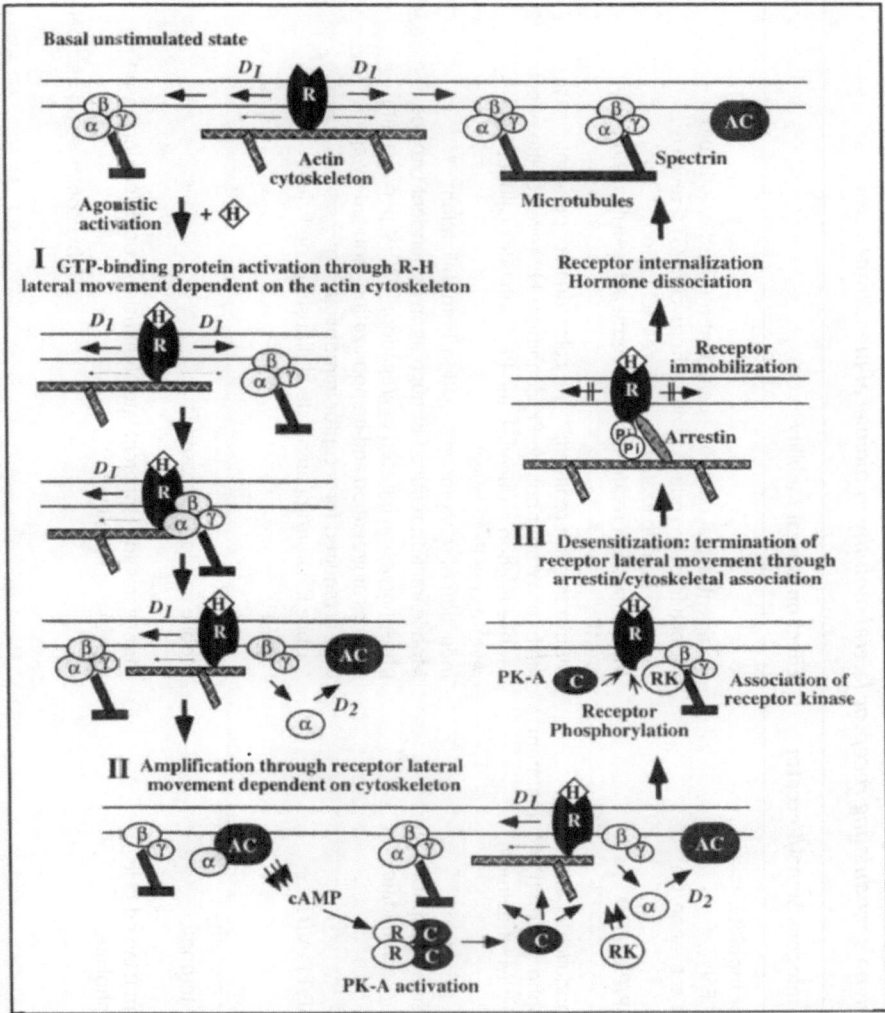

Fig. 8.3. A speculative scheme for the role of cytoskeletal elements in regulating a GTP-binding protein activating receptor signal transduction system through modulation of plasma membrane protein lateral movement. Receptors are linked to actin filaments which facilitate movement, while GTP-binding proteins are immobilized through linkage to microtubules, probably through molecules like spectrin/fodrin (see text). Diffusion controlled events within the plane of the membrane and cytosol are indicated by D_1 and D_2 respectively; see Figures 5.5 and 6.8 for details of the role of receptor lateral mobility in hormonal stimulation of signal transduction and the desensitization of response subsequent to hormonal stimulation, respectively. R, receptor; H, hormone; α,β,γ, trimeric GTP-binding protein subunits; AC, adenylate cyclase; RK, receptor kinase; PK-A, cAMP-dependent protein kinase; R,C, PK-A regulatory and catalytic subunits; Pi, inorganic phosphate.

Table 8.2. Examples of heterologous signals modulating receptor lateral mobility in biological membranes.

Protein	Heterologous Signal/Receptor	Effect on Lateral Mobility
A. Signal transduction decreasing receptor mobility		
MHC Class I	TNF/IFNγ	Apparent lateral diffusion coefficient reduced c. four-fold[97]
CD2	T-cell activation/Ca^{2+}	Mobile fraction reduced;[44,47]x ionomycin/EGTA reverse receptor immobilization upon T cell activation[44]
	cAMP/PK-A (??)	Mobile fraction reduced by the permeant cAMP analog DB-cAMP[44]
CD4	Calmodulin-dependent phosphorylation (PMA treatment)	Mobile fraction reduced;[79]* the calmodulin-dependent kinase inhibitor W7 but not the PK-C inhibitor H7 reverses the effect
Fcγ RIIa*	PK-C (PMA treatment)	Mobile fraction reduced;[55]+ the PK-C inhibitor calphostin reverses the PMA effect
CR3 (C3b receptor)^	PMA treatment	Induction of receptor aggregation/immobilization[68,70,71]+
acetylcholine receptor (nicotinic)	Tyrosine phosphorylation Agrin/Pervanadate$/$Ca^{2+}$	Mobile fraction reduced through agrin-dependent tyrosine phosphorylation-induced aggregation;[78,119,122]+ pervanadate$ reduces aggregation and immobilization induced by agrin[78,119] Ca^{2+} is required for receptor immobilization[150]
vasopressin V_2-receptor (renal)	cAMP/PK-A (??)	Mobile fraction/apparent lateral diffusion coefficient reduced[129]++
B. Signal transduction increasing receptor mobility		
epithelial Na+ channel (amiloride-sensitive)	aldosterone	Mobile fraction increased[77]+
PDGF receptor	Serum (starved cells)	Mobile fraction/apparent lateral diffusion coefficient increased[36]#
CR3 (C3b receptor)^	Lymphokines	Increase in mobility[66,67]+

Table continues on next page

Table 8.2. (coontinued)

CD2	calmodulin/calcineurin	Calcineurin phosphatase inhibitor cyclosporin A reverses receptor immobilization[48] upon T-cell activation[44,47+]
	cAMP/PK-A (??)	The permeant cAMP analog DB-cAMP reduces Ca^{2+}-dependent immobilization upon T-cell activation[44]
HLA I β2-microglobulin	Differentiation@	Increase in mobile fraction (by almost two-fold) and apparent lateral diffusion coefficient[151]
Fcγ RIIIB*	PK-C (staurosporine°)	Staurosporine reduces immobilization (mobile fraction is increased) by reducing binding to CR3[96]
acetylcholine receptor (nicotinic)	PK C	PMA inhibits agrin-induced receptor aggregation[150] Staurosporine (but not H7)§ inhibits agrin-induced receptor aggregation[152+]

MHC, major histocompatibility complex; TNF, tumor necrosis factor; IFNγ, interferon γ; EGTA, ethyleneglycol-bis(β-aminoethyl)-N,N,N',N'-tetraacetic acid; PK-A, cAMP-dependent protein kinase; DB-cAMP, N6,2'-O-dibutryladenosine-cAMP; PMA, phorbol-12-myristate-13-acetate; PK-C, Ca2+, phospholipid dependent kinase; PDGF, platelet-derived growth factor. +Cytoskeletal linkage implicated (see Table 8.1). ^C3b complement receptor (integrin CD11b/CD18 or Mac-1). *Low-affinity IgG Fc receptor. $Tyrosine phosphatase inhibitor. ^C3b complement receptor (integrin CD11b/CD18 or Mac-1). *See Fig. 8.4. *Low-affinity IgG Fc receptor. $Tyrosine phosphatase subsequent to treatment with the permeant cAMP analog 8-bromoadenosine-cAMP indicate a reduced mobile fraction and apparent lateral diffusion coefficient compared to untreated cells, consistent with the observation that a nonreceptor-mediated increase in intracellular cAMP levels leads to receptor endocytosis in the absence of ligand.129 In contrast, preliminary results for short-term stimulation with salmon calcitonin, a heterologous adenylate cyclase stimulating hormone, indicate the lack of an effect on V2-receptor lateral movement (Jans DA, unpublished). #See Table 3.3. @Differentiation was induced by replacing glucose with galactose in the medium.[151] °General serine/threonine and tyrosine kinase inhibitor. §The serine kinase inhibitor H7 does not block receptor aggregation, although it reduces receptor phosphorylation.152

Fig. 8.4. Dependence of CD2 immobilization on Ca^{2+} mobilization in Jurkat T cells activated by paired antibodies (see chapter 7, section D). Data points are from ref. 44, representing results for the CD2 mobile fraction subsequent to treatments with activating antibodies in the presence of the Ca^{2+} ionophore, ionomycin, the Ca^{2+} chelator, ethyleneglycol-bis(β-aminoethyl)-N,N,N′,N′-tetraacetic acid (EGTA), EGTA and ionomycin (together), the permeant cAMP analog $N^6,2'$-O-dibutryladenosine-cAMP, the actin filament perturber cytochalasin D, the microtubule perturbing agent, colchicine, and butyric acid, a physiologically produced short chain fatty acid which inhibits Ca^{2+} release from intracellular stores in certain biological systems. The regression coefficient (r) is indicated.

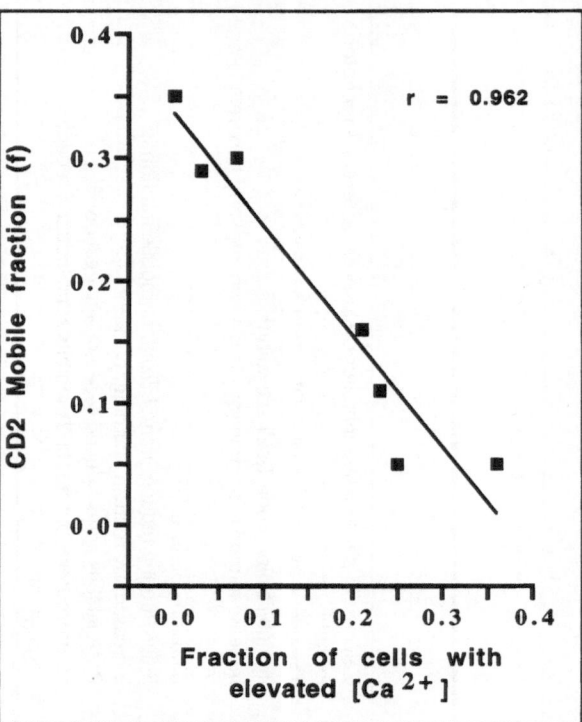

receptor induced by serum in starved cells,[36] and the reduction of MHC class I lateral movement by the cytokines tumor necrosis factor and interferon-γ (see Tables 3.3A, 8.2).[97,98] Lymphokines analogously appear to increase lateral movement of CR3,[66,67] whereas PMA treatment in the absence of ligand leads to receptor aggregation (see chapter 7, section E).[68,70,71] As discussed in chapter 3 (section G) and illustrated in Table 3.3, the stage of the cell cycle has also been shown to affect plasma membrane protein lateral mobility, impinging on cellular responsiveness to extracellular signals in terms of heterologous control of receptor lateral mobility.[97]

In most of the examples shown in Table 8.2, regulation of receptor lateral mobility through the cytoskeleton is strongly implicated (see also Table 8.1). A possible exception, at least in terms of a direct link to the cytoskeleton, appears to be the low-affinity IgG receptor Fcγ RIIIB, whose immobilizing interaction with CR3 is reduced by the protein kinase inhibitor staurosporine,[96] implying that protein-protein interactions within the plane of the membrane, able to be regulated by extracellular signals and phosphorylation, may also modulate membrane protein mobility.

In summary, it is clear that stimulation of one receptor can lead to modulation of the lateral mobility properties of other receptors. This stems from the fact that multiple signals result in the activation of common signaling components such as kinases/phosphatases or second messenger molecules. This is of particular significance since it means that multiple signaling cascades can be regulated at the level of the membrane through modulating the primary response of heterologous receptors to ligand stimulation. By modifying the lateral mobility of receptors in heterologous receptor signaling pathways, the rate-limiting step of signal transduction of those pathways can be controlled. Modulation of the lateral mobility of heterologous receptors also serves as a mechanism of integrating responses mediated by a variety of different receptors and signaling pathways in order to effect long-term phenotypic changes such as in immune responses, transformation and differentiation.

Clearly, receptor movement within the plane of the membrane, the cytoskeleton and signal transduction can be seen as three interacting components, each one of which can regulate and modulate the activity or function of the others (see Fig. 8.5). Phenomena such as anchorage modulation (see chapter 3, section D), where locally bound lectin can have long-range inhibitory effects on capping in B-lymphocytes,[153-156] may be able to be understood in terms of this concept.

E. POTENTIAL PHARMACOLOGICAL APPLICATIONS OF THE MOBILE RECEPTOR HYPOTHESIS

The importance of receptor lateral mobility in integrated response to stimuli under physiologically relevant conditions, and particularly the fact that receptor mobility can clearly be modulated by heterologous extracellular signals as exemplified by Figure 8.5 implies that plasma membrane lateral movement should be a potential site of pharmacological intervention in relevant clinical situations. One could, for example, inhibit cell growth or response through preventing receptor lateral mobility, which, as seen in chapters 5-7, plays a central role in signal transduction. For example, since receptors relying on lateral movement and immobilization are integral to adhesion-dependent processes, such as phagocytosis, cell-cell recognition in immune responses, and cell migration/metastasis, manipulation of the lateral mobility of these receptors would be a way to control and regulate these processes. Anti-adhesion therapy to prevent metastasis could thus take the form of either inhibiting the lateral movement of

integrins to focal adhesion sites, or preventing their immobilization at those sites. This could be an alternative to more conventional approaches of using antibodies to block recognition or sense/antisense oligonucleotides to modulate the expression of surface molecules (see ref. 157). Crosslinking and hence immobilizing antibodies (rather than antibodies which only block binding) could be used therapeutically, either to activate cells in a given context (e.g., Fc receptors) or to inhibit signal transduction by preventing receptor lateral movement (e.g., of GTP-binding protein activating receptors). As already alluded to in chapter 4 (section E), the possibility that receptor antagonists of GTP-binding protein activating receptor systems might function through blocking receptor lateral movement and hence signal transduction (see ref. 158) has been shown to be incorrect, since antagonist-occupied vasopressin V_2- and N-formyl-peptide-receptors are clearly mobile (see chapter 6, section G).[2,8,118]

The various cold/phospholipase/poly-L-lysine/cationized ferritin treatments used to modulate membrane fluidity/protein mobility discussed in previous chapters (e.g., Figs. 3.3, 5.3 and 8.2)[3,5,7,9-12] seem to have little direct application in clinical situation. There is a body of knowledge, however, regarding agents which can modulate cellular phospholipase activities in living cells (e.g., phorbol esters, arachidonic acid, etc., and see below), and these could at least theoretically be utilized in this context in a considered fashion. Further, since stimulation of one receptor and/or its second messenger pathway is able to effect immobilization or increase the mobility of other receptors (see section D above), and since receptor mobility/immobilization in the presence of ligand is integral to signal transduction in a variety of receptor systems, it is clear that cellular response could be able to be controlled through the use of agonists modulating heterologous receptor lateral mobility. Anti-adhesion therapy in this context might take the form of treatments modulating intracellular Ca^{2+}, activating or

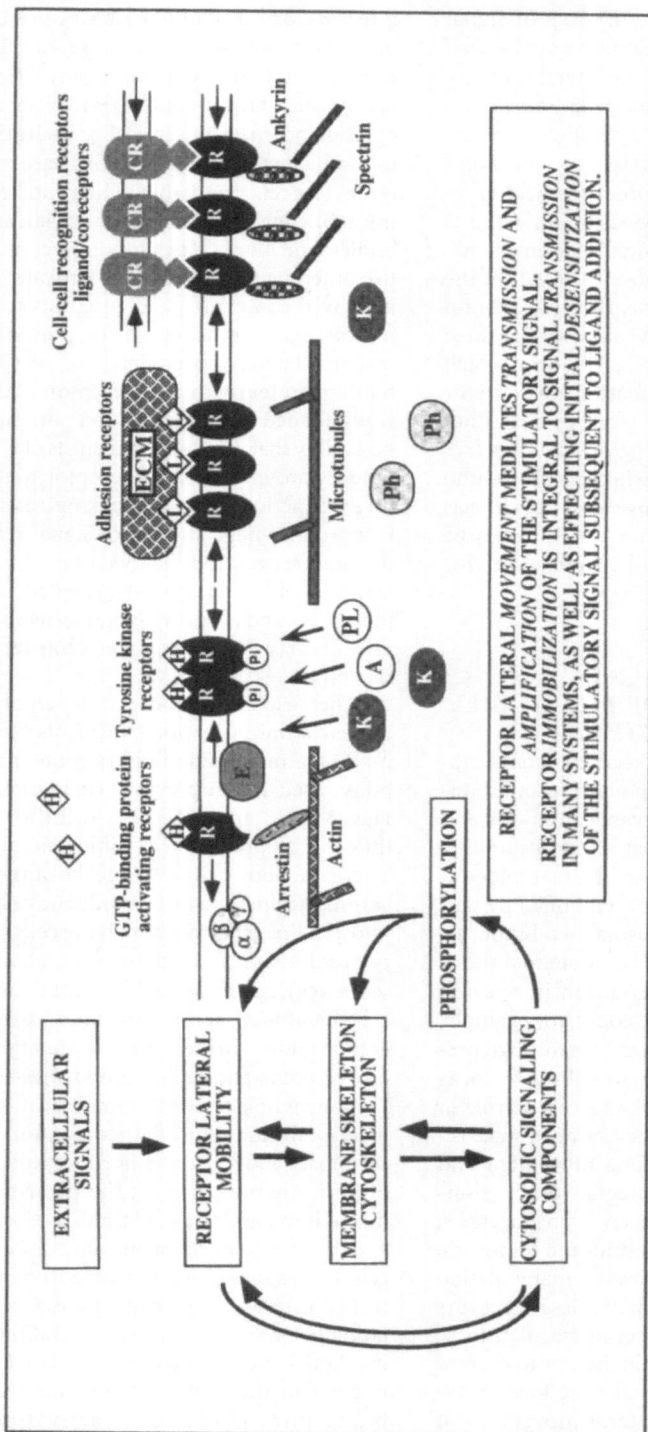

Fig. 8.5. The central role of membrane receptor lateral movement, the membrane skeleton/cytoskeleton and cytosolic signaling components in mediating cellular response. Each component interacts with and modifies the function and activity of the others. Phenomena such as anchorage modulation[153-156] may be able to be understood in these terms (see text; see also chapter 3, section D). H, hormone; R, receptor; CR, coreceptor; α,β,γ, trimeric GTP-binding protein subunits; ECM, extracellular matrix; K, kinase; E, effector enzyme (e.g., adenylate cyclase or phospholipase Cβ); A, adaptor molecule; Ph, phosphatase; PL, phospholipase Cγ; Pi, inorganic phosphate

inhibiting specific kinases, such as PK-A or PK-C or phosphatases such as calcineurin, or perhaps even modifying the cytoskeleton. In this latter respect, it is of interest that cytoskeletal perturbing agents, such as cytochalasin D, taxol and vinblastin, exhibit anti-tumor activity,[159] while kinase inhibitors, such as staurosporine, calphostin C and H7, can induce apoptotic cell death in transformed cell lines.[160,161] To take an example, it is clear that a variety of reagents could at least theoretically be used to modulate CD2's mobile fraction through intracellular Ca^{2+} (see Tables 8.1 and 8.2), and thereby its biological activity in cell-cell recognition, which could be useful in a specific set of circumstances or medical context. For approaches to modulating intracellular Ca^{2+} and PK-C activity in T cells in a therapeutic setting through antibodies, interleukin-2, etc., see ref. 162.

It should also be remembered that a number of adhesion/cell recognition receptors also serve as receptors for viruses, being essential for cellular recognition and subsequent entry.[163] These include ICAM-1 which is recognized by human rhinoviruses (see ref. 163), MHC class I molecules which are docking sites for the alphavirus Semliki Forest virus,[164,165] and CD4, which is a high-affinity receptor for human immunodeficiency viruses HIV-1 and -2. One possible approach to prevent the spread of infection might be to use treatments (e.g., phorbol esters) which immobilize and downregulate the receptor in the absence of ligand.[79] Although problems in terms of limiting the effects of many of the cytoskeletal-perturbing or kinase activating/inhibiting reagents mentioned to specific target cells might be envisaged, such concerns essentially apply to all anti-cancer and other therapies. The approach of using such agents or similar ones in conjunction with cell-specific (e.g., antigen or receptor-specific) drug delivery systems (e.g., refs. 166-168) merits consideration, and it seems that the idea of targeting receptor lateral movement as a site of pharmacological action should not be discounted on such grounds alone.

An additional aspect to such considerations is that different forms of plasma membrane integral receptors exist which vary in their lateral mobility properties. An example is the CD2 ligand/coreceptor LFA-3, which has two isoforms with radically different lateral mobility properties (see chapter 7, section D).[45] The transmembrane anchored form is essentially immobile and has severely reduced adhesion strengthening activity, in contrast to the highly mobile glycosyl-phosphoinositol-(GPI) linked form.[45] Understanding of the gene regulation and post-transcriptional modification determining the relative amounts of these isoforms in the cell membrane could lead to possibilities in terms of modulating the cellular LFA-3 lateral mobility and accordingly cell-adhesion/cell-cell recognition properties. Similar possibilities may exist in the case of the CR3 integrin, where PMA and various other treatments can mobilize intracellular stores of receptor and bring them to the cell surface (see chapter 7 section E).[70,71,169-172]

In the case of many enveloped virus-cell recognition interactions, lateral movement within the plane of the membrane to the coated pit structures is required for the initial endocytotic uptake process to take place (see refs. 164 and 165).[173,174] It is conceivable that this could be used as a possible target for pharmacological intervention using crosslinking antibodies or signal transduction reagents. Once endocytosed by the host cell, enveloped viruses fuse with the endosomal membrane, a step essential for viral infection to proceed. Intriguingly, several studies[175-177] indicate that virus component-mediated membrane fusion is dependent on the lateral mobility of membrane integral viral components. This has been shown for the influenza virus HA (hemagglutinin)[175] and the Sendai virus F (the fusion protein) and HN (hemagglutinin-neuraminidase)[176] proteins. Low pH treatments inhibiting both lateral mobility and fusogenic activity were used on HA viral strain variants to establish a clear correlation between loss of mobility and inactivation of fusogenic activity, in accord with the notion that lateral motion of the HA proteins is required for fusion.[175]

Henis et al[176] used a crosslinking immobilizing antibody/concanavalin A combination to block the lateral mobility of F and/or HN and found that the mobility of both components was necessary for Sendai virus fusion with cells. Consistent with these results, Volsky et al[177] used cationized ferritin (see also chapter 5, section C) to crosslink and immobilize membrane proteins and found concomitant inhibition of Sendai virus-mediated fusion. The implication from these and other similar studies is that membrane protein lateral movement is essential for membrane fusion, and since the latter is required for viral infection to proceed in the case of enveloped viruses, such as influenza virus, reoviruses, retroviruses and Rubella virus, one can speculate that targeting the fusogenic stage through inhibiting viral protein lateral mobility might be a pharmacological approach of potential usefulness to protect against or combat viral disease.

Finally, as touched upon in chapter 3 (section E), GPI-linked proteins (e.g., LFA-3) are generally highly laterally mobile and in some cases appear to be able to insert spontaneously into membranes in the absence of detergent which may have application in drug delivery.[178] Kooyman et al[179] showed that GPI-linked proteins are capable of undergoing intermembrane transfer between cell types in vivo, whereby GPI-linked proteins (in this case, complement restricting factors) expressed on the surface of transgenic mouse red blood cells could be transferred in a functional form to vascular endothelial cells. This is a most interesting observation in terms of the possibility of delivering therapeutic proteins or other molecules to target tissues.[179] To what extent the high lateral mobility of the GPI-linked proteins is a key factor in this transfer is unknown, but based on the results above, particularly with respect to membrane fusion effected by viral proteins, it seems reasonable to postulate that high mobility may be essential to ensure access

of the requisite proteins to the cell-cell intermembrane contact site(s). It thus seems that high protein lateral mobility might itself be able to be exploited to effect drug delivery.

F. CONCLUDING REMARKS

As indicated in the preface of this book, one reason for its conception was to encourage consideration of the membrane in two-dimensional rather than one-dimensional terms; that is, that the membrane is a lattice of sites for interaction within its plane, rather than a simple barrier between the extracellular medium, the source of hormones/growth factors and other ligands, and the cytosol where the signaling events of phosphorylation, regulation of second messenger concentrations, etc. occur. Twenty years after formulation of the Mobile Receptor Hypothesis, and about 25 years since conception of the fluid mosaic model of biological membranes, it is to be hoped that this endeavour has not been entirely in vain.

The evidence expounded in this book represents a reasonable case for acceptance of the hypothesis's basic tenet that receptor lateral movement in the plane of the membrane is essential for signal transduction. Receptor movement as well as immobilization is required in a variety of signaling systems as enumerated at the beginning of the chapter. Lateral movement of hormone-occupied receptor is integral to hormonal stimulation and signal amplification in systems where membrane-associated trimeric GTP-binding proteins mediate signaling and where receptor dimerization/aggregation is required for signal transduction as in the case of tyrosine kinase receptors, while lateral movement of unliganded receptor to the sites of interaction is required in various receptor-mediated adhesion interactions. Immobilization of ligand-occupied receptor movement is the initial event in abrogation of the stimulatory signal mediated by GTP-binding protein activating receptors, while it is essential to communication of the stimulatory signal in systems such as those of receptors/coreceptors mediating adhesion/cell-recognition. Receptor immobilization also appears to be a prerequisite for internalization in all receptor systems. As also discussed above, both the rate of diffusion (determining the rate of signal transduction) and particularly the fraction of mobile receptors (determining the amplitude of the stimulatory response) are critical parameters in determining the nature of the signal transduced into the cytosol. Since diffusion-mediated events in the cytosol are orders of magnitude faster than those within the plane of the membrane, the former determine the overall rate of both signal transduction.

As discussed at the end of section D and illustrated in Figure 8.5, plasma membrane protein lateral mobility appears to be a critical control or integration point for signals from both outside and within the cell. Modulation of lateral movement through dynamic interactions with the membrane skeleton and cytoskeleton and intracellular signals, such as phosphorylation, enables cellular response to be integrated; receptor lateral movement determines the initial cellular response, which in turn regulates receptor (including heterologous receptor) lateral mobility. In this context, the fact that receptor lateral movement is so central to signal transduction and cellular response opens the way for potential pharmacological applications (section E above). Further understanding of integral plasma membrane protein and particularly receptor lateral mobility, and of the precise signals and mechanisms that regulate it, may enable the tenets of the Mobile Receptor Hypothesis to be exploited in a clinical setting in the not too distant future.

REFERENCES

1. Jans DA. The mobile receptor hypothesis revisited: a mechanistic role for hormone receptor lateral mobility in signal transduction. Biochim Biophys Acta 1992; 1113:271-276.
2. Jans DA, Pavo I. A mechanistic role for polypeptide hormone receptor lateral mobility in signal transduction. Amino Acids 1995; 9:93-109.
3. Jans DA, Peters R, Zsigo J et al. The adenylate cyclase-coupled vasopressin V_2-receptor is highly laterally mobile in membranes of LLC-PK$_1$ renal epithelial cells at physiological temperature. EMBO J 1989; 8(9):2431-2438.
4. Jans DA, Peters R, Fahrenholz F. Lateral mobility of the phospholipase-C-activating vasopressin V_1-type receptor in A7r5 smooth muscle cells: a comparison with the adenylate cyclase-coupled V_2-receptor. EMBO J 1990; 9(9):2693-2699.
5. Jans DA, Peters R, Jans P et al. Ammonium chloride affects receptor number and lateral mobility of the vasopressin V_2-type receptor in the plasma membrane of LLC-PK$_1$ renal epithelial cells: role of the cytoskeleton. Exper Cell Res 1990; 191:121-128.
6. Jans DA, Peters R, Fahrenholz F. An inverse relationship between receptor internalization and the fraction of laterally mobile receptors for the vasopressin renal-type V_2-receptor; an active role for receptor immobilization in down-regulation? FEBS Lett 1990; 274:223-226.
7. Jans DA, Peters R, Jans P et al. Vasopressin V_2-receptor mobile fraction and ligand-dependent adenylate cyclase-activity are directly correlated in LLC-PK$_1$ renal epithelial cells. J Cell Biol 1991; 114(1):53-60.
8. Pavo I, Jans DA, Peters R et al. A vasopressin antagonist that binds to the V_2-receptor of LLC-PK$_1$ renal epithelial cells is highly laterally mobile but does not effect ligand-induced receptor immobilization. Biochim Biophys Acta 1994; 1223:240-246.
9. Zakharova OM, Rosenkranz AA, Sobolev AS. Modification of fluid lipid and mobile protein fractions of reticulocyte plasma membranes affects agonist-stimulated adenylate cyclase. Application of the percolation theory. Biochim Biophys Acta 1995; 1236:177-184.
10. Sobolev AS, Rosenkranz AA, Kazarov AR. Interaction of proteins of the adenylate cyclase complex: area-limited mobility of movement along the whole membrane? Analysis with the application of the percolation theory. Biosc Rep 1984; 4:897-902.

11. Sobolev AS, Kazarov AR, Rosenkranz AA. Application of percolation theory principles to the analysis of interaction of adenylate cyclase complex proteins in cell membranes. Mol Cell Biochem 1988; 81(1):19-28.

12. Atlas D, Volsky DJ, Levitzki A. Lateral mobility of beta-receptors involved in adenylate cyclase activation. Biochim Biophys Acta 1980; 597(1):64-69.

13. Tolkovsky AM, Levitzki A. Mode of coupling between the β-adrenergic receptor and adenylate cyclase in turkey erythrocytes. Biochemistry 1978; 17:3795-3810.

14. Tolkovsky AM, Levitzki, A. Coupling of a single adenylate cyclase to two receptors: adenosine and catecholamine. Biochemistry 1978; 17:3811-3817.

15. Tolkovsky A, Braun S, Levitzki A. Kinetics of interaction between β-receptors, GTP-proteins and the catalytic subunit of turkey erythrocyte adenylate cyclase. Proc Natl Acad Sci USA 1982; 79:213-217.

16. Hanski E, Rimon G, Levitzki A. Adenylate cyclase activation by the β-adrenergic receptor as a diffusion controlled process. Biochemistry 1979; 18:846-853.

17. Rimon G, Hanski E, Levitzki A. Temperature dependence of beta receptor, adenosine receptor, and sodium fluoride stimulated adenylate cyclase from turkey erythrocytes. Biochemistry 1980; 19:4451-4460.

18. Orly J, Schramm M. Fatty acids as modulators of membrane functions: catecholamine-activated adenylate cyclase of the turkey erythrocyte. Proc Natl Acad Sci USA 1975; 72:3433-3437.

19. Bergman RN, Hechter H. Neurophyseal hormone-responsive renal adenylate cyclase IV. A random-hit matrix model for coupling in a hormone-sensitive adenylate cyclase system. J Biol Chem 1978; 253:3238-3250.

20. Kwon G, Axelrod D, Neubig RR. Lateral mobility of tetramethylrhodamine (TMR) labelled G protein alpha and beta gamma subunits in NG 108-15 cells. Cell Signal 1994; 6(6):663-679.

21. Neubig RR. Membrane organization in G-protein mechanisms. FASEB J 1994; 8:939-946.

22. Neubig RR, Sklar LA. Subsecond modulation of formyl peptide-linked guanine nucleotide-binding proteins by guanosine 5'-O-(3-thio)triphosphate in permeabilized neutrophils. Mol Pharmacol 1993; 43(5):734-740.

23. Brandt DR, Ross EM. Catecholamine-stimulated GTPase cycle; multiple sites of regulation by β-adrenergic receptor and Mg^{2+} studied in reconstituted receptor-G_s vesicles. J Biol Chem 1986; 261:1656-1664.

24. Ransnas LA, Insel PA. Subunit dissociation is the mechanism for hormonal activation of the Gs protein in native membranes. J Biol Chem 1988; 263(33):17239-17242.

25. Alousi AA, Jasper JR, Insel PA et al. Stoichiometry of receptor-Gs-adenylate cyclase interactions. FASEB J 1991; 5:2300-2303.

26. Mueller H, Weingarten R, Ransnas LA et al. Differential amplification of antagonistic receptor pathways in neutrophils. J Biol Chem 1991; 266(20):12939-12943.

27. Rees AR, Gregoriou M, Johnson P et al. High affinity epidermal growth factor receptors on the surface of A-431 cells have restricted lateral diffusion. EMBO J 1984; 3:1843-1847.

28. Ballotti R, Lammers R, Scimeca IC et al. Intermolecular transphosphorylation between insulin receptors and EGF-insulin receptor chimerae. EMBO J 1989; 8:3303-3309.

29. Gadella TW Jr, Jovin TM. Oligomerization of epidermal growth factor receptors on A431 cells studied by time-resolved fluorescence imaging microscopy. A stereochemical model for tyrosine kinase receptor activation. J Cell Biol 1995; 129(6):1543-1558.

30. Gilboa L, Ben Levy R, Yarden Y et al. Roles for a cytoplasmic tyrosine and tyrosine kinase activity in the interactions of Neu receptors with coated pits. J Biol Chem 1995; 270(13):7061-7067.

31. Weiner DB, Lui J, Cohen JA et al. A point mutation in the neu oncogene mimics ligand induction of receptor aggregation. Nature 1989; 339:230-231.

32. Heffetz D, Zick Y. Receptor aggregation is necessary for activation of the soluble insulin receptor kinase. J Biol Chem 1986; 261:889-894.

33. Giugni TD, Braslau DL, Haigler HT. Electric field-induced redistribution and postfield relaxation of epidermal growth factor receptors on A431 cells. J Cell Biol 1987; 104(5):1291-1297.

34. Venkatakrishnan G, McKinnon CA, Pilapil CG et al. Nerve growth factor receptors are preaggregated and immobile on responsive cells. Biochemistry 1991; 30(11):2748-2753.

35. Ljungquist-Hoeddelius P, Lirvall M, Wasteson A et al. Lateral diffusion of PDGF-β receptor in human fibroblasts. Bioscience Reports 1991; 11(1):43-52.

36. Ljungquist P, Wasteson A, Magnusson KE. Lateral diffusion of plasma membrane receptors labelled with either platelet-derived growth factor (PDGF) or wheat germ agglutinin (WGA) in human leukocytes and fibroblasts. Bioscience Reports 1989; 9:63-73.

37. Yarden Y, Schlessinger J. Epidermal growth factor induces rapid, reversible aggregation of the purified epidermal growth factor receptor. Biochem 1987; 26:1443-1451.

38. Posner RG, Subramanian K, Goldstein B et al. Simultaneous cross-linking by two nontriggering bivalent ligands causes synergistic signaling of IgE Fc epsilon RI complexes. J Immunol 1995; 155(7):3601-3609.

39. Menon AK, Holowka D, Webb WW et al. Clustering, mobility, and triggering activity of small oligomers of immunoglobulin E on rat basophilic leukemia cells. J Cell Biol 1986; 102:534-540.

40. Menon AK, Holowka D, Webb WW et al. Cross-linking of receptor-bound IgE to aggregates larger than dimers leads to rapid immobilization. J Cell Biol 1986; 102:541-550.

41. Schlessinger J, Webb WW, Elson EL. Lateral motion and valence of Fc receptors on rat peritoneal mast cells. Nature 1976; 264:550-552.

42. Mao SY, Varin-Blank N, Edidin M et al. Immobilization and internalization of mutated IgE receptors in transfected cells. J Immunol 1991; 146(3):958-966.

43. McCloskey MA, Liu ZY, Poo MM. Lateral electromigration and diffusion of Fcε receptors on rat basophilic leukemia cells: effects of IgE binding. J Cell Biol 1984; 99(3):778-787.

44. Liu SJ, Hahn WC, Bierer BE et al. Intracellular mediators regulate CD2 lateral diffusion and cytoplasmic Ca^{2+} mobiliza-

tion upon CD2-mediated T cell activation. Biophys J 1995; 68(2):459-470.

45. Chan PY, Lawrence MB, Dustin ML et al. Influence of receptor lateral mobility on adhesion strengthening between membranes containing LFA-3 and CD2. J Cell Biol 1991; 115(1):245-255.

46. Carpen O, Dustin ML, Springer TA et al. Motility and ultrastructure of large granular lymphocytes on lipid bilayers reconstituted with adhesion receptors LFA-1, ICAM-1, and two isoforms of LFA-3. J Cell Biol 1991; 115(3):861-871.

47. Liu SJ, Golan DE. Mobilization of intracellular free Ca^{2+} upon T cell activation via TCR/CD3 is required to induce a decrease in CD2 lateral mobility in Jurkat T cell membranes. Biophys J 1994; 66:149a.

48. Liu SJ, Golan DE. Regulation of CD2 lateral mobility by calmodulin and calcineurin in Jurkat T cell membranes. Biophys J 1994; 66:149a.

49. Duband JL, Nuckolls GH, Ishihara A et al. Fibronectin receptor exhibits high lateral mobility in embryonic locomoting cells but is immobile in focal contacts and fibrillar streaks in stationary cells. J Cell Biol 1988; 107:1385-1396.

50. Myohanen HT, Stephens RW, Hedman K et al. Distribution and lateral mobility of the urokinase-receptor complex at the cell surface. J Histochem Cytochem 1993; 41(9):1291-1301.

51. Panaotou G, Waterfield MD. The assembly of signalling complexes by receptor tyrosine kinases. Bioessays 1993; 15(3):171-177.

52. Ullrich A, Schlessinger J. Signal transduction by receptors with tyrosine kinase activity. Cell 1990; 61:203-212.

53. Noh DY, Shin SH, Rhee SG. Phosphoinositide-specific phospholipase C and mitogenic signaling. Biochim Biophys Acta 1995; 1242(2):99-113.

54. Yamashita T, Mao SY, Metzger H. Aggregation of the high-affinity IgE receptor and enhanced activity of p53/56[lyn] protein-tyrosine kinase. Proc Natl Acad Sci USA 1994; 91(23):11251-11255.

55. Zhang F, Yang B, Odin JA et al. Lateral mobility of Fcγ RIIa is reduced by protein kinase C activation. FEBS Lett 1995; 376(1-2):77-80.

56. Sarmay G, Rozsnyay Z, Koncz G et al. Interaction of signaling molecules with human Fcγ RIIb1 and the role of various Fcγ RIIb isoforms in B-cell regulation. Immunol Lett 1995; 44(2-3):125-131.

57. Apgar JR. Association of the crosslinked IgE receptor with the membrane skeleton is independent of the known signaling mechanisms in rat basophilic leukemia cells. Cell Regul 1991; 2(3):181-191.

58. Paolini R, Numerof R, Kinet JP. Phosphorylation/dephosphorylation of high-affinity IgE receptors: a mechanism for coupling/uncoupling a large signaling complex. Proc Natl Acad Sci USA 1992; 89(22):10733-10737.

59. Kent UM, Mao SY, Wofsy C et al. Dynamics of signal transduction after aggregation of cell-surface receptors: studies on the type I receptor for IgE. Proc Natl Acad Sci USA 1994; 91(8):3087-3091.

60. Schlessinger J. Signal transduction by allosteric receptor oligomerization. Trends Biochem Sci 1988; 13:443-447.

61. Schlessinger J. The epidermal growth factor receptor as a multifunctional allosteric protein. Biochemistry 1989; 27:3119-3123.

62. Yarden Y, Schlessinger J. Self-phosphorylation of epidermal growth factor: evidence for a model of intermolecular allosteric activation. Biochem 1987; 26:1434-1442.

63. Becker KE, Ishizaka T, Metzger H et al. Surface IgE on human basophils during histamine release. J Exp Med 1973; 138:394-409.

64. Lawson D, Fewtrell C, Gomperts B et al. Anti-immunoglobulin-induced histamine secretion by rat peritoneal mast cells studied by immunoferritin electron microscopy. J Exp Med 1975; 142(2):391-402.

65. Magro AM, Alexander A. Histamine-release - in vitro studies of inhibitory region of dose-response curve. J Immunol 1974; 112:1762-1765.

66. Griffin FM Jr, Mullinax PJ. Effects of differentiation in vivo and of lymphokine treatment in vitro on the mobility of C3 receptors of human and mouse mononuclear phagocytes. J Immunol 1985; 135(5):3394-3397.

67. Griffin Jnr FM, Mullinax PJ. Augmentation of macrophage complement receptor function in vitro. III. C3b receptors that promote phagocytosis migrate within the plane of macrophage plasma membrane. J Med Exp 1981; 154:291-305.

68. Hermanowski-Vosatka A, Detmers PA, Gotze O et al. Clustering of ligand on the surface of a particle enhances adhesion to receptor-bearing cells. J Biol Chem 1988; 263(33):17822-17827.

69. Ross GD, Reed W, Dalzell JG et al. Macrophage cytoskeletal association with CR3 and CR4 regulates receptor mobility and phagocytosis of iC3b-opsonized erythrocytes. J Leukoc Biol 1992; 51(2):109-117.

70. Detmers PA, Wright SD, Olsen E et al. Aggregation of complement receptors on human neutrophils in the absence of ligand. J Cell Biol 1987; 105:1137-1145.

71. Pryzwansky KB, Wyatt T, Reed W et al. Phorbol ester induces transient focal concentrations of functional, newly expressed CR3 in neutrophils at sites of specific granule exocytosis. Eur J Cell Biol 1991; 54(1):61-75.

72. Goldstein B, Griego R, Wofsy C. Diffusion-limited forward rate constants in two dimensions. Application to the trapping of cell surface receptors by coated pits. Biophys J 1984; 46(5):573-586.

73. Vlodavsky I, Fielding PE, Johnson LK et al. Inhibition of low density lipoprotein uptake in confluent endothelial cell monolayers correlates with a restricted surface receptor redistribution. J Cell Physiol 1979; 100(3):481-495.

74. Thatte HS, Bridges KR, Golan DE. Microtubule inhibitors differentially affect translational movement, cell surface expression, and endocytosis of transferrin receptors in K562 cells. J Cell Physiol 1994; 160:345-357.

75. Thatte HS, Bridges KR, Golan DE. ATP depletion causes translational immobilization of cell surface transferrin receptors in K562 cells. J Cell Physiol 1996; 166:446-452.

76. Tank DW, Fredericks WJ, Barak LS et al. Electric field-induced redistribution and postfield relaxation of low density lipoprotein receptors on cultured human fibroblasts. J Cell Biol 1985; 101(1):148-157.

77. Smith PR, Stoner JC, Viggiano SC et al. Effects of vasopressin and aldosterone on the lateral mobility of eptihelial Na⁺ channels in A6 epithelial cells. J Memb Biol 1995; 147(2):195-205.

78. Meier T, Perez GM, Wallace BG. Immobilization of nicotinic acetylcholine receptors on mouse C2 myotubes by agrin-induced protein tyrosine phosphorylation. J Cell Biol 1995; 131(2):441-451.

79. Grebenkamper K, Tosi PF, Lazarte JE et al. Modulation of CD4 lateral mobility in intact cells by an intracellularly applied antibody. Biochem J 1995; 312(1):251-259.

80. Fire E, Zwart DE, Roth MG et al. Evidence from lateral mobility studies for dynamic interactions of a mutant influenza hemagglutinin with coated pits. J Cell Biol 1991; 115:1585-1594.

81. Ferns M, Deiner M, Hall Z. Agrin-induced acetylcholine receptor clustering in mammalian muscle requires tyrosine phosphorylation. J Cell Biol 1996; 132(5):937-944.

82. Roess DA, Niswender GD, Barisas BG. Cytochalasins and colchicine increase the lateral mobility of human chorionic gonadotropin-occupied luteinizing hormone receptors on ovine luteal cells. Endocrinol 1988; 122:261-269.

83. Roettger BF, Rentsch RU, Hadac EM et al. Insulation of a G protein-coupled receptor on the plasmalemmal surface of the pancreatic acinar cell. J Cell Biol 1995; 130:579-590.

84. Srinivasan Y, Elmer L, Davis J et al. Ankyrin and spectrin associate with voltage-dependent sodium channels in brain. Nature 1988; 333(6164):177-180.

85. Woda BA, Woodin MB. The interaction of lymphocyte membrane proteins proteins with the lymphocyte membrane proteins with the lymphocyte cytoskeletal matrix. J Immunol 1984; 133(5):27767-2772.

86. Bennett N, Sitaramayya A. Inactivation of photoexcited rhodopsin in retinal rods: the roles of rhodopsin kinase and 48-kDa protein (arrestin). Biochemistry 1988; 27:1710-1715.

87. Mao SY, Alber G, Rivera J et al. Interaction of aggregated native and mutant IgE receptors with the cellular skeleton. Proc Natl Acad Sci USA 1992; 89(1):222-226.

88. Robertson D, Holowka D, Baird B. Cross-linking of immunoglobulin E-receptor complexes induces their interaction with the cytoskeleton of rat basophilic leukemia cells. J Immunol 1986; 136:4565-4572.

89. Chopra H, Hatfield JS, Chang YS et al. Role of tumor cell cytoskeleton and membrane glycoprotein IRgpIIb/IIIa in platelet adhesion to tumor cell membrane and tumor cell-induced platelet aggregation. Cancer Res 1988; 48:3787-3800.

90. Jaffe SH, Friedlander F, Matsuzaki F et al. Differential effects of the cytoplasmic domains of cell adhesion molecules on cell aggregation and sorting-out. Proc Natl Acad Sci USA 1990; 87:3589-3593.

91. Carpen O, Pallai P, Staunton DE et al. Association of intercellular adhesion molecule-1 (ICAM-1) with actin-containing cytoskeleton and alpha-actinin. J Cell Biol 1992; 118:1223-1234.

92. Burridge K. Substrate adhesions in normal and transformed fibroblasts: organization and regulation of cytoskeletal, membrane, and extracellular matrix components at focal contacts. Cancer Rev 1987; 4:18-78.

93. Kupfer A, Singer SJ. The specific interaction of helper T-cells and antigen-presenting B cells. IV. Membrane and cytoskeletal reorganizations in the bound T cell as a function of antigen dose. J Exp Med 1989; 170:1697-1713.

94. Freed BM, Lempert N, Lawrence DA. The inhibitory effects of N-ethylmaleimide, colchicine and cytochalasins on human T-cell functions. Int J Immunopharmacol 1989; 11:459-465.

95. Barak LS, Webb WW. Diffusion of low density lipoprotein-receptor complex on human fibroblasts. J Cell Biol 1982; 95:846-852.

96. Poo H, Krauss JC, Mayo-Bond L et al. Interaction of Fc gamma receptor type IIIB with complement receptor type 3 in fibroblast transfectants: evidence from lateral diffusion and resonance energy transfer studies. J Mol Biol 1995; 247(4):597-603.

97. Stolpen AH, Golan DE, Pober JS. Tumor necrosis factor and immune interferon act in concert to slow the lateral diffu-

sion of proteins and lipids in human endothelial cell membranes. J Cell Biol 1988; 107:781-789.

98. Polgar K, Yacono PW, Golan DE et al. Immune interferon gamma inhibits translational diffusion of a plasma membrane protein in preimplantation stage mouse embryos: a T-helper 1 mechanism for immunologic reproductive failure. Am J Obstet Gynecol 1996; 174(1/1):282-287.

99. Behrisch A, Dietrich C, Noegel AA et al. The actin-binding protein hisactophilin binds in vitro to partially charged membranes and mediates actin coupling to membranes. Biochemistry 1995; 34:15182-15190.

100. Goldmann WH, Niggli V, Kaufmann S et al. Probing actin and liposome interaction of talin and talin-vinculin complexes: a kinetic, thermodynamic and lipid labeling study. Biochemistry 1992; 31:7665-7671.

101. Kaufmann S, Kas J, Goldmann WH et al. Talin anchors and nucleates actin filaments at lipid membranes. A direct demonstration. FEBS Lett 1992; 314:203-205.

102. Isenberg G. Actin binding proteins—lipid interactions. J Muscle Res Cell Motil 1991; 12:136-144.

103. Paccaud JP, Reith W, Johansson B et al. Role of internalization signals and receptor mobility. J Biol Chem 1993; 268(31):23191-23196.

104. Ishijima SA, Asakura H, Suzuta T. Participation of cytoplasmic organelles in E-rosette formation. Immunol Cell Biol 1991; 69:403-409.

105. Horwitz A, Duggan K, Buck C et al. Interaction of plasma membrane fibronectin receptor with talin. A transmembrane linkage. Nature 1986; 320:531-532.

106. Burridge K, Molony L, Kelly T. Adhesion plaques: sites of transmembrane interaction between the extracellular matrix and the actin cytoskeleton. J Cell Sci Suppl 1987; 8:211-229.

107. Schlessinger J, Axelrod D, Koppel DE et al. Lateral transport of a lipid probe and labeled proteins on a cell membrane. Science 1977; 195:307-309.

108. Schlessinger J, Koppel DE, Axelrod D et al. Lateral transport on cell membranes:

mobility of concanavalin A receptors on myoblasts. Proc Natl Acad Sci USA 1976; 73:2409-2413.

109. Pribluda VS, Metzger H. Transmembrane signaling by the high-affinity IgE receptor on membrane preparations. Proc Natl Acad Sci USA 1992; 89(23):11446-11450.

110. Henning SW, Meuer SC, Samstag Y. Serine phosphorylation of a 67-kDa protein in human T lymphocytes represents an accessory receptor-mediated signaling event. J Immunl 1994; 152:4804-4815.

111. Graeser D, Neubig RR. Compartmentation of receptors and guanine nucleotide-binding proteins in NG108-15 cells: lack of cross-talk in agonist binding among the alpha 2-adrenergic, muscarinic, and opiate receptors. Mol Pharmacol 1993; 43(3):434-443.

112. Robinson MS. The role of clathrin, adaptors and dynamin in endocytosis. Curr Opin Cell Biol 1994: 6(4):538-544.

113. Fischer-von-Mollard-G, Stahl B, Li C et al. Rab proteins in regulated exocytosis. Trends Biochem Sci 1994; 19(4):164-168.

114. Klueppelberg UG, Gates LK, Gorelick FS et al. Agonist regulated phosphorylation of the pancreatic cholecystokinin receptor. J Biol Chem 1991; 266:2403-2408.

115. Lutz MP, Pinon DI, Gates LK et al. Control of cholecystokinin receptor dephosphorylation in pancreatic acinar cells. J Biol Chem 1993; 268:12136-12142.

116. Jesaitis AJ, Bokoch GM, Tolley JO et al. Lateral segregation of neutrophil chemotactic receptors into actin- and fodrin-rich plasma membrane microdomains depleted in guanyl nucleotide regulatory proteins. J Cell Biol 1988; 107:921-928.

117. Jesaitis AJ, Tolley JO, Allen RA. Receptor-cytoskeleton interactions and membrane traffic may regulate chemoattractant-induced superoxide production in human granulocytes. J Biol Chem 1986; 261:13662-13669.

118. Johansson B, Wymann MP, Holmgren-Peterson K et al. N-formyl peptide receptors in human neutrophils display distinct membrane distribution and lateral mobility when labeled with agonist and antagonist. J Cell Biol 1993; 121:1281-1289.

119. Wallace BG. Regulation of the interaction of nicotinic acetylcholine receptors with the cytoskeleton by agrin-activated protein tyrosine kinase. J Cell Biol 1995; 128:1121-1129.

120. Bloch RJ, Morrow JS. An unusual beta-spectrin associated with clustered acetylcholine receptors. J Cell Biol 1989; 108:481-494.

121. Tinsley JM, Blake DJ, Zuellig RA et al. Increasing complexity of the dystrophin-associated protein complex. Proc Natl Acad Sci USA 1994; 91:8307-8313.

122. Ferns M, Deiner M, Hall Z. Agrin-induced acetylcholine receptor clustering in mammalian muscle requires tyrosine phosphorylation. J Cell Biol 1996; 132(5):937-944.

123. Nelson WJ, Hammerton RW. A membrane-cytoskeletal complex containing Na^+,K^+-ATPase, ankyrin, and fodrin in Madin-Darby canine kidney (MDCK) cells. Implications for the biogenesis of epithelial cell polarity. J Cell Biol 1989; 108:893-902.

124. Nelson WJ, Veshnock PJ. Ankyrin binding to Na^+,K^+-ATPase and implications for the organization of membrane domains of polarized cells. Nature 1987; 328:533-536.

125. Paller MS. Lateral mobility of Na,K-ATPase and membrane lipids in renal cells. Importance of cytoskeletal integrity. J Membr Biol 1994; 142(1):127-135.

126. Golan DE, Veatch W. Lateral mobility of band 3 in the human erythrocyte membrane studied by fluorescence photobleaching recovery: evidence for control by cytoskeletal interactions. Proc Natl Acad Sci USA 1980; 77:2537-2541.

127. Tsuji A, Ohnishi, S. Restriction of the lateral motion of Band 3 in the erythrocyte membrane by the cytoskeletal network: dependence on spectrin association state. Biochemistry 1986; 25:6133-6139.

128. Tsuji A, Kawasaki S, Ohnishi, S et al. Regulation of band 3 mobilities in erythrocyte ghost membrane by protein association and cytoskeletal network. Biochemistry 1988; 27:7447-7452.

129. Jans DA, Hemmings BA. cAMP-dependent protein kinase activation affects vasopressin V_2-receptor number and internalization in $LLC-PK_1$ renal epithelial cells. FEBS Lett 1991; 281:267-271.

130. Sohlemann-P, Hekman-M, Puzicha-M et al. Binding of purified recombinant β-arrestin to guanine-nucleotide-binding-protein-coupled receptors. Eur J Biochem 1995; 232(2):464-472.

131. Lohse MJ, Andexinger S, Pitcher J et al. Receptor-specific desensitization with purified proteins. Kinase dependence and receptor specificity of β-arrestin and arrestin in the $\beta2$-adrenergic receptor and rhodopsin systems. J Biol Chem 1992; 267(12):8558-8564.

132. Benovic JC, Kuhn H, Weyand I et al. Functional desensitization of the isolated β-adrenergic receptor by the β-adrenergic receptor kinase: potential role of an analog of the retinal protein arrestin (48-kDa protein). Proc Natl Acad Sci USA 1987; 84:8879-8882.

133. Johnson GL, Dhanasekaran N. The G-protein family and their interactions with receptors. Endocrine Reviews 1992; 10(3):317-331.

134. Walter RJ, Berlin RD, Pfeiffer JR et al. Polarization of endocytosis and receptor topography on cultured macrophages. J Cell Biol 1980; 86(1):199-211.

135. Raff MC, De Petris S. Ligand-induced redistribution of membrane macromolecules: some possible implications for cancer. J Clin Pathol Suppl R Coll Pathol 1974; 7:31-34.

136. Allison AC, Davies P, De Petris S. Role of contractile microfilaments in macrophage movement and endocytosis. Nature New Biol 1971; 232(31):153-155.

137. Dustin P. Microtubules. KG. Berlin: Springer-Verlag GmbH & Co. 1978.

138. Sarndahl E, Bokoch GM, Stendahl O et al. Stimulus-induced dissociation of alpha subunits of heterotrimeric GTP-binding proteins from the cytoskeleton of human neutrophils. Proc Natl Acad Sci USA 1993; 90:6552-6556.

139. Leiber D, Jasper JR, Alousi AA et al. Alteration in Gs-mediated signal transduction in S49 lymphoma cells treated with inhibitors of microtubules. J Biol Chem 1993; 268:3833-3837.

140. Macias MJ, Musacchio A, Ponstingl H et al. Structure of the pleckstrin homology

domain from β-spectrin. Nature 1994; 369:675-677.

141. Wang DS, Shaw R, Winkelmann JC et al. Binding of PH domains of β-adrenergic receptor kinase and β-spectrin to WD40/β-transducin repeat containing regions of the β-subunit of trimeric G-proteins. Biochem Biophys Res Commun 1994; 203(1):29-35.

142. Salim K, Gout I, Guruprasad L et al. Crystal structure of the pleckstrin homology domain from dynamin. Nat Struct Biol 1994; 1(11):782-788.

143. Musacchio A, Gibson T, Rice P et al. The PH domain: a common piece in the structural patchwork of signalling proteins. Trends Biochem Sci 1993; 18(9):343-348.

144. Shaw G. Identification of novel pleckstrin homology (PH) domains provides a hypothesis for PH domain function. Biochem Biophys Res Commun 1993; 195(2): 1145-1151.

145. Gesemann M, Cavalli V, Denzer AJ et al. Alternative splicing of agrin alters its binding to heparin, dystroglycan, and the putative agrin receptor. Neuron 1996; 16(4):755-767.

146. Denzer AJ, Gesemann M, Schumacher B et al. An amino-terminal extension is required for the secretion of chick agrin and its binding to extracellular matrix. J Cell Biol 1995; 131(6/1):1547-1560.

147. Namba Y, Ito M, Zu Y et al. Human T-cell L-plastin bundles actin filaments in a calcium-dependent manner. J Biochem (Tokyo) 1992; 112:503-507.

148. Pacaud M, Derancourt I. Purification and further characterization of macrophage 70-kDa protein, a calcium-regulated, actin-binding protein identical to L-plastin. Biochemistry 1993; 32:3448-3455.

149. Hausdorff WP, Caron MG, Lefkowitz RJ. Turning off the signal: desensitisation of the β-adrenergic receptor function. FASEB J 1990; 4:2881-2889.

150. Wallace BG. Regulation of agrin-induced acetyl choline receptor aggregation by Ca²⁺ and phorbol ester. J Cell Biol 1988; 107:267-278.

151. Magnusson KE, Gustafsson M, Holmgren K et al. Small intestinal differentiation in human colon carcinoma HT29 cells has distinct effects on the lateral diffusion of lipid (ganglioside GM1) and proteins (HLA class 1, HLA class 2, and neoplastic epithelial antigens) in the apical cell membrane. J Cell Physiol 1990; 143(2):381-390.

152. Wallace BG. Staurosporine inhibits agrin-induced acetyl choline receptor phosphorylation and aggregation. J Cell Biol 1994; 125:661-668.

153. Yahara I, Edelman GM. Modulation of lymphocyte receptor mobility by locally bound concanavalin A. Proc Natl Acad Sci USA 1975; 72:1579-1583.

154. Schlessinger J, Elson EL, Webb WW et al. Receptor diffusion on cell surfaces modulated by locally bound concanavalin A. Proc Natl Acad Sci USA 1977; 74(3):1110-1114.

155. Henis YI, Elson EL. Inhibition of the mobility of mouse lymphocyte surface immunoglobulins by locally bound concanavalin A. Proc Natl Acad Sci USA 1981; 78:1072-1076.

156. Edelman GM, Yahara I, Wang JL. Receptor mobility and receptor-cytoplasmic interactions in lymphocytes. Proc Natl Acad Sci USA 1973; 70:1442-1446.

157. Sharar S, Wim RK, Harlan JM. The adhesion cascade and anti-adhesion therapy: an overview. Springer Semin Immunopathol 1995; 16:359-378.

158. Henis YI, Hekman M, Elson EL et al. Lateral diffusion of β-receptors in membranes of cultured liver cells. Proc Natl Acad Sci USA 1982; 79:2907-2911.

159. Merck Index 11, 2796, 9049, 9887, respectively.

160. Bruno S, Ardelt B, Skierski JS et al. Different effects of staurosporine, an inhibitor of protein kinases, on the cell cycle and chromatin structure of normal and leukemic lymphocytes. Cancer Res 1992; 52(2):470-473.

161. Jarvis WD, Turner AJ, Povirk LF et al. Induction of apoptotic DNA fragmentation and cell death in HL-60 human promyelocytic leukemia cells by pharmacological inhibitors of protein kinase C. Cancer Res 1994; 54(7):1707-1714.

162. Graves J, Lucas S, Cantrell D. The regulation of protein kinase C in T lymphocytes. In: Rees RC, ed. The biology and

clinical applications of interleukin-2. Oxford, New York, Tokyo: IRL Press at Oxford University Press, 1990:82-89.

163. Geleziunas R, Bour S, Wainberg MA. Human immunodeficiency virus type 1-associated CD4 downmodulation. Adv Virus Res 1994; 44:203-266.

164. Kielian M. Membrane fusion and the alphavirus life cycle. Adv Virus Res 1995; 45:113-151.

165. Helenius A, Morein B, Fries E et al. Human (HLA-A and HLA-B) and murine (H-2K and H-2D) histocompatibility antigens are cell surface receptors for Semliki Forest virus. Proc Natl Acad Sci USA 1978; 75(8):3846-3850.

166. Rosenkranz AA, Yachmenev SV, Jans DA et al. Receptor-mediated endocytosis and nuclear transport of a transfecting DNA construct. Exper Cell Res 1992; 199:323-329.

167. Akhlynina TV, Yachmenev SV, Jans DA et al. The use of internalizable derivatives of chlorin e6 for increasing its photosensitizing activity. Photochem Photobiol 1993; 58:45-48.

168. Akhlynina TV, Rosenkranz AA, Jans DA et al. Insulin-mediated intracellular targeting enhances the photodynamic activity of chlorin e6. Cancer Res 1995; 55:1014-1019.

169. Griffin FM Jr, Mullinax PJ. High concentrations of bacterial lipopolysaccharide, but not microbial infection-induced inflammation, activate macrophage C3 receptors for phagocytosis. J Immunol 1990; 145(2):697-701.

170. Detmers PA, Powell DE, Walz A et al. Differential effects of neutrophil-activating peptide 1/IL-8 and its homologues on leukocyte adhesion and phagacytosis. J Immunol 1991; 147(12):4211-4217.

171. Vedder NB, Harlan JM. Increased surface expression of CD11b/CD18 (Mac-1) is not required for stimulated neutrophil adherence to cultured endothelium. J Clin Invest 1988; 81:676-682.

172. Siu K, Detmers PA, Levin SM et al. Transient adhesion of neutrophils to endothelium. J Exp Med 1989; 169:1779-1793.

173. Helenius A, Kartenbeck J, Simons K et al. On the entry of Semliki Forest virus into BHK-21 cells. J Cell Biol 1980; 84:404-420.

174. Marsh M, Helenius A. Adsorptive endocytosis of Semliki Forest virus. J Mol Biol 1980; 142:439-454.

175. Gutman O, Danieli T, White JM et al. Effects of exposure to low pH on the lateral mobility of influenza hemagglutinin expressed at the cell surface: correlation between mobility inhibition and inactivation. Biochemistry 1993; 32(1):101-106.

176. Henis YI, Herman-Barhom Y, Aroeti B et al. Lateral mobility of both envelope proteins (F and HN) of Sendai virus in the cell membrane is essential for cell-cell fusion. J Biol Chem 1989; 264:17119-17125.

177. Volsky DJ, Loyter A. Inhibition of membrane fusion by suppression of lateral movement of membrane proteins. Biochim Biophys Acta 1978; 514:213-224.

178. Zhang F, Schmidt WG, Hou Y et al. Spontaneous incorporation of the glycosyl-phosphatidylinositol-linked protein Thy-1 into cell membranes. Proc Natl Acad Sci USA 1992; 89:5231-5235.

179. Kooyman DL, Byrne GW, McClellan S. In vivo transfer of GPI-linked complement restriction factors from erythrocytes to the endothelium. Science 1995; 269(5220):89-92.

INDEX

A

α-actinin. 176, 181, 194-195
A431 cells, 29, 84-85, 94, 96, 100, 144, 166
A7r5 cells, 22, 86, 90, 144
Actin binding protein (ABP), 171, 176, 195
  hisactophilin, 193, 199
  microfilaments, 5, 171, 176, 195
  polymerization, 55, 194
  stress fibers (F-actin), 27, 56-57, 180
  talin, 176, 181, 193-195, 199
  vinculin, 182, 193, 199
Adaptin, 96, 147
Adaptor protein complex (AP-2), 96, 147
Adenylate cyclase, 3, 5, 6, 7, 40, 67, 88, 98,
  101, 106, 118-126, 128-132, 139, 149,
  154, 198, 201, 203, 206
  interaction with Gsa. 98, 106, 118, 126,
  128-130
Adhesion, 10, 12, 52, 63, 83, 139, 165, 173,
  174, 176-182, 192, 194, 200, 205, 207,
  208
  assay, 177
  receptors, 10, 56, 178, 181, 192, 193, 207
  strengthening, 176, 177, 179, 182, 207
ADP ribosylation, 5
Adrenergic receptors, 100
  α₁β-adrenergic receptor, 156
  α₂-adrenergic receptor, 98, 108, 119, 128
Agonist, 36, 84, 86, 89, 90, 97, 100-101,
  118, 119, 126, 130-131, 140-141,
  149-156, 200, 205
Agrin, 69, 71, 73, 195, 199-200, 202
Amplification, 6, 129-131, 139, 182, 192,
  193, 208
Anchorage modulation, 56-57, 67, 205-206
Ankyrin, 52, 55, 66, 67, 73, 193, 195
Antagonist, 32, 84, 100, 109, 150-156, 205
Anti-adhesion therapy, 205
Antibodies, 1, 3, 10, 17, 18, 25-26, 36,
  51-53, 56, 62, 64, 70-72, 84, 87, 90, 94,
  96, 106, 109, 147, 166-168, 170-171,
  174-178, 180-182, 194, 204-205,
  207-208
  divalent. 1, 3, 62, 84, 87
  Fab, 53, 62, 70, 84, 87, 96, 166
  (Fab)2, 53
  monovalent, 3, 51, 53, 62, 84, 87, 96

Apparent lateral diffusion, 22, 25-27, 30,
  36, 38-40, 49-55, 62-67, 70-72, 74,
  87-92, 94, 100, 103, 108-109, 120,
  122-123, 128, 140, 142-145, 147-148,
  151-152, 168, 174, 178, 180-181,
  193-195, 198, 200, 202-203
Arrestin, 149, 154, 156, 194, 199
Artificial membrane. 7, 17, 26, 37, 39-40,
  49-50, 72. See also Membrane.
Asialoglycoprotein receptor, 33, 53
Autofluorescence, 17-18, 84, 89
Azide, 57, 87
  sodium, 87, 143, 145, 171

B

β-adrenergic receptor, 8, 32, 67, 100, 106,
  118-121, 126-130, 199
Bacteriorhodopsin, 40, 49-50, 72, 102
Band 3, 33, 39, 40, 51-53, 55-56, 61, 64, 67,
  72-73, 196
Blebs, 56, 66

C

c-ErbB-2. 92, 94, 166. See also Neu
  receptor.
Ca2+, 39, 70-71, 74, 91, 96-98, 100, 102,
  127, 140, 149, 167-168, 170, 174-176,
  195, 199, 200, 202, 204, 207
Calcineurin, 176, 202, 207
Calmodulin, 27, 70-71, 74, 100, 140, 174,
  176, 200, 202
cAMP. 3, 68, 98, 120, 123-125, 129, 130,
  133, 139-140, 177, 200, 203. See also
  Second messenger molecule.
  -dependent protein kinase, 100, 102,
  140, 154, 200-202
  permeant analog, 71, 154, 176, 202, 204
Capping. 1, 3, 52, 56, 57, 194, 205. See also
  Patching.
CD antigens
  CD11b, 170, 194
  CD11b/CD18. 36, 52, 74, 182, 203. See
  also CR3, Mac-1.
  CD18, 170
  CD2, 10, 34, 52, 58, 63, 68, 71, 73, 173-
  179, 181, 192, 194, 200, 202, 204,
  207. See also LFA-2.
  CD3, 173-174

CD32, 35, 60, 64, 69, 71, 200. *See also* Fc receptor.
CD4, 34, 60, 64, 68, 71, 73, 173-174, 176, 202, 207
CD51/CD29, 180
CD58, 174. *See also* LFA-3.
CD8, 34, 173, 174. *See also* Fibronectin, receptor; VLA-4.
Cell fusion, 1, 4
Cell-cell recognition, 1, 10, 12, 139, 165, 173-180, 192, 194, 205, 207
cGMP-phosphodiesterase, 98
Chinese hamster ovary (CHO), 29, 34, 36, 58, 62, 85, 150
cells, 30, 70, 71, 87, 156, 174
Cholecystokinin type A (CCKA) receptor, 87, 92, 100, 104, 122, 144, 145, 148-150, 195
Cholera toxin, 5, 26, 30, 118, 126
Cholesterol, 37, 40, 53, 54, 120. *See also* Lipid.
Coated pits, 57, 62, 64, 73, 140-145, 147, 148, 193-194, 207
Colchicine, 30, 36, 55, 57, 129, 147, 148, 176, 194, 195, 198, 204
Collision coupling theory, 7–10, 130
Concanavalin A, 26, 31, 52, 54-57, 208. *See also* Lectin, wheat germ agglutinin.
CR1 (C1 complement receptor), 36, 61, 64, 140
CR3 (C3b complement receptor), 36, 52, 67, 71, 72, 170, 182, 194, 200, 202, 204, 207. *See also* Mac-1, CD antigens.
Cytochalasin, 36, 55-57, 148, 150, 171, 173, 176, 182, 194-196, 198, 200, 204, 207
Cytokine receptor, 8, 9, 12, 87-88, 109, 110, 167, 192. *See also* Interleukin.
Cytoplasmic domain, 57, 59, 62-65, 71, 73, 167, 171, 172, 176, 200
Cytoskeleton, 7, 26, 39, 50, 51, 55-57, 64, 66, 67, 72, 73, 83, 87, 108, 125, 128, 129, 140, 148, 150, 154, 156, 170-171, 173, 176, 179, 181-183, 191, 193-195, 197-199, 204-207, 209

D

Deglycosylation, 148
Deoxyglucose, 145, 171
Dephosphorylation, 140, 149, 176. *See also* Phosphatase.
Desensitization, 83, 127, 139-163, 171, 200-201

Diacyl glycerol (DAG), 97, 102
Dimerization, 97. *See also* Tyrosine kinase receptor.
Domain structure, 7, 53, 65-67, 73, 83
Downregulation, 65, 127, 129, 130, 139, 149, 156, 176, 182, 192, 200
Dynamin, 129, 150, 195, 198

E

Effector enzyme, 5, 6, 98, 109, 118, 121, 126, 128, 156, 157, 206
EGTA, 68, 70-71, 174-175, 200, 202, 204
Emission, 17, 18. *See also* Fluorescence.
Endocytosis, 36, 71, 87, 96, 140-141, 143, 145-150, 154, 156, 173, 192, 203
Epidermal growth factor (EGF), 7, 9, 84, 87, 92, 94, 95-97, 103, 104, 140, 142, 145, 156, 166
receptor, 59, 64, 73, 84-85, 90, 94-97, 100, 103, 144, 147-148, 166
Epinephrine, 119, 121, 156
Extracellular domain, 26, 57, 61, 65, 103, 106, 109, 167, 180
Extracellular matrix, 10, 51, 53, 56-57, 65, 67, 71, 73, 173, 179-180, 192, 206

F

Fc receptor, 8-10, 35, 69, 74, 140, 145, 167, 171, 173-174, 177, 179, 192-194, 203, 205. *See also* CD antigen, CD32.
Fcε RI, 10, 35, 41, 60, 64, 167-173, 177, 195
Fcγ RIIA, 35, 60, 64, 69, 74, 200, 202
Fcγ RIIIB, 35, 39, 52, 71-72, 170, 203-204
Feedback loop, 156
Ferritin, 27, 120, 121, 205, 208
Fibroblast growth factor (FGF), 67, 103, 107
Fibroblasts, 27, 29, 31, 34, 39, 41, 50, 59, 66, 67, 69, 71, 85, 90-91, 100, 119, 143-144, 148, 180-181
Fibronectin, 10, 37, 51, 65, 73, 180-182
receptor, 10, 15, 35, 39, 51, 66, 73, 173, 180-182, 194, 195. *See also* VLA-4, CD antigens.
Fluid mosaic model, 1-3, 12, 65, 83, 208
Fluorescein isothiocyanate (FITC), 17-18, 26, 31, 36

Fluorescence, 17, 18, 20-25, 84, 88-89
emitted, 19, 20, *See also* Emission.
excitation. *See also* Excitation, 18
microphotolysis,17
nanovid method, 40
photobleaching recovery (FBR), 12,
17-19, 21, 24-26, 40-41, 49, 54, 83,
84, 88-89, 91, 94, 96, 121-123, 128,
140, 142, 150, 167-168, 170, 177,
198
polarization measurements, 121, 147
recovery after photobleaching (FRAP),
17, 27-28, 31, 37, 50, 58, 85, 108
resonance energy transfer, 166
resonance energy transfer microscopy,
94
spectroscopy measurement methods, 40
Focal contact sites, 10, 180, 182, 193
Fodrin. *See also* Spectrin, 129, 198, 201
Forskolin, 118, 123, 125, 154

G

Ganglioside GM1, 5, 26
Glucagon receptor, 87, 98, 103, 106, 125
Glycophorin, 37, 39
A, 36, 52, 56, 65, 72
B, 65
C, 36, 39, 53
Glycoprotein, 10, 26, 32, 49, 52, 58, 63,
65-66, 71, 74, 102, 171, 174
Glycosylation, 57, 61, 65, 148
Glycosyl phosphoinositol (GPI) anchor, 37,
57-58, 62-63, 73, 173, 177, 181
Granulocyte-macrophage colony stimulating factor (GM-CSF), 8-9, 109
GTP analog, 119, 198
Gpp(NH)p, 119
GTP-binding protein, 1, 5, 6, 8, 12, 36, 49,
83–138, 140, 149, 151, 156, 157, 191,
192, 195, 198, 199, 201, 206, 208
activating receptor, 86, 88, 91-92, 94,
98–104, 106, 109, 110, 117, 122,
139, 147-150, 156, 170, 172, 182,
191-194, 199-201, 205, 208
lateral movement, 199
activation, 98, 102, 118, 126-128, 139,
149, 154, 165, 167, 171, 192
coupling receptor, 5, 9, 12, 55, 56, 90-91,
94, 100-103, 106, 108, 120, 166
domain sequences, 103, 106
Gα, 5, 98-99, 103, 106, 108-109, 118,
126-127, 129-130, 154
Gαo, 98-99, 108

G$\beta\gamma$, 5, 98-99, 106, 108-109, 126-127,
129, 131, 149, 154, 198
Gi, 98, 101, 119, 128-129
Giα, 98, 118, 126-127, 129
Go, 70, 127
Gq, 98, 102, 127
Gs, 5, 88, 100-101, 118, 128, 130, 140,
198
Gsa, 98, 118, 126, 131, 198
Gt, 128
Gtbg, 126-127
GTPase, 97-98, 107, 150
lateral movement, 56, 107-109, 128
monomeric, 96
stimulating system, 98, 170
trimers, 9, 102, 108-109, 119, 126, 154,
156, 199
Guanine nucleotides, 5, 97-99, 126, 129,
150

H

Hemagglutinin, 32, 64, 140, 207
Her2, 92, 94, 103, 166. *See also* Neu
receptor; c-ErbB-2.
Heterologous desensitization, 149, 200
Homologous desensitization, 149, 200
Human chorionic gonadotropin receptor, 6
Human leukocyte antigen (HLA), 34, 69,
72, 74, 203
Hydrophilicity plots, 102-104

I

ICAM-1, 174, 176-177, 207
ICAM-2, 174
IgE, 8, 10, 35, 41, 60, 64, 145, 167-172, 177,
195
IgG, 27, 35, 60, 64, 74, 95, 103, 167, 170,
171, 200, 203-204
Immobilization, 8-10, 12, 49-53, 73, 83, 96,
100, 106, 120, 123, 125, 129-130,
139-163, 165-195, 200, 202, 204-205,
208
Immunoglobulin (Ig), 8, 27, 36-37, 52, 95,
167, 192
domain, 103
Inositol trisphosphate (IP$_3$), 97, 102
Insulin, 6, 51, 58, 63, 73, 84, 85, 87, 91,
94-96, 103-104, 140, 144-145, 147,
156, 166. *See also* Second messenger
molecule.
Integral membrane protein, 1, 26, 55, 65,
67

Interleukin (IL) receptors
    IL-2, 8-9, 87, 90, 109
       tac peptide, 87
    IL-3, 8-9, 109
    IL-3/5, 8
    IL-4, 8-9, 109
    IL-5, 9, 109
    IL-6, 8
Internalization, 5, 12, 36, 62, 64, 100, 106,
    110, 122, 139-150, 154, 156, 165,
    171-173, 176, 192-195, 209. *See also*
    Receptor.
Ionophore
    A23187, 182
    ionomycin, 71, 174-175, 200, 204
Isoprotenerol, 5, 54, 100, 120-121, 126

K

K_D *See also* Receptor, 131

L

L-plastin, 176, 195, 199-200
Lateral mobility measurement, 17, 20, 40,
    83, 84, 88-90, 100, 107-109, 122, 148,
    168
Lectin, 26, 31, 52, 56, 94, 205. *See also*
    Concanavalin A, Wheat germ
    agglutinin.
Ligand, 8, 10, 17-18, 20, 23, 51-52, 64,
    72-73, 83-84, 88, 94-96, 98-100, 106,
    109, 125, 129-131, 140, 146-151, 154,
    156, 166-189, 192-195, 200, 203-205,
    207-208
    bivalent, 95, 168, 174
    crosslinking, 10, 170
    dimeric, 9, 10
    monomeric, 60, 64, 169-172
Lipid, 1-3, 7, 12, 17, 21, 25-26, 37, 39-41,
    49-50, 53-57, 63, 65-67, 73, 90, 108,
    110, 118-121, 128, 130
    boundary, 67
    cholesterol, 53
    fluid, 54, 67, 120-121
    fluidity, 53, 72, 73, 119-120, 125
    probe, 17-18, 20, 21, 26, 28, 37, 39-41,
       49-50, 52, 54, 56, 67, 72, 84, 88,
       100, 121, 141
LLC-PK$_1$ cells, 21-23, 28, 86, 88-89, 91,
    122-125, 128, 141-142, 144, 197-198
Low density lipoprotein (LDL) receptor,
    34, 56, 61, 64, 73, 192

Luteinizing hormone (LH). *See also*
    Lutropin, 92, 94, 102, 147, 196
    receptor, 73, 140, 147-148, 195
Lymphocyte function-associated antigen
    (LFA)
    -1, 52, 174, 176, 177
    -2, 1, 174, *See also* CD2.
    -3, 10, 37, 40, 52, 57-58, 63, 73, 173-174,
       176-181, 192, 207-208, *See also*
       CD58.

M

Mac-1, 36, 52, 74, 170, 182, 194, 203. *See
    also* CR3, CD antigen.
Macrodomain, 65
Macrophages, 8, 10, 27, 35, 60, 71, 167,
    173, 174, 182
Major histocompatability complex (MHC)
    antigens, 39, 52, 55, 65, 69, 71
Mast cells, 8, 10, 29, 35, 167, 170, 173
Membrane, 26–39, 49, 51, 53-54, 58, 72,
    83, 90, 98, 108, 110, 117-138, 139,
    140, 147, 150, 154, 157, 165-189, 202
    artificial, 7, 17, 26, 37, 39-40, 49, 50, 72.
       See also Artificial membrane.
    basolateral, 21, 24, 29, 36, 66-67
    cholesterol, 120
    fluidity, 1-3, 53-55, 73, 119-121, 130,
       147, 205
    isolated, 7, 37-40, 50, 55, 67, 94, 119,
       127, 150
    mitochondrial, 49
    plasma, 3, 4, 7, 12, 17-18, 26, 28, 40,
       49-81, 83, 85, 96, 108, 110,
       117-118, 120-121, 140, 144, 173,
       182, 199, 204-205, 207, 209
    skeleton, 7, 40, 51, 53, 55-57, 64, 66-67,
       73, 148, 150, 171, 191, 193, 195,
       198-199, 206, 209
    T cell, 36
Metastasis, 180, 182, 191, 205
Methylamine, 36, 87, 145, 146
Microtubule, 5, 26, 52, 55-57, 73, 128, 140,
    147-148, 150, 176, 193-196, 198-199,
    201, 204. *See also* Tubulin.
    associated protein-2 (MAP-2), 55
    motor protein, 129
Mo-1, 170, 194
Mobile fraction, 22, 25-27, 30, 36, 38-39,
    49-51, 193-195, 198, 200, 202, 204,
    207

Mobile receptor hypothesis, 1-15, 41, 83, 118, 130, 182, 191–217
Model membrane, 37. *See also* Artificial membrane.
Motility, 39, 180-182, 191
Muscarinic receptor, 120
  M1, 98, 127
  M2, 98, 120
  M4, 156

N

N-formyl-peptide receptor, 100, 128, 150, 152, 153
Na+ channel, 33, 66, 69, 71, 73, 194, 196, 202
Nanovid, 40
Nerve growth factor (NGF), 84, 87, 92, 104, 140, 144-146, 166
  receptor, 96, 142, 144-146, 166
Neu receptor (p185 ^{neu}), 59, 73, 86, 94, 96, 106, 147, 166. *See also* C-ErbB-2, Her2.
Neuraminidase, 32, 207
Neutrophil, 8, 10, 76, 87, 126, 128-130, 150, 173, 182, 198
NIH 3T3 fibroblasts, 33, 59, 86
Nonspecific binding, 18, 20, 23, 84, 89, 141

P

p53/56^{lyn}, 167, 170
p59^{fyn}, 96, 109, 167
Patching, 1, 57. *See also* Capping.
Percolation model, 67
Peripheral membrane protein, 1, 39, 108
Pervanadate, 69-71, 202
Phagocytosis, 10, 71, 167, 173, 182, 205
Phalloidin, 27, 197
Phorbol ester, 64, 173, 176, 205, 207
Phorbol-12 myristate-13 acetate (PMA), 68, 71, 74, 182, 200, 202, 204, 207
Phosphatase, 7, 32, 58, 63, 70, 71, 149, 157, 176, 202, 204, 206, 207. *See also* Dephosphorylation.
Phospholipase, 102, 120, 205
  A$_2$, 53, 54, 98, 120-122, 127, 167
  C, 3, 98, 149
  Cβ, 98, 102, 106, 126-127, 206
  Cγ, 7, 96, 97, 107, 165, 167, 206
Phospholipid probes, 54

Phosphorylation, 2, 7, 9, 55, 57, 62-64, 70, 71, 74, 94, 96, 100, 103, 107, 109, 127, 129, 139-140, 148-150, 156, 165, 167, 170, 176, 183, 193-196, 198-200, 202, 204, 208-209
Platelet-derived growth factor (PDGF), 7, 9, 70, 74, 84, 87, 92, 94-96, 103-104, 166, 203
  receptor, 72, 74, 86, 95-96, 103, 166, 200, 202
Pleckstrin homology domain, 129, 198-199
Polylysine, 120, 122
Profilin, 199-200
Prostaglandin E$_1$, 119

R

Receptor
  aggregation, 87, 96, 140-142, 166-167, 173, 182, 202-204
  dissociation constant, 125. *See also* K$_D$.
  internalization, 106, 110, 139, 140, 144, 145, 154, 171, 192. *See also* Internalization.
  kinase, 94, 96, 106, 127, 129, 140, 149, 154, 166, 199, 201
  negative mutant, 156
  phosphorylation. *See* Phosphorylation.
Restriction of lateral movement, 17, 39, 41, 49, 52, 65-67, 83, 90, 109, 173, 194, 198
Rhodamine, 30, 70, 84, 87, 151, 168-169, 197
Rhodopsin, 32, 37, 39-40, 49-50, 98, 100, 102, 127-129, 149, 199
Rotational mobility, 87, 170

S

Saffman-Delbruck equation, 25-26, 50, 53
Second messenger molecule, 3, 100, 204, *See also* cAMP, IP$_3$.
Selectin, 173
Shc, 96, 167
Sheep erythrocyte rosetting (SER), 176
Signal-to-noise-ratio, 17
Signal transduction, 2, 5, 7-10, 12, 17, 41, 53, 55-56, 65, 67-72, 74, 83-84, 96-101, 108-140, 150, 156-157, 165-189, 191-194, 199-205, 207-209
Somatostatin, 98, 100, 101

Spectrin, 26, 30, 36, 38, 51, 52, 55, 64, 66, 67, 73, 129, 193-196, 198, 199, 201, *See also* Fodrin.
Spherocytes, 26, 28, 31, 39, 55
Staurosporine, 71, 156, 203, 207
Steroidogenesis, 6
Stress fibers. *See* Actin.

T

Taxol, 147, 207
Techniques to measure lateral mobility, 40
Therapeutic applications. *See* Anti-adhesion therapy.
Thermotropic separation, 120
Tight junction, 67
Transducin (Gt), 98, 126
Transfected cell lines, 29, 32, 57-58, 61, 70, 71, 85, 150, 171, 174
Transferrin receptor, 34, 41, 55, 73, 147, 194-195
Transmembrane domain, 57-58, 63, 65, 94, 102-104, 109, 167, 173, 177, 180
Tubulin, 27, 55, 129, 147, 195, 199. *See also* Microtubule.
 polymerization, 55, 147, 176

Tyrosine kinase receptor, 9, 10, 12, 91-98, 100, 102-104, 107, 109-110, 117, 122, 139, 143, 145, 148, 156, 165-167, 172, 182, 192, 193, 208
 dimerization, 7, 94, 95, 117, 129, 166, 173. *See also* Dimerization.

V

Vasopressin, 5, 9, 20, 22, 69, 71, 84, 89, 90, 102, 123-125, 140-142, 150, 152, 154
 V_1 receptor, 8, 22, 90, 92, 94, 98, 100, 102, 144-145, 148, 151, 154
 V_2 receptor, 8, 20, 22, 88-92, 98, 100, 122-125, 130, 139-141, 144-145, 148-154, 177, 194, 196, 198, 202, 205
Vasotocin, 154
Vesicular stomatitis virus (VSV), 32, 49
 G spike glycoprotein, 58, 74
Vinblastin, 55, 129, 147, 196, 198, 207
Vinculin. *See* Actin binding protein
Viscosity, 25, 26, 40, 41, 49, 50, 56, 65, 121
VLA-4, 180. *See also* CD antigens; Fibronectin, receptor.
VLA-5, 180

W

Wheat germ agglutinin, 26, 31, 56. *See also* Concanavalin A.